COMPLEX ANALYSIS

COMPLEX ANALYSIS
A Modern First Course
in Function Theory

Jerry R. Muir, Jr.
The University of Scranton

Published by John Wiley & Sons, Inc., Hoboken, New Jersey
Published simultaneously in Canada

For general information on our other products and services or for technical support, please contact our Customer Care
Department within the United States at (800) 762-2974, outside the United States at (317) 572-3993 or fax (317) 572-
4002.

Wiley also publishes its books in a variety of electronic formats. Some content that appears in print
may not be available in electronic formats. For more information about Wiley products, visit our web site at www.wiley.com.

Library of Congress Cataloging-in-Publication Data:

Muir, Jerry R.
   Complex analysis : a modern first course in function theory / Jerry R.
Muir, Jr.
      pages cm
   Includes bibliographical references and index.
   ISBN 978-1-118-70522-3 (cloth)
 1. Functions of complex variables. 2. Mathematical analysis. I. Title.
   QA331.M85 2014
   515–dc23
                                             2014035668

10 9 8 7 6 5 4 3 2 1

*To Stacey,
proofreader, patron,
and partner*

CONTENTS

PREFACE

This unfortunate name, which seems to imply that there is something unreal about these numbers and that they only lead a precarious existence in some people's imagination, has contributed much toward making the whole subject of complex numbers suspect in the eyes of generations of high school students.

– Zeev Nehari [21] on the use of the term *imaginary number*

In the centuries prior to the movement of the 1800s to ensure that mathematical analysis was on solid logical footing, complex numbers, those numbers algebraically generated by adding $\sqrt{-1}$ to the real field, were utilized with increasing frequency as an ever-growing number of mathematicians and physicists saw them as useful tools for solving problems of the time. The 19th century saw the birth of complex analysis, commonly referred to as function theory, as a field of study, and it has since grown into a beautiful and powerful subject.

The functions referred to in the name "function theory" are primarily analytic functions, and a first course in complex analysis boils down to the study of the complex plane and the unique and often surprising properties of analytic functions. Familiar concepts from calculus – the derivative, the integral, sequences and series – are ubiquitous in complex analysis, but their manifestations and interrelationships are novel in this setting. It is therefore possible, and arguably preferable, to see these topics addressed in a manner that helps stress these differences, rather than following the same ordering seen in calculus.

Complex Analysis: A Modern First Course in Function Theory, First Edition. Jerry R. Muir, Jr.
© 2015 John Wiley & Sons, Inc. Published 2015 by John Wiley & Sons, Inc.

This text grew from course notes I developed and tested on many unsuspecting students over several iterations of teaching undergraduate complex analysis at Rose-Hulman Institute of Technology and The University of Scranton. The following characteristics, rooted in my personal biases of how best to think of function theory, are worthy of mention.

- Complex analysis should never be underestimated as simply being calculus with complex numbers in place of real numbers and is distinguished from being so at every possible opportunity.

- Series are placed front and center and are a constant presence in a number of proofs and definitions. Analyticity is defined using power series to emphasize the difference between analytic functions and the differentiable functions studied in calculus. There is an intuitive symmetry between analyzing zeros using power series and singularities using Laurent series. The early introduction of power series allows the complex exponential and trigonometric functions to be defined as natural extensions of their real counterparts.

- Many properties of analytic functions seem counterintuitive (perhaps unbelievable) to students recently removed from calculus, and seeing these as early as possible emphasizes the distinctive nature of complex analysis. In service of this, Liouville's theorem, factorization using zeros, the open mapping theorem, and the maximum principle are considered prior to the more-involved Cauchy integral theory.

- Analytic function theory is built upon the trinity of power series, the complex derivative, and contour integrals. Consequently, the Cauchy–Riemann equations, an alternative expression of analyticity tied to differentiability in two real variables, are naturally partnered with the conformal mapping theorem at the end of the line of properties of analytic functions. Harmonic functions, also strongly reliant on this multivariable calculus topic, are the subject of the final chapter, allowing their study to benefit from the full theory of analytic functions.

- The geometric mapping properties of planar functions give intuition that was easily provided by the graphs of functions in calculus and help to tie geometry to function theory. In particular, linear fractional (Möbius) transformations are developed in service to this principle, prior to the introduction of analyticity or conformal mapping.

- The study of any flavor of analysis requires a box of tools containing basic geometric and topological facts and the related properties of sequences. These topics, in the planar setting, are addressed up front for easy reference, so as not to interrupt the subsequent presentation of function theory.

When faced with the choice of glossing over some details to more quickly "get to the good stuff" or ensuring that the development of topics is logically complete and consistent, I opted for the latter, leaving the reader the freedom to decide how to approach the text. This was done (with one caveat[1]) subject to the constraint that all material encountered in the typical undergraduate sequence of calculus courses is assumed without proof. This includes results that may not be proved in those courses but whose proofs are part of a standard course in real analysis. This helps to streamline the presentation while reducing overlap with other courses. For example, complex sequences are shown to converge if and only if their real and imaginary components converge. Then the assumed algebraic rules for convergent real sequences imply the same rules for complex sequences. A similar decomposition into real and imaginary parts readily provides familiar rules for derivatives and integrals of complex-valued functions of a real variable. It is important to clarify that material proved in a real analysis course that is not considered in calculus, such as aspects of topology or convergence of sequences of functions, is dealt with here.

It is my hope that this text provides a clear, concise exposition of function theory that allows the reader to observe the development of the theory without "losing the forest through the trees." To wit, connections between complex analysis and the sciences and engineering are noted but not explored. Although interesting and important, presenting applications to areas with which the reader is not assumed familiar while maintaining the previously mentioned commitment to logical completeness would have significantly lengthened the text and interrupted the flow of ideas.

The apex of a first complex analysis course is the residue theorem and its application to the evaluation of real integrals (Sections 5.3 and 5.4), and a number of pedagogical options are available for reaching or surpassing that point depending on the background of the students, the goals of the instructor, and the amount of rigor desired. The minimal route to get to Section 5.4 without skipping material used at some point along the way is: Sections 1.2–1.7, 2.1–2.4, 2.7–2.9, 3.1–3.4, 4.1–4.3, 5.1–5.4. It has been my experience that Section 5.4 can be reached in a fourteen-week-semester course taught to students with a multivariable calculus background with the possible omission of only one or two of the latter sections of Chapter 3, if necessary. The completeness of the text leaves the decision in the instructor's hands of what results could be accepted without proof or covered lightly. For instance, one willing to assert that properties of complex sequences or continuous functions behave like their real counterparts, assume that integrals may be exchanged with series as needed eschewing the details of uniform convergence, or postpone the presentation of geometric mapping properties and linear fractional transformations should find the text suitably flexible, as should an instructor whose students already have comfort with the material covered in the first chapter or so. An instructor venturing into the last chapter on harmonic functions will want to have covered Sections 3.5 and 3.6 at its start and Section 2.6 later on.

Each section of the text concludes with remarks, under the heading "Summary and Notes," that review the main points of the section, note who discovered them and when, place them in context either within the whole of the text or within the broader mathematical world, and/or note interesting connections to the sciences. No

---

[1]We assume the Jordan curve theorem, that every simple closed planar curve has an inside and outside. See Remark 4.1.4 for discussion.

one will ever accuse me of being well versed in the history of mathematics, but I have always found the topic interesting. I hope the tidbits provided within these notes whet the reader's appetite to learn more about it than I know and find the history-related references in the bibliography to be as educational as I did.

Exercises of both a computational and theoretical nature are given. Many of the problems asking students to "prove" or "show" something require little more than a calculation, while others have more depth. Those marked with the symbol "▷" contain results that are referenced later in the text. Of note, there are several exercises that either provide an alternative definition of, or point of view on, a concept or introduce a complex analysis topic that is not included in the body of the text. Examples include the definition of compactness using open covers, the definition of contour integrals using limits of Riemann sums, Hadamard products of analytic functions, a proof of Cauchy's theorem for cycles, the argument principle, the application of the residue theorem to sum certain numerical series, the uniqueness of the inverse Laplace transform up to continuity, and Fejér's theorem for the Cesàro summability of Fourier series. These would serve as interesting projects for good students, as would a thorough reading of the epilogue on the Riemann mapping theorem.

My undergraduate professors at SUNY Potsdam were fond of saying that a mathematics book should always be read with paper and pencil in hand. I hope that any detail lost to the tug of war between brevity and lucidity can be recovered with a bit of work on the side, to the reader's benefit.

Function theory has been around for a long time, and I am grateful for the instruction I received as a student and the excellent complex analysis books listed in the bibliography. My love of, and point of view on, the subject surely germinated during the privileged time I spent as a student of Ted Suffridge at the University of Kentucky and from the books [7, 26] from which I learned and referenced during those years. I have no doubt their influence is present throughout the following pages in places even I would not expect.

I thank my students over the years, especially those who weren't shy about pointing out flaws when this work was in its infancy, and Stephen Aldrich, Michael Dorff, and Stacey Muir, who used early drafts and offered valuable feedback. Stacey's unlimited help and support deserves special recognition – for the last decade, she has been contractually obligated not to run away when I'm at my wit's end with a project.

Finally, I could not be more appreciative of the team at Wiley for their able support during the preparation of this text, for allowing me the freedom to write the book I imagined, and for believing in the first place that the course notes of a crackpot professor could become a reasonable book. That said, my fingers did the typing, and, alas, any misteaks are mine.

JERRY R. MUIR, JR.

Scranton, Pennsylvania

CHAPTER 1

THE COMPLEX NUMBERS

In this chapter, we introduce the complex numbers and their interpretation as points in a number plane, an analog to the real number line. We develop the algebraic, geometric, and topological properties of the set of complex numbers, many of which mirror those of the real numbers. These properties, especially the topological ones, are connected to sequences, and thus we conclude the chapter by studying the basic nature of sequences and series. At the conclusion of the chapter, we will possess the tools necessary to begin the study of functions of a complex variable.

1.1 Why?

Our work in this text can best be understated as follows: Let's throw $\sqrt{-1}$ into the mix and see what happens to the calculus. The result is a completely different flavor of analysis, a separate field distinguished from its real-variable sibling in some striking ways.

The use of $\sqrt{-1}$ as an intermediate step in finding solutions to real-variable problems goes back centuries. In the Renaissance, Italian mathematicians used complex numbers as a tool to find *real* roots of cubic equations. The algebraic use of complex

Complex Analysis: A Modern First Course in Function Theory, First Edition. Jerry R. Muir, Jr.
© 2015 John Wiley & Sons, Inc. Published 2015 by John Wiley & Sons, Inc.

numbers became much more mainstream due to the work of Leonhard Euler in the 18th century and later, Carl Friedrich Gauss. Euler and Jean le Rond d'Alembert are generally credited with the first serious considerations of functions of a complex variable – the former considered such functions as an intermediate step in the calculation of certain *real* integrals, while the latter saw these functions as useful in his study of fluid mechanics.

Introducing complex numbers as a stepping stone to solve real problems is a common historical theme, and it is worth recalling how other familiar systems of numbers can be viewed to solve particular algebraic and analytic problems. The natural numbers, integers, rational numbers, and real numbers satisfy the set containments $\mathbb{N} \subseteq \mathbb{Z} \subseteq \mathbb{Q} \subseteq \mathbb{R}$, but each subsequent set has characteristics not present in its predecessor. Where the natural numbers $\mathbb{N} = \{1, 2, \dots\}$ are closed under addition and multiplication, extending to the integers $\mathbb{Z} = \{\dots, -2, -1, 0, 1, 2, \dots\}$ provides an additive identity and inverses. The set of rational numbers $\mathbb{Q}$, consisting of all fractions of integers, has multiplicative inverses of its nonzero elements and hence is an algebraic *field* under addition and multiplication.

The move from $\mathbb{Q}$ to the real numbers $\mathbb{R}$ is more analytic than algebraic. Although $\mathbb{Q}$ is a field, it is not *complete*, meaning there are "holes" that need to be filled. For instance, consider the equation $x^2 = 2$. Since $1^2 = 1$ and $2^2 = 4$, it seems that a solution to the equation should exist and lie somewhere in between. Further analysis reveals that a solution should lie between $5/4$ and $3/2$. Successive subdivisions may be used to target where a solution should lie, but that point is not in $\mathbb{Q}$. Beyond unsolvable algebraic equations lies the number π, the ratio of a circle's circumference to its diameter, which can also be shown not to lie in $\mathbb{Q}$. The alleviation of these problems comes by allowing the set $\mathbb{R}$ of real numbers to be the *completion* of $\mathbb{Q}$. In satisfyingly imprecise terms, $\mathbb{R}$ is equal to $\mathbb{Q}$ with the "holes filled in." This is done so that the *axiom of completeness* (i.e. the least upper bound property) holds. See Appendix B for more detail. The result is that $\mathbb{R}$ is a *complete ordered field*.

The upgrade from the real numbers to the complex numbers has both algebraic and analytic motivation. The real numbers are not *algebraically* complete, meaning there are polynomial equations such as $x^2 = -1$ with no solutions. The incorporation of $\sqrt{-1}$ mentioned earlier is a direct response to this. But the work of Euler and d'Alembert shows how moving outside $\mathbb{R}$ facilitates analytic methods as well. While their work did much for bringing credibility to the use of complex numbers, it was during the 19th century, in the movement to deliver rigor to mathematical analysis, that complex function theory gained its footing as a separate subject of mathematical study, due largely to the work of Augustin-Louis Cauchy, Bernhard Riemann, and Karl Weierstrass.

Function theory is the study of the calculus of complex-valued functions of a complex variable. The analysis of functions on this new domain will quickly distinguish itself from real-variable calculus. As the reader will soon see, by combining the algebra and geometry inherent in this new setting, we will be able to perform a great deal of analysis that is not available on the real domain. Such analysis will include some intuition-bending results and techniques that solve problems from calculus that are not easily accessible otherwise.

Before setting out on our study of complex analysis, we must agree on a starting point. We assume that the reader is familiar with the fundamentals of differential, integral, and multivariable calculus. The language of sets and functions is freely used; the unfamiliar reader should examine Appendix A.

Lastly, to whet our appetites for what is to come, here are a handful of exercises appearing later in the text the statements of which are understandable from calculus, but whose solutions are either made possible or much simpler by the introduction of complex numbers.

Forthcoming Exercises

1. Derive triple angle identities for $\sin 3\theta$ and $\cos 3\theta$. [Section 2.4, Exercise 6]

2. Find a continuous one-to-one planar transformation that maps the region lying inside the circles $(x - 1)^2 + y^2 = 4$ and $(x + 1)^2 + y^2 = 4$ onto the upper half-plane $y \geq 0$. [Section 2.6, Exercise 13]

3. Find the radius of convergence of the Taylor series expansion of the function

$$f(x) = \frac{\sin x}{1 + x^4}$$

   about $a = 2$. [Section 3.2, Exercise 3]

4. Verify the summation identity, where $c \in \mathbb{R}$ is a constant.

$$\sum_{n=1}^{\infty} \frac{1}{n^2 + c^2} = \begin{cases} \dfrac{\pi^2}{6} & \text{if } c = 0, \\[2ex] \dfrac{\pi}{2c} \coth \pi c - \dfrac{1}{2c^2} & \text{if } c \neq 0 \end{cases}$$

   [Section 5.3, Exercise 11]

5. Evaluate the following integrals, where $n \in \mathbb{N}$ and $a, b > 0$.

   (a) $\displaystyle\int_0^{2\pi} \cos^n t \, dt$,   [Section 2.8, Exercise 4]

   (b) $\displaystyle\int_{-\infty}^{\infty} \frac{\sin ax}{x(x^2 + b^2)} \, dx$,   [Section 5.4, Exercise 7]

   (c) $\displaystyle\int_0^{\infty} \frac{\sqrt[n]{x}}{x^2 + a^2} \, dx$,   [Section 5.4, Exercise 10]

6. Find a real-valued function u of two variables that satisfies

$$\frac{\partial^2 u}{\partial x^2} + \frac{\partial^2 u}{\partial y^2} = 0$$

   inside the unit circle and continuously extends to equal 1 for points on the circle with $y > 0$ and 0 for points on the circle with $y < 0$. [Section 6.2, Exercise 3]

1.2   The Algebra of Complex Numbers

As alluded to in Section 1.1, we desire to expand from the set of real numbers in a way that provides solutions to polynomial equations such as $x^2 = -1$. One may be tempted to simply define a number that solves this equation. The drawback to doing so is that the negative of this number would also be a solution, and this could cause some ambiguity in the definition. We therefore choose a different method.

1.2.1 Definition. A *complex number* is an ordered pair of real numbers. The set of complex numbers is denoted by $\mathbb{C}$.

By definition, any $z \in \mathbb{C}$ has the form $z = (x, y)$ for numbers $x, y \in \mathbb{R}$. What distinguishes complex numbers from their counterparts, the two-dimensional vectors in $\mathbb{R}^2$, is their algebra – specifically, their multiplication.

1.2.2 Definition. If (a, b) and (c, d) are complex numbers, then we define the algebraic operations of *addition* and *multiplication* by

$$(a, b) + (c, d) = (a + c, b + d)$$
$$(a, b)(c, d) = (ac - bd, ad + bc).$$

Clearly, $\mathbb{C}$ is closed under both of these operations. (Adding or multiplying two complex numbers results in another complex number.)

Notice that if $a, b \in \mathbb{R}$, then $(a, 0) + (b, 0) = (a + b, 0)$ and $(a, 0)(b, 0) = (ab, 0)$. Therefore $a \mapsto (a, 0)$ is a natural algebraic embedding of $\mathbb{R}$ into $\mathbb{C}$. Accordingly, it is natural to write a for the complex number $(a, 0)$, and in this way, we consider $\mathbb{R} \subseteq \mathbb{C}$.

For any complex number $z = (x, y)$,

$$z = (x, 0) + (0, 1)(y, 0) = x + (0, 1)y.$$

In other words, each complex number can be written uniquely in terms of its two real components and the complex number $(0, 1)$. This special complex number gets its own symbol.

1.2.3 Definition. The *imaginary unit* is the complex number $i = (0, 1)$. A complex number z expressed as

$$z = x + iy \tag{1.2.1}$$

is said to be in *rectangular form*.

Because every complex number can be written uniquely as above, we (usually) refrain from using the ordered pair notation in favor of using the rectangular form. Notice that i is a solution to the equation $z^2 = -1$.

It is left as an exercise to verify that 0 is the additive identity and 1 is the multiplicative identity, every member of $\mathbb{C}$ has an additive inverse, both operations are associative and commutative, multiplication distributes over addition, and if $z \neq 0$ is written as in (1.2.1), then it has the multiplicative inverse

$$z^{-1} = \frac{1}{z} = \frac{x}{x^2 + y^2} + i\left(\frac{-y}{x^2 + y^2}\right) \tag{1.2.2}$$

in $\mathbb{C}$. This shows that $\mathbb{C}$ is a *field*.

1.2.4 Definition. For a complex number z written as in (1.2.1), we call the real numbers x and y the *real part* and *imaginary part* of z, respectively, and use the symbols $x = \operatorname{Re} z$ and $y = \operatorname{Im} z$. If $\operatorname{Re} z = 0$, then z is called *imaginary* (or *purely imaginary*). The *conjugate* of z is the complex number $\overline{z} = x - iy$. The *modulus* (or *absolute value*) of z is the nonnegative real number $|z| = \sqrt{x^2 + y^2}$.

It is a direct calculation to verify the relationship

$$|z|^2 = z\overline{z} \tag{1.2.3}$$

for all $z \in \mathbb{C}$. Other useful identities involving moduli and conjugates of complex numbers are left to the exercises.

1.2.5 Example. The identity (1.2.3) is useful for finding the rectangular form of a complex number. For instance, consider the quotient

$$z = \frac{1 + 2i}{2 + i}.$$

To find the expressions from Definition 1.2.4, we multiply by the conjugate of the denominator over itself,

$$z = \frac{1 + 2i}{2 + i} \frac{2 - i}{2 - i} = \frac{4 + 3i}{5},$$

to get a positive denominator. We see that $\operatorname{Re} z = 4/5$, $\operatorname{Im} z = 3/5$, $|z| = 1$, and $\overline{z} = (4 - 3i)/5$.

Summary and Notes for Section 1.2.

The set of complex numbers $\mathbb{C}$ consists of ordered pairs of real numbers. The real numbers are the those complex numbers of the form $(x, 0)$ for $x \in \mathbb{R}$, and $i = (0, 1)$ is the imaginary unit. Algebraic operations are defined to make $\mathbb{C}$ a field and so that $i^2 = -1$. We write the complex number $z = (x, y)$ in the rectangular form $z = x + iy$, and the conjugate of z is $\overline{z} = x - iy$.

In their attempts to find real solutions to cubic equations, Italian mathematicians found it necessary to manipulate complex numbers. Perhaps the first to consider them was Gerolamo Cardano in the 16th century, who named them "fictitious numbers." Rafael Bombelli introduced the algebra of complex numbers shortly thereafter. At that time, the square roots of negative numbers were just manipulated as a means to an end. The ordered pair definition can be traced to William Rowan Hamilton almost three centuries later.

Exercises for Section 1.2.

1. For the following complex numbers z, calculate $\operatorname{Re} z$, $\operatorname{Im} z$, $|z|$, and $\overline{z}$.

   (a) $z = 3 + 2i$

   (b) $z = \dfrac{1 + i}{i}$

(c) $z = \dfrac{2 - i}{1 + i} + i$

(d) $z = (4 + 2i)\overline{(3 + i)}$

2. Verify the following algebraic properties of $\mathbb{C}$.

   (a) The complex numbers 0 and 1 are the additive and multiplicative identities of $\mathbb{C}$, respectively.

   (b) Each $z \in \mathbb{C}$ has an additive inverse.

   (c) Addition and multiplication of complex numbers is associative. In other words, $z + (w + v) = (z + w) + v$ and $z(wv) = (zw)v$ for all $z, w, v \in \mathbb{C}$.

   (d) Addition and multiplication of complex numbers is commutative. That is, $z + w = w + z$ and $zw = wz$ for all $z, w \in \mathbb{C}$.

   (e) Multiplication of complex numbers distributes over addition. That is, $a(z + w) = az + aw$ for all $a, z, w \in \mathbb{C}$.

   (f) If $z \in \mathbb{C}$ is nonzero, then its multiplicative inverse is as given in (1.2.2).

3. $\triangleright$ Verify the following identities involving the conjugate.

   (a) For each $z \in \mathbb{C}, \overline{\overline{z}} = z$.

   (b) For each $z \in \mathbb{C}$,
   $$\operatorname{Re} z = \frac{z + \overline{z}}{2}, \qquad \operatorname{Im} z = \frac{z - \overline{z}}{2i}.$$

   (c) For all $z, w \in \mathbb{C}$,
   $$\overline{z + w} = \overline{z} + \overline{w}, \qquad \overline{zw} = \overline{z}\,\overline{w}.$$

4. $\triangleright$ Verify the following identities involving the modulus. For each, let $z, w \in \mathbb{C}$.

   (a) $|zw| = |z||w|$

   (b) $|z/w| = |z|/|w|$ if $w \neq 0$

   (c) $|\overline{z}| = |z|$

   (d) $-|z| \leq \operatorname{Re} z \leq |z|$

5. $\triangleright$ Prove that for all $z, w \in \mathbb{C}$,
$$|z + w|^2 = |z|^2 + 2 \operatorname{Re} z\overline{w} + |w|^2.$$

6. Let $p\colon \mathbb{C} \to \mathbb{C}$ be a polynomial. That is, $p(z) = \sum_{k=0}^{n} a_k z^k$ for $a_0, \dots, a_n \in \mathbb{C}$ and $a_n \neq 0$. A *root* (or *zero*) of p is a number $r \in \mathbb{C}$ such that $p(r) = 0$. Show that if $a_0, \dots, a_n \in \mathbb{R}$, then $\overline{r}$ is a root of p whenever r is a root of p.

7. Let $a, b \in \mathbb{R}$. Consider the function $T\colon \mathbb{R}^2 \to \mathbb{R}^2$ given by
$$T(x, y) = (\operatorname{Re}[(a + ib)(x + iy)], \operatorname{Im}[(a + ib)(x + iy)]).$$

   Prove that T is a linear transformation on $\mathbb{R}^2$, and determine a 2×2 matrix form for T. What does T represent?

8. For ordered pairs of real numbers (a, b) and (c, d), what drawbacks are there to defining multiplication of complex numbers by $(a, b)(c, d) = (ac, bd)$?

1.3 The Geometry of the Complex Plane

The real number line is the geometric realization of the set of real numbers and accordingly is a useful tool for conceptualization. Since complex numbers are defined to be ordered pairs of real numbers, it is only natural to visualize the set of complex numbers as the points in the Cartesian coordinate plane $\mathbb{R}^2$. This geometric interpretation is essential to the analysis of complex functions.

1.3.1 Definition. When its points are considered to be complex numbers, the Cartesian coordinate plane is referred to as the *complex plane* $\mathbb{C}$. The x- and y-axes in the plane are called the *real and imaginary axes*, respectively, in $\mathbb{C}$.

Because addition of complex numbers mirrors addition of vectors in $\mathbb{R}^2$, we use vectors to geometrically interpret addition in terms of parallelograms. Continuing this line of thought, we see that the value $|z|$, as the distance from the point $z \in \mathbb{C}$ to 0, is the length (or magnitude) of the vector z. If $z, w \in \mathbb{C}$, then $z - w$, in vector form, is the vector pointing from w to z. Therefore $|z - w|$ is equal to the distance between z and w. Lastly, we note that the operation of complex conjugation is realized geometrically as reflection in the real axis. See Figure 1.1.

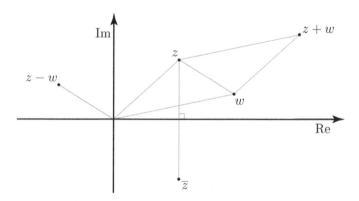

Figure 1.1 $z + w$, $z - w$, and $\overline{z}$ for some $z, w \in \mathbb{C}$

Another geometric consequence of the parallelogram interpretation is the triangle inequality, which gives that if $z, w \in \mathbb{C}$, then the distance from 0 to $z + w$ is never greater than the sum of the distances from 0 to z and 0 to w. We prove it as follows.

1.3.2 Triangle Inequality. *If $z, w \in \mathbb{C}$, then*

$$|z + w| \leq |z| + |w|. \tag{1.3.1}$$

Proof. We use Exercises 4 and 5 of Section 1.2 to calculate

$$|z + w|^2 = |z|^2 + 2\,\mathrm{Re}\,z\overline{w} + |w|^2$$
$$\leq |z|^2 + 2|z\overline{w}| + |w|^2$$

$$= |z|^2 + 2|z||w| + |w|^2$$
$$= (|z| + |w|)^2.$$

Taking the square root of both sides completes the proof.   □

If $a \in \mathbb{C}$ and $r > 0$, then the circle in $\mathbb{C}$ centered at a of radius r is the set of all points whose distance from a is r. From our above observation, this circle can be described in set notation by $\{z \in \mathbb{C} : |z - a| = r\}$. The inside of a circle, called a *disk*, is a commonly used object and is denoted by

$$D(a; r) = \{z \in \mathbb{C} : |z - a| < r\}. \tag{1.3.2}$$

The most prominently used disk in the plane is the unit disk – the disk centered at 0 of radius 1. For this, we use the special symbol

$$\mathbb{D} = D(0; 1). \tag{1.3.3}$$

1.3.3 Example. Let us consider the geometry of the set

$$E = \{z \in \mathbb{C} : |1 + iz| < 2\}$$

in two ways. First, write $z = x + iy$ to see that the condition defining E is equivalent to $|1 - y + ix| < 2$ or $\sqrt{x^2 + (y - 1)^2} < 2$. This describes all planar points of distance less than 2 from $(0, 1) = i$. Hence $E = D(i; 2)$.

In this circumstance, there is an advantage to eschewing real and imaginary parts. Note that
$$|1 + iz| = |i(-i + z)| = |i||z - i| = |z - i|.$$
Thus $E = \{z \in \mathbb{C} : |z - i| < 2\} = D(i; 2)$.

Summary and Notes for Section 1.3.

Since the complex numbers are ordered pairs of real numbers, the set $\mathbb{C}$ of complex numbers is geometrically realized as the plane $\mathbb{R}^2$. The addition and modulus of complex numbers parallel the addition and magnitude of planar vectors. The triangle inequality gives an important bound on sums.

In 1797, a Norwegian surveyor named Caspar Wessel was the first of many to consider the geometric interpretation of the complex numbers, but his work was largely unknown as was that of Jean-Robert Argand in 1806. (The complex plane is often referred to as the *Argand plane*.) The work of Carl Friedrich Gauss in the first half of the 19th century brought the concept to the masses.

Exercises for Section 1.3.

1. Geometrically illustrate the parallelogram rule for the complex numbers $2 + i$ and $1 + 3i$.

2. Geometrically illustrate the relationship between the complex number $-1 + 2i$ and its conjugate.

3. Describe the following sets geometrically. Sketch each.

(a) $\{z \in \mathbb{C} : |z - 1 + i| = 2\}$

(b) $\{z \in \mathbb{C} : |z - 1|^2 + |z + 1|^2 \leq 6\}$

(c) $\{z \in \mathbb{C} : \text{Im}\, z > \text{Re}\, z\}$

(d) $\{z \in \mathbb{C} : \text{Re}(iz + 1) < 0\}$

4. Provide a geometric description of complex multiplication for nonzero $z, w \in \mathbb{C}$. It is helpful to write $z = r(\cos\theta + i\sin\theta)$ and $w = \rho(\cos\varphi + i\sin\varphi)$ and use trigonometric identities.

5. For which pairs of complex numbers is equality attained in the triangle inequality? Prove your answer.

6. ▷ Prove that for all $z, w \in \mathbb{C}$, $\|z| - |w\| \leq |z - w|$.

7. Show that for all $z \in \mathbb{C}$, $|\text{Re}\, z| + |\text{Im}\, z| \leq \sqrt{2}\,|z|$.

8. This exercise concerns lines in $\mathbb{C}$.

   (a) Let $a, b \in \mathbb{C}$ with $b \neq 0$. Prove that the set

   $$L = \left\{z \in \mathbb{C} : \text{Im}\left(\frac{z - a}{b}\right) = 0\right\}$$

   is a line in $\mathbb{C}$. Explain the role of a and b in the geometry of L. (*Hint*: Recall the vector form of a line from multivariable calculus.)

   (b) Let C be a circle in $\mathbb{C}$ with center c and radius $r > 0$. If a lies on C, write the line tangent to C at a in the form given in part (a).

1.4 The Topology of the Complex Plane

The *topology* of a certain space (in our case $\mathbb{C}$) gives a useful alternative to traditional geometry to describe relationships between points and sets. The key concepts of limits and continuity from calculus are tied to the topology of the real line, as we will see is also true in the plane. That connection just scratches the surface of how powerful a tool we will find planar topology to be for analyzing functions.

We begin by observing that with respect to a given subset of $\mathbb{C}$, each point of $\mathbb{C}$ is of one of three types.

1.4.1 Definition. Let $A \subseteq \mathbb{C}$ and $a \in \mathbb{C}$. Then a is an *interior point* of A if A contains a disk centered at a, a is an *exterior point* of A if it is an interior point of the complement $\mathbb{C} \setminus A$, and a is a *boundary point* of A if it is neither an interior point of A nor an exterior point of A. (See Figure 1.2.)

These points form the following sets.

1.4.2 Definition. Let $A \subseteq \mathbb{C}$. The set of interior points of A is called the *interior* of A and is denoted A°. The set of boundary points of A is called the *boundary* of A and is denoted ∂A.

Note that the set of exterior points of A is the interior of $\mathbb{C} \setminus A$, and so we need not define a new symbol for this set. We have that $\mathbb{C}$ can be decomposed into the disjoint union

$$\mathbb{C} = A^\circ \cup \partial A \cup (\mathbb{C} \setminus A)^\circ.$$

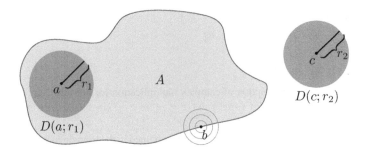

Figure 1.2 a is an interior point of A, b is a boundary point of A, and c is an exterior point of A

The set A contains all of its interior points, none of its exterior points, and none, some, or all of its boundary points. The extremal cases are special.

1.4.3 Definition. Let $A \subseteq \mathbb{C}$. If $\partial A \cap A = \varnothing$, then A is *open*. If $\partial A \subseteq A$, then A is *closed*.

The following properties can be deduced from the above definitions. Their proofs are left as an exercise.

1.4.4 Theorem. *The following hold for $A \subseteq \mathbb{C}$.*

(a) *The set A is closed if and only if its complement $\mathbb{C} \setminus A$ is open.*

(b) *The set A is open if and only if for every $a \in A$, there exists $r > 0$ such that $D(a;r) \subseteq A$.*

(c) *A point $a \in \mathbb{C}$ is in ∂A if and only if $D(a;r) \cap A \neq \varnothing$ and $D(a;r) \setminus A \neq \varnothing$ for all $r > 0$.*

1.4.5 Example. We study the disk $D(a;r)$ for some $a \in \mathbb{C}$ and $r > 0$. If $z_0 \in D(a;r)$, then let $\rho = r - |z_0 - a|$. Then $0 < \rho \leq r$. For all $z \in D(z_0;\rho)$,

$$|z - a| = |z - z_0 + z_0 - a| \leq |z - z_0| + |z_0 - a| < \rho + (r - \rho) = r$$

by the triangle inequality, showing $z \in D(a;r)$. Therefore $D(z_0;\rho) \subseteq D(a;r)$. This implies that $D(a;r)$ is an open set.

It is left as an exercise (using an argument quite similar to the one just presented) to show that the exterior points of $D(a;r)$ form the set $\{z \in \mathbb{C} : |z - a| > r\}$. Therefore the boundary of the disk is

$$\partial D(a;r) = \{z \in \mathbb{C} : |z - a| = r\}, \tag{1.4.1}$$

which is exactly the circle of radius r centered at a.

1.4.6 Definition. Given any set $A \subseteq \mathbb{C}$, the set

$$\overline{A} = A \cup \partial A \tag{1.4.2}$$

is called the *closure* of A.

Many of the properties of the closure are addressed in the exercises.

1.4.7 Example. If $D(a; r)$ is the disk in Example 1.4.5, then its closure is the *closed disk*

$$\overline{D}(a; r) = \overline{D(a; r)} = \{z \in \mathbb{C} : |z - a| \leq r\}. \tag{1.4.3}$$

1.4.8 Definition. Let $A \subseteq \mathbb{C}$ and $a \in \mathbb{C}$. We say that a is a *limit point* of A provided that $D(a; r) \cap A \setminus \{a\} \neq \varnothing$ for every $r > 0$.

In other words, a is a limit point of A if every disk centered at a intersects A at a point *other than* a. This leads to another useful characterization of closed sets.

1.4.9 Theorem. *A set $E \subseteq \mathbb{C}$ is closed if and only if E contains all of its limit points.*

Proof. Suppose that E is closed and that a is a limit point of E. Were a an exterior point of E, we would have $D(a; r) \subseteq \mathbb{C} \setminus E$ for some $r > 0$. Since this contradicts that a is a limit point of E, it must be that a is an interior point or boundary point of E. Either way, $a \in E$.

Conversely, assume that E contains all of its limit points. Suppose that $a \in \partial E \setminus E$. For any $r > 0$, $D(a; r) \cap E \neq \varnothing$ by Theorem 1.4.4. Since $a \notin E$, $D(a; r) \cap E \setminus \{a\} \neq \varnothing$ for all $r > 0$, and thus a is a limit point of E. This shows that $a \in E$, a contradiction. Thus $\partial E \subseteq E$, and hence E is closed. $\qquad\square$

We continue with two more definitions.

1.4.10 Definition. A set $A \subseteq \mathbb{C}$ is *bounded* if $A \subseteq D(0; R)$ for some $R > 0$.

1.4.11 Definition. A set $K \subseteq \mathbb{C}$ is *compact* if K is closed and bounded.

One must be careful not to be misled by the simplicity of the above definition and underestimate the importance of compact sets to the study of analysis. In fact, many properties of complex functions depend on compactness.

A reader with some previous exposure to topological concepts may have seen compactness defined in terms of "open covers." This definition is of great importance to the study of topology, but does not serve our purpose in this text. That our definition is equivalent is the content of the Heine–Borel theorem. An outline of the proof of this theorem is included in the exercises.

We now consider our final topological concept.

1.4.12 Definition. Nonempty sets $A, B \subseteq \mathbb{C}$ are *separated* if $\overline{A} \cap B = A \cap \overline{B} = \varnothing$. A nonempty set $E \subseteq \mathbb{C}$ is *connected* if E is not equal to the union of separated sets. Otherwise, E is *disconnected*.

While this definition may seem complicated, it should bring the reader comfort that the intuitive notion of connectedness matches the rigorous definition. The following observation is key to a method of detecting connectedness that will be sufficient for most circumstances we will encounter.

1.4.13 Lemma. *Let $A, B \subseteq \mathbb{C}$ be separated sets, $a \in A$, $b \in B$, and L be the line segment with endpoints a and b. Then $L \nsubseteq A \cup B$.*

Proof. Suppose $L \subseteq A \cup B$. If $u = (b - a)/|b - a|$, then $L = \{a + tu : 0 \leq t \leq |b - a|\}$. Set

$$t_0 = \sup\{t \in [0, |b - a|] : a + tu \in A\}, \qquad c = a + t_0 u \in L.$$

(See Appendix B for properties of the supremum.)

If $c \in A$, then $c \notin \overline{B}$, and hence $D(c; r) \cap B = \varnothing$ for some $r > 0$. Furthermore, $t_0 < |b - a|$, and if $t_0 < t < \min\{t_0 + r, |b - a|\}$, then $a + tu \in L \setminus A \subseteq B$. But $|(a + tu) - c| = t - t_0 < r$, a contradiction.

If $c \in B$, then $c \notin \overline{A}$, and so $D(c; r) \cap A = \varnothing$ for some $r > 0$. But $t_0 > 0$, and there must exist $\max\{t_0 - r, 0\} < t < t_0$ such that $a + tu \in A$. (See Theorem B.4.) But $|(a + tu) - c| = t_0 - t < r$, a contradiction. Hence $L \nsubseteq A \cup B$. $\qquad\square$

We now see that if a nonempty set $E \subseteq \mathbb{C}$ is such that the line segment connecting two arbitrary points in E lies in E, then E is connected. *For instance, all open and closed disks are connected, as are all lines, rays, and line segments.* This can be taken a step further. The proof of the following theorem and related results are considered in the exercises. See Figure 1.3.

1.4.14 Theorem. *Let $E \subseteq \mathbb{C}$ be nonempty. If for all $a, b \in E$, there are $a_0, \ldots, a_n \in E$ such that $a_0 = a$, $a_n = b$, and for each $k = 1, \ldots, n$, the line segment with endpoints a_{k-1} and a_k lies in E, then E is connected.*

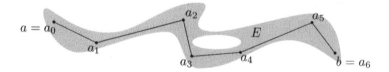

Figure 1.3   An illustration of Theorem 1.4.14 with $n = 6$

In our upcoming work, the most important connected sets are also open and warrant their own name.

1.4.15 Definition. A connected open subset of $\mathbb{C}$ is called a *domain*.

1.4.16 Theorem. *Let $\Omega \subseteq \mathbb{C}$ be a nonempty open set. Then Ω is a domain if and only if it is not equal to the union of disjoint nonempty open sets.*

Proof. It is equivalent to show that Ω is disconnected if and only if $\Omega = A \cup B$, where both A and B are open, nonempty, and $A \cap B = \varnothing$.

If Ω is disconnected, then let $\Omega = A \cup B$, where A and B are separated. Let $a \in A$. Since a is an exterior point of B, there is $r_1 > 0$ such that $D(a; r_1) \cap B = \varnothing$. Since $a \in \Omega$, there is $r_2 > 0$ such that $D(a; r_2) \subseteq \Omega$. If $r = \min\{r_1, r_2\} > 0$, then $D(a; r) \subseteq \Omega \setminus B = A$, showing A is open. A symmetric argument shows B is open.

Conversely, suppose $\Omega = A \cup B$ for disjoint nonempty open sets A, B. Every point of A is an exterior point of B and hence $A \cap \overline{B} = \varnothing$. Similarly, $\overline{A} \cap B = \varnothing$, showing A and B are separated. Thus Ω is disconnected. $\qquad\square$

We conclude this section with one more definition.

1.4.17 Definition. A maximal connected subset of a set $E \subseteq \mathbb{C}$ is called a *component* of E.

This means that $A \subseteq E$ is a component of E if A is connected and for any connected set $B \subseteq E$ such that $A \subseteq B$, it must be that $A = B$. It is an exercise to show that every point of E lies in a component of E (and hence components exist).

1.4.18 Example. We know that intervals in $\mathbb{R}$ are connected. If $E = (-1, 0) \cup (0, 1)$, then E is disconnected using the separated sets $A = (-1, 0)$ and $B = (0, 1)$. In fact, A and B are components of E.

One may easily decompose any set with at least two elements into the union of two nonempty disjoint subsets, showing the importance of $\overline{A}$ and $\overline{B}$ in Definition 1.4.12. Observing that $\overline{A} \cap \overline{B} = \{0\} \neq \varnothing$ in this example shows why only one closure is considered at a time.

1.4.19 Example. We conclude by analyzing a set with regard to all concepts introduced in this section. Filling in the details of the statements made is left as an exercise. Let

$$E = \{z \in \mathbb{C} : |\operatorname{Im} z| < |\operatorname{Re} z|\}.$$

(See Figure 1.4.) Each point $a \in E$ is an interior point of E, and hence E is open. Indeed, one may show that $D(a; r) \subseteq E$, where $r = (|\operatorname{Re} a| - |\operatorname{Im} a|)/2$, using the triangle inequality. Likewise, if $a \in \mathbb{C}$ is such that $|\operatorname{Re} a| < |\operatorname{Im} a|$, then $D(a; r) \subseteq \mathbb{C} \setminus E$ if $r = (|\operatorname{Im} a| - |\operatorname{Re} a|)/2$, showing a is an exterior point of E. We also have that ∂E consists of those $a \in \mathbb{C}$ for which $|\operatorname{Re} a| = |\operatorname{Im} a|$ since for such a and $r > 0$, at least one of $a \pm r/2$ lies in E and a lies in $\mathbb{C} \setminus E$. We conclude from this reasoning that the limit points of E are precisely the points in $\overline{E}$. Since $\partial E \nsubseteq E$, E is not closed. Moreover, E is not bounded, as $(0, \infty) \subseteq E$. Hence E fails both conditions required of compactness. Lastly, we note that $A = \{z \in E : \operatorname{Re} z < 0\}$ and $B = \{z \in E : \operatorname{Re} z > 0\}$ are connected because any pair of points in one set is connected by a line segment contained within that set. The above logic shows A and B are open, and hence E is disconnected by Theorem 1.4.16. It follows that A and B are connected components of E and are each domains.

Summary and Notes for Section 1.4.

We have defined basic topological concepts such as open, closed, compact, bounded, and connected sets in the complex plane $\mathbb{C}$.

The field of topology is vast and deep and comes in many flavors. Point set topology is the study of abstract topological spaces where open sets are defined by a set of axioms. Despite its importance, it is a relatively young area, only coming into its own in the early 20th century. This particular flavor of topology was motivated

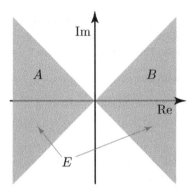

Figure 1.4   The set E and its components A and B from Example 1.4.19

by problems in abstract analysis. Indeed, John L. Kelley, in the preface to his classic book on general topology [14], wrote, "I have, with difficulty, been prevented by my friends from labeling [this book]: What Every Young Analyst Should Know."

Exercises for Section 1.4.

1. For each of the following sets E, determine whether E is open, closed, bounded, compact, connected, a domain. In addition, identify E°, ∂E, $\overline{E}$, the collection of limit points of E, and the components of E. Do not include proofs.

   (a) $E = \{z \in \mathbb{C} : \operatorname{Re} z > 0\}$

   (b) $E = \{z \in \mathbb{C} : 0 < |z| < 1\}$

   (c) $E = \{z \in \mathbb{C} : |\operatorname{Re} z| \leq 1,\ |z| > 2\}$

   (d) $E = \bigcup_{n \in \mathbb{Z}} D(n; 1/2)$

   (e) $E = \{z \in \mathbb{C} : \operatorname{Re} z \neq |z|\}$

   (f) $E = \{z \in \mathbb{C} : \operatorname{Re} z, \operatorname{Im} z \in \mathbb{Q}\}$

   (g) $E = \{1/n + i/m : n, m \in \mathbb{N}\}$

2. Justify the statements made in Example 1.4.19.

3. Let $a \in \mathbb{C}$ and $r > 0$. Show that the set of exterior points of the disk $D(a; r)$ is $\{z \in \mathbb{C} : |z - a| > r\}$.

4. Prove Theorem 1.4.4.

5. Prove the following for a set $E \subseteq \mathbb{C}$.

   (a) The set $\overline{E}$ is a closed set.

   (b) The set E is closed if and only if $E = \overline{E}$.

6. Let $A \subseteq \mathbb{C}$ and $a \in \mathbb{C}$. Prove that a is a limit point of A if and only if $D(a; r) \cap A$ is infinite for all $r > 0$.

7. Let $A \subseteq \mathbb{C}$. Show that if $B \subseteq A$, then $\overline{B} \subseteq \overline{A}$.

8. $\triangleright$ Prove that a set $A \subseteq \mathbb{C}$ is open if and only if for every $a \in A$, there is a closed disk $\overline{D}(a; r) \subseteq A$ for some $r > 0$.

9. Let $\mathcal{U} = \{U_\alpha : \alpha \in I\}$ be a collection of open subsets of $\mathbb{C}$. (Here, I is an index set. See Appendix A.)

    (a) Prove that $\bigcup_{\alpha \in I} U_\alpha$ is open.

    (b) Prove that $\bigcap_{\alpha \in I} U_\alpha$ is open if I is finite.

    (c) Give an example of an infinite collection $\mathcal{U}$ where the intersection in part (b) is not open.

10. Let $\mathcal{E} = \{E_\alpha : \alpha \in I\}$ be a collection of closed subsets of $\mathbb{C}$, where I is an index set.

    (a) Prove that $\bigcap_{\alpha \in I} E_\alpha$ is closed.

    (b) Prove that $\bigcup_{\alpha \in I} E_\alpha$ is closed if I is finite.

    (c) Give an example of an infinite collection $\mathcal{E}$ where the union in part (b) is not closed.

11. Find an example of a disconnected set whose closure is connected.

12. Suppose that $E, F \subseteq \mathbb{C}$ are connected. Is $E \cap F$ necessarily connected if $E \cap F \neq \varnothing$? Provide a proof or a counterexample.

13. Use the following steps to show that if $E \subseteq \mathbb{C}$, then each $a \in E$ lies in a component of E.

    (a) Suppose I is an index set and $E_\alpha \subseteq \mathbb{C}$ is connected for each $\alpha \in I$. Show that if $\bigcap_{\alpha \in I} E_\alpha \neq \varnothing$, then $\bigcup_{\alpha \in I} E_\alpha$ is connected.

    (b) Let F be the union of all connected subsets of E containing a. Show that F is a component.

14. Show that every component of an open set is a domain. (*Hint*: Part (a) of Exercise 13 is helpful.)

15. Which subsets of $\mathbb{C}$ are both open and closed? Prove your answer.

16. Prove Theorem 1.4.14. Show that the converse holds if E is open using the following strategy: Let $a \in E$ and A consist of all $b \in E$ such that a and b are connected by a sequence of line segments as in the statement of Theorem 1.4.14. Show A and $E \setminus A$ are open.

17. Let $U \subseteq \mathbb{C}$ be a nonempty open set. Prove that U has an *exhaustion* by compact sets. That is, show that there are compact sets $K_n \subseteq U$ for each $n \in \mathbb{N}$ such that $K_n \subseteq K_{n+1}$ for every n and $U = \bigcup_{n=1}^\infty K_n$.

18. In this exercise, we consider the abstract topological definition of compactness. Let $K \subseteq \mathbb{C}$. An *open cover* of K is a collection $\mathcal{U} = \{U_\alpha : \alpha \in I\}$ of open subsets of $\mathbb{C}$ such that $K \subseteq \bigcup_{\alpha \in I} U_\alpha$. A subcollection $\{U_{\alpha_1}, \ldots, U_{\alpha_n}\} \subseteq \mathcal{U}$ is said to be a *finite subcover* of K if $K \subseteq \bigcup_{k=1}^n U_{\alpha_k}$. The *Heine–Borel theorem* states that a set K is compact if and only if every open cover of K contains a finite subcover of K. Prove the Heine–Borel theorem using the following steps.

    (a) If every open cover of K contains a finite subcover of K, show that K is bounded.

    (b) If every open cover of K contains a finite subcover of K, show that K is closed.

    (c) Let $S \subseteq \mathbb{C}$ be a closed square with sides parallel to the axes. Show that every open cover of S contains a finite subcover of S. (*Hint*: Look ahead to the subdivision argument in the proof of the Bolzano–Weierstrass theorem [Theorem 1.6.14] and use something similar in a proof by contradiction.)

(d) Show that if K is compact, then every open cover of K contains a finite subcover of K by choosing a square S that contains K and extending the open cover of K to an open cover of S by adding the set $\mathbb{C} \setminus K$.

1.5 The Extended Complex Plane

In analysis, we frequently need to deal with infinite limits. Even in calculus, one sees limits approaching $\pm\infty$. Unfortunately, there is a multitude of directions in $\mathbb{C}$ in which a limit can approach "infinity." To escape this problem, we use a clever topological identification that will relate the complex plane to a sphere.

As is learned in multivariable calculus, $\mathbb{R}^3$ is the set (space) of all points of the form (x_1, x_2, x_3), where $x_1, x_2, x_3 \in \mathbb{R}$. The *unit sphere* in $\mathbb{R}^3$ is the set of all points of distance 1 from the origin; in other words, it is the set

$$S = \{(x_1, x_2, x_3) \in \mathbb{R}^3 : x_1^2 + x_2^2 + x_3^2 = 1\}.$$

We put the elements of $\mathbb{C}$ into a one-to-one correspondence with the elements of S with the exception of the "north pole" $N = (0, 0, 1)$. To do so, we first associate the plane $\mathbb{C}$ to the plane in $\mathbb{R}^3$ described by the equation $x_3 = 0$. This is done in the natural way by writing $z \in \mathbb{C}$ as $x_1 + ix_2$ and identifying this with the point $(x_1, x_2, 0)$. If we let L be the line in $\mathbb{R}^3$ through the points $(x_1, x_2, 0)$ and N, then $L \cap S \setminus \{N\}$ contains exactly one point, which we call Z. This establishes the desired one-to-one correspondence between $z \in \mathbb{C}$ and $Z \in S \setminus \{N\}$. Topologically, this correspondence "wraps" the plane onto the sphere, leaving only the north pole uncovered. Note that $\mathbb{D}$ is sent to the "lower hemisphere," $\partial\mathbb{D}$ is fixed, and $\mathbb{C} \setminus \overline{\mathbb{D}}$ is sent to the "upper hemisphere." See Figure 1.5.

1.5.1 Definition. The one-to-one correspondence described above is called the *stereographic projection* of $\mathbb{C}$ onto $S \setminus \{N\}$.

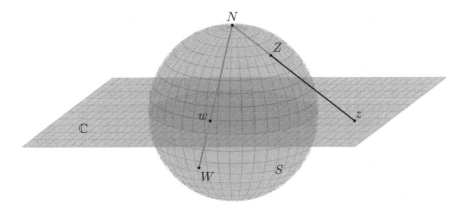

Figure 1.5   Stereographic projection of $z \in \mathbb{C} \setminus \overline{\mathbb{D}}$ onto Z and $w \in \mathbb{D}$ onto W

As the north pole is the lone point on the sphere left uncovered by the stereo-graphic projection, it is natural to identify this point with ∞.

1.5.2 Definition. The *extended complex plane* is the set $\mathbb{C}_\infty = \mathbb{C} \cup \{\infty\}$, where ∞ inherits all of the natural properties from the north pole of the sphere through the stereographic projection. Because of this correspondence, the extended complex plane is often called the *Riemann sphere.*

This method of adding the point ∞ to $\mathbb{C}$ is alternatively referred to as the *one-point compactification of* $\mathbb{C}$.

Summary and Notes for Section 1.5.

Through the stereographic projection, we can identify the points of the plane $\mathbb{C}$ with the points of a sphere without its "north pole." This then allows the north pole to be identified with ∞. Unlike on the real line, where we consider signed infinities, the complex ∞ is a single geometric point, and the full sphere is called the extended complex plane, denoted by $\mathbb{C}_\infty$.

The stereographic projection is certainly not a new idea, and it was not motivated by complex analysis. Ancient Greeks knew of the projection, and it was used for centuries for making both celestial and geographic maps.

Exercises for Section 1.5.

1. Which sets in S correspond to the real and imaginary axes in $\mathbb{C}$ under the stereographic projection?

2. Let $z = x + iy \in \mathbb{C}$ be given. Calculate the coordinates of the point $Z = (x_1, x_2, x_3) \in S$ corresponding to z under the stereographic projection.

3. Let $Z = (x_1, x_2, x_3) \in S$ be given. Calculate the complex number z corresponding to Z under the stereographic projection.

4. For any $z, w \in \mathbb{C}_\infty$ define the *spherical distance* between z and w, denoted $d(z, w)$, to be the distance in $\mathbb{R}^3$ between the points $Z, W \in S$ corresponding to z and w under the stereographic projection. Use Exercises 2 and 3 to prove the following.

   (a) If $z, w \in \mathbb{C}$, then
   $$d(z, w) = \frac{2|z - w|}{\sqrt{(1 + |z|^2)(1 + |w|^2)}}.$$

   (b) If $z \in \mathbb{C}$, then
   $$d(z, \infty) = \frac{2}{\sqrt{1 + |z|^2}}.$$

5. $\triangleright$ Show that circles on S correspond to circles and lines in $\mathbb{C}$. (*Hint:* Recall the following facts from analytic geometry: Every circle in the xy-plane can be expressed by an equation of the form $x^2 + y^2 + ax + by + c = 0$ for some $a, b, c \in \mathbb{R}$. Every circle on S is the intersection of S with a plane in $\mathbb{R}^3$ described by the equation $Ax_1 + Bx_2 + Cx_3 = D$ for some $A, B, C, D \in \mathbb{R}$. Then use Exercise 2.)

Figure 1.6 $z_n \to a$

1.6   Complex Sequences

Sequences are a fundamental tool in analysis for the purpose of approximation. They are closely related to the topology of $\mathbb{C}$ and essential to the study of series, which lie at the core of function theory.

1.6.1 Definition. A function of the form $z \colon \{m, m+1, \ldots\} \to \mathbb{C}$ for some $m \in \mathbb{Z}$ is called a *sequence* of complex numbers.

The function value $z(n)$ for some $n \geq m$ is denoted by z_n. This allows a sequence to be described, in a more familiar manner, by a list $\{z_m, z_{m+1}, \ldots\}$ or $\{z_n\}_{n=m}^{\infty}$. The subscripts are the *indices* of the sequence. Although indices can begin at any integer m, we will typically consider sequences with $m = 1$ for simplicity. This does not affect generality, however, since any sequence can be *reindexed*. In other words, if a sequence begins at $n = m$ for some $m \in \mathbb{Z}$, then we could replace each index n by $n - m + 1$ so that the sequence begins with the index $n = 1$ or vice versa. In most circumstances, such a maneuver is only used to provide some sort of algebraic simplification.

1.6.2 Definition. A sequence $\{z_n\}_{n=1}^{\infty}$ of complex numbers *converges* to a number $a \in \mathbb{C}$, called the *limit* of $\{z_n\}$, provided that for any $\varepsilon > 0$, there is some $N \in \mathbb{N}$ such that $|z_n - a| < \varepsilon$ whenever $n \geq N$. A sequence *diverges* if it fails to converge.

Convergence is typically denoted by writing $z_n \to a$ as $n \to \infty$ (or just $z_n \to a$ when the context is clear) or by the expression

$$\lim_{n \to \infty} z_n = a. \tag{1.6.1}$$

Geometrically, $z_n \to a$ if for any $\varepsilon > 0$, there exists $N \in \mathbb{N}$ such that for all $n \geq N$, $z_n \in D(a; \varepsilon)$. See Figure 1.6.

Because of our understanding of ∞ from Section 1.5, we can deal simply with infinite limits, a special type of divergence.

1.6.3 Definition. The sequence $\{z_n\}_{n=1}^{\infty}$ of complex numbers *diverges to* ∞, denoted by $z_n \to \infty$ (as $n \to \infty$) or $\lim_{n \to \infty} z_n = \infty$, if given any $R > 0$, there is $N \in \mathbb{N}$ such that $|z_n| > R$ for all $n \geq N$.

Exercise 4 of Section 1.5 can be used to show that the set $\{z \in \mathbb{C} : |z| > R\}$ corresponds to the set of points on the sphere S of distance less than $\varepsilon = 2/\sqrt{1 + R^2}$

from the north pole N under the stereographic projection, and therefore this definition of an infinite limit naturally corresponds to a sequence on S converging to N.

We now consider some intuitive properties of sequences.

1.6.4 Theorem. *A convergent sequence has a unique limit.*

Proof. Suppose that $\{z_n\}_{n=1}^{\infty}$ converges to both a and b in $\mathbb{C}$. If $a \neq b$, then let $\varepsilon = |a-b|/2 > 0$. For some $N_1, N_2 \in \mathbb{N}$, $n \geq N_1$ implies $|z_n - a| < \varepsilon$ and $n \geq N_2$ implies $|z_n - b| < \varepsilon$. But if $n \geq \max\{N_1, N_2\}$, then by the triangle inequality,

$$2\varepsilon = |a - b| = |a - z_n + z_n - b| \leq |a - z_n| + |z_n - b| < 2\varepsilon,$$

a contradiction. Thus $a = b$. $\qquad\qquad\qquad\qquad\qquad\qquad\qquad\qquad\qquad\square$

1.6.5 Definition. A sequence $\{z_n\}_{n=1}^{\infty}$ of complex numbers is *bounded* if there exists $R > 0$ such that $|z_n| \leq R$ for all $n \in \mathbb{N}$.

1.6.6 Theorem. *If a sequence converges, then it is bounded.*

Proof. Suppose that the sequence $\{z_n\}_{n=1}^{\infty}$ of complex numbers converges to some $a \in \mathbb{C}$. Then for some $N \in \mathbb{N}$, $|z_n - a| < 1$ whenever $n \geq N$. Fix

$$R = \max\{|z_1|, \ldots, |z_N|, 1 + |a|\}.$$

Clearly, if $n \leq N$, then $|z_n| \leq R$. If $n > N$, then by the triangle inequality,

$$|z_n| = |z_n - a + a| \leq |z_n - a| + |a| < 1 + |a| \leq R.$$

Hence the sequence is bounded. $\qquad\qquad\qquad\qquad\qquad\qquad\qquad\qquad\quad\square$

Sequences of real numbers are familiar from calculus. The following theorem shows how our study of complex sequences can rely on their real counterparts.

1.6.7 Theorem. *Let $\{z_n\}_{n=1}^{\infty}$ be a sequence of complex numbers. For each $n \in \mathbb{N}$, write $z_n = x_n + iy_n$. Then $\{z_n\}$ converges if and only if both $\{x_n\}$ and $\{y_n\}$ converge. In this case,*

$$\lim_{n\to\infty} z_n = \lim_{n\to\infty} x_n + i \lim_{n\to\infty} y_n. \tag{1.6.2}$$

Proof. Suppose $x_n \to a$ and $y_n \to b$ for some $a, b \in \mathbb{R}$. Let $\varepsilon > 0$. There are $N_1, N_2 \in \mathbb{N}$ such that $|x_n - a| < \varepsilon/2$ for all $n \geq N_1$ and $|y_n - b| < \varepsilon/2$ for all $n \geq N_2$. If $N = \max\{N_1, N_2\}$, then for all $n \geq N$,

$$|z_n - (a + ib)| = |x_n - a + i(y_n - b)| \leq |x_n - a| + |y_n - b| < \frac{\varepsilon}{2} + \frac{\varepsilon}{2} = \varepsilon$$

using the triangle inequality. This shows $\{z_n\}$ converges to $a + ib$, verifying (1.6.2).

Conversely, suppose $z_n \to c$ for some $c \in \mathbb{C}$. Let $\varepsilon > 0$. There is $N \in \mathbb{N}$ such that $|z_n - c| < \varepsilon$ for all $n \geq N$. But for such n,

$$|x_n - \operatorname{Re} c| = |\operatorname{Re}(z_n - c)| \leq |z_n - c| < \varepsilon$$

using Exercise 4 from Section 1.2. Hence $\{x_n\}$ converges. A similar argument gives that $\{y_n\}$ converges. $\qquad\square$

We know from calculus that convergent real sequences satisfy the following algebra rules. The complex versions can be proved by resorting to real and imaginary parts and using Theorem 1.6.7 and are left as exercises.

1.6.8 Theorem. *Suppose that $\{z_n\}_{n=1}^{\infty}$ and $\{w_n\}_{n=1}^{\infty}$ are sequences of complex numbers such that $z_n \to a$ and $w_n \to b$ for some $a, b \in \mathbb{C}$ as $n \to \infty$. Furthermore, let $c \in \mathbb{C}$. Then*

(a) $\lim_{n\to\infty} c z_n = ca$,

(b) $\lim_{n\to\infty}(z_n + w_n) = a + b$,

(c) $\lim_{n\to\infty} z_n w_n = ab$, and

(d) $\lim_{n\to\infty} z_n/w_n = a/b$ *if $w_n \neq 0$ for all $n \in \mathbb{N}$ and $b \neq 0$.*

From calculus, we have techniques, such as l'Hôpital's rule, for dealing with limits of real sequences. Theorem 1.6.7 gives us one route to apply these real techniques to find limits of complex sequences. Another method, relying on the modulus, is the following.

1.6.9 Theorem. *Let $\{z_n\}_{n=1}^{\infty}$ be a sequence of complex numbers, $\{c_n\}_{n=1}^{\infty}$ be a sequence of nonnegative real numbers, and $a \in \mathbb{C}$.*

(a) *If $c_n \to 0$ and $|z_n - a| \leq c_n$ for all $n \in \mathbb{N}$, then $z_n \to a$.*

(b) *If $c_n \to \infty$ and $|z_n| \geq c_n$ for all $n \in \mathbb{N}$, then $z_n \to \infty$.*

We leave the proof of this theorem as an exercise but consider the following example of its helpfulness.

1.6.10 Example. Consider the sequence of complex numbers

$$\left\{ \frac{n+i}{(1-2i)^n} \right\}_{n=1}^{\infty}.$$

By the triangle inequality, we have

$$\left| \frac{n+i}{(1-2i)^n} \right| \leq \frac{n+1}{|1-2i|^n} = \frac{n+1}{5^{n/2}}.$$

An application of l'Hôpital's rule shows that the right-hand side converges to 0. Therefore our sequence converges to 0 by Theorem 1.6.9. Note that we do not have a version of l'Hôpital's rule for complex sequences to directly apply in these circumstances. (Nor will we.)

1.6.11 Definition. Let $\{z_n\}_{n=1}^{\infty}$ be a sequence of complex numbers and $\{n_k\}_{k=1}^{\infty}$ be a strictly increasing sequence of positive integers. The sequence $\{z_{n_k}\}_{k=1}^{\infty}$ is called a *subsequence* of $\{z_n\}$.

In looser terms, a subsequence of a sequence is a sequence formed by taking terms of the original sequence, in order.

1.6.12 Theorem. *A sequence $\{z_n\}_{n=1}^{\infty}$ of complex numbers converges to a number $a \in \mathbb{C}$ if and only if every subsequence of $\{z_n\}$ converges to a.*

Proof. Suppose that $z_n \to a$ as $n \to \infty$ and that $\{z_{n_k}\}_{k=1}^{\infty}$ is a subsequence of $\{z_n\}$. Then for any $\varepsilon > 0$, there exists $N \in \mathbb{N}$ such that $|z_n - a| < \varepsilon$ whenever $n \geq N$. Since $\{n_k\}_{k=1}^{\infty}$ is strictly increasing, $n_k \geq k$ for all $k \in \mathbb{N}$. Therefore $|z_{n_k} - a| < \varepsilon$ whenever $k \geq N$, and hence $z_{n_k} \to a$ as $k \to \infty$.

The converse follows because a sequence is a subsequence of itself. $\square$

We now present some vital connections between sequences and the topology of $\mathbb{C}$. The first classifies closed sets in terms of sequences.

1.6.13 Theorem. *A set $E \subseteq \mathbb{C}$ is closed if and only if every convergent sequence of elements of E has its limit in E.*

Proof. Suppose that E is closed. Let $\{z_n\}_{n=1}^{\infty}$ be a sequence of elements of E that converges to some $a \in \mathbb{C}$. If $a \notin E$, then a is an exterior point of E. There exists $\varepsilon > 0$ such that $D(a; \varepsilon) \cap E = \varnothing$. It follows that $|z_n - a| \geq \varepsilon$ for all $n \in \mathbb{N}$, a contradiction. Thus $a \in E$.

Conversely, suppose that every convergent sequence in E converges to a point in E. Let $a \in \partial E$. Then for each $n \in \mathbb{N}$, there exists $z_n \in D(a; 1/n) \cap E$ by Theorem 1.4.4. Since $|z_n - a| < 1/n$ for all $n \in \mathbb{N}$, $z_n \to a$ by Theorem 1.6.9. Therefore $a \in E$, showing E is closed. $\square$

We now come to an important theorem that relates compactness to sequences. Its power is in the impact that a set being closed and bounded has on sequences within the set.

1.6.14 Bolzano–Weierstrass Theorem. *A set $K \subseteq \mathbb{C}$ is compact if and only if every sequence of elements of K has a subsequence that converges to an element of K.*

Proof. Suppose that K is compact, and hence bounded, and let $\{z_n\}_{n=1}^{\infty}$ be a sequence in K. There is a square $S = \{z \in \mathbb{C} : a \leq \operatorname{Re} z \leq b, \ c \leq \operatorname{Im} z \leq d\}$, where $a, b, c, d \in \mathbb{R}$ and $b - a = d - c = \alpha$ for some $\alpha > 0$, such that $K \subseteq S$. If S is divided symmetrically into four closed subsquares, then (at least) one subsquare contains infinitely many terms of $\{z_n\}$. Call this subsquare S_1, and let $n_1 \in \mathbb{N}$ be such that $z_{n_1} \in S_1$. Note that $S_1 = \{z \in \mathbb{C} : a_1 \leq \operatorname{Re} z \leq b_1, \ c_1 \leq \operatorname{Im} z \leq d_1\}$, where $b_1 - a_1 = d_1 - c_1 = \alpha/2$.

Now continue this process inductively to generate a collection of closed squares $S_k = \{z \in \mathbb{C} : a_k \leq \operatorname{Re} z \leq b_k, \ c_k \leq \operatorname{Im} z \leq d_k\}$, where $b_k - a_k = d_k - c_k = \alpha/2^k$, $S_k \subseteq S_{k-1}$ for all $k \geq 2$, and each S_k contains infinitely many terms of $\{z_n\}$. (See Figure 1.7.) At each step, we choose $n_k \in \mathbb{N}$ such that $n_k > n_{k-1}$ and $z_{n_k} \in S_k$. Evidently, $a_k < b_j$ for all $j, k \in \mathbb{N}$. Set $x_0 = \sup\{a_k : k \in \mathbb{N}\}$. Then $a_k \leq x_0 \leq b_k$ for all k. Likewise, set $y_0 = \sup\{c_k : k \in \mathbb{N}\}$ so that $c_k \leq y_0 \leq d_k$ for all k. If $z_0 = x_0 + iy_0$, then $z_0 \in S_k$ for all k.

Let $\varepsilon > 0$. Choose $N \in \mathbb{N}$ such that $\alpha/2^N < \varepsilon/\sqrt{2}$. Then the distance between any two points in S_N is less than ε, proving $S_N \subseteq D(z_0; \varepsilon)$. It immediately follows that $|z_{n_k} - z_0| < \varepsilon$ for all $k \geq N$, showing $z_{n_k} \to z_0$. Because $\{z_{n_k}\}_{k=1}^{\infty} \subseteq K$ and K is closed, $z_0 \in K$ by Theorem 1.6.13. Thus $\{z_n\}$ has a subsequence that converges to an element of K.

Conversely, if every sequence of members of K has a subsequence converging to an element of K, then every convergent sequence of elements of K must converge to a member of K by Theorem 1.6.12. Thus K is closed by Theorem 1.6.13. Were K unbounded, for each $n \in \mathbb{N}$, there would exist $z_n \in K$ such that $|z_n| \geq n$. But any subsequence of $\{z_n\}_{n=1}^{\infty}$ would be unbounded and hence divergent by Theorem 1.6.6. Therefore K is compact. □

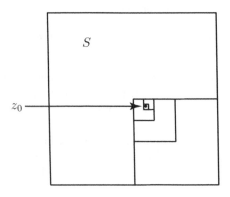

Figure 1.7   Possible inductive steps in the proof of Theorem 1.6.14

We immediately use the sequential notion of compactness to prove the following valuable theorem.

1.6.15 Theorem. *Suppose that for each $n \in \mathbb{N}$, $K_n \subseteq \mathbb{C}$ is a nonempty compact set and that $K_{n+1} \subseteq K_n$ for all n. Then*

$$\bigcap_{n=1}^{\infty} K_n \neq \varnothing.$$

Proof. For each $n \in \mathbb{N}$, let $z_n \in K_n$. Then $\{z_n\}_{n=1}^{\infty}$ is a sequence of elements of K_1 and hence there is a subsequence $\{z_{n_k}\}_{k=1}^{\infty}$ that converges to some $a \in K_1$ by the Bolzano–Weierstrass theorem.

Let $n \in \mathbb{N}$. Since $n_k \geq n$ when $k \geq n$, $\{z_{n_k}\}_{k=n}^{\infty}$ is a sequence in K_n that converges to a. Because K_n is closed, $a \in K_n$ by Theorem 1.6.13. But n was arbitrarily chosen, so $a \in \bigcap_{n=1}^{\infty} K_n$. □

We conclude this section with some remarks about Cauchy sequences.

1.6.16 Definition. A sequence $\{z_n\}_{n=1}^{\infty}$ of complex numbers is a *Cauchy sequence* if for every $\varepsilon > 0$, there is some $N \in \mathbb{N}$ such that $|z_n - z_m| < \varepsilon$ whenever $m, n \geq N$.

That every Cauchy sequence of real numbers is convergent is a characteristic called the completeness of $\mathbb{R}$. We have a similar result in $\mathbb{C}$. Its proof is left as an exercise.

1.6.17 Completeness of the Complex Numbers. *A sequence of complex numbers converges if and only if it is a Cauchy sequence.*

Summary and Notes for Section 1.6.

Sequences of complex numbers and their limits are defined in exactly the same way as their real counterparts are in calculus, and the algebra of convergent sequences holds as we would expect, as seen by considering the real and imaginary parts of their terms. Sequences are ubiquitous in analysis; they are a basic tool for developing complicated ideas.

Closed sets and compact sets can be characterized using sequences which makes these sets easier to study. In general topological spaces, these characterizations do not hold, which leads to calling certain sets sequentially closed or sequentially compact.

Exercises for Section 1.6.

1. Find limits of the following sequences, or explain why they diverge.

   (a) $\left\{ \dfrac{i2^n - n}{(3 + i)^n} \right\}_{n=1}^{\infty}$

   (b) $\left\{ \dfrac{n + i^n}{n} \right\}_{n=1}^{\infty}$

   (c) $\left\{ \dfrac{1 - in^2}{n(n + 1)} \right\}_{n=1}^{\infty}$

   (d) $\left\{ \dfrac{(1 + i)^n}{n} \right\}_{n=1}^{\infty}$

2. Prove Theorem 1.6.8 by using the real and imaginary parts of $\{z_n\}$ and $\{w_n\}$ and Theorem 1.6.7.

3. Prove Theorem 1.6.8 using Definition 1.6.2. Theorem 1.6.6 is helpful.

4. Prove Theorem 1.6.9.

5. ▷ Let $\{z_n\}_{n=1}^{\infty}$ be a sequence of complex numbers. Show that if $z_n \to a$ for some $a \in \mathbb{C}$, then $|z_n| \to |a|$.

6. A sequence $\{x_n\}_{n=1}^{\infty}$ of real numbers is *monotone* if it is either increasing or decreasing. The *monotone convergence theorem* from calculus states that a bounded monotone sequence must converge. Prove it.

7. ▷ Let $a \in \mathbb{C}$.

   (a) Prove that if $|a| < 1$, then $a^n \to 0$.

   (b) Prove that if $|a| > 1$, then $a^n \to \infty$.

8. Let $\{z_n\}_{n=1}^{\infty}$ and $\{w_n\}_{n=1}^{\infty}$ be sequences of complex numbers.

(a) Show that if $\{z_n\}$ is bounded and $w_n \to \infty$, then $(z_n + w_n) \to \infty$.

(b) If $z_n \to \infty$ and $w_n \to \infty$, does it follow that $(z_n + w_n) \to \infty$?

9. ▷ Let $\{x_n\}_{n=1}^{\infty}$ and $\{y_n\}_{n=1}^{\infty}$ be convergent sequences of real numbers such that $x_n \leq y_n$ for all $n \in \mathbb{N}$. Show that

$$\lim_{n \to \infty} x_n \leq \lim_{n \to \infty} y_n.$$

10. Let $\{z_n\}_{n=1}^{\infty}$ and $\{w_n\}_{n=1}^{\infty}$ be sequences in $\mathbb{C} \setminus \{0\}$. If $\{1/z_n\}$ is bounded and $w_n \to 0$, show that $z_n/w_n \to \infty$.

11. ▷ Let $A \subseteq \mathbb{C}$ and $a \in \mathbb{C}$. Prove that $a \in \overline{A}$ if and only if there is a sequence of elements of A converging to a.

12. Let $\{z_n\}_{n=1}^{\infty}$ be a sequence of complex numbers, and suppose that a is a limit point of $\{z_n : n \in \mathbb{N}\}$. Show that there is a subsequence of $\{z_n\}$ converging to a.

13. ▷ Let $a \in \mathbb{C}$ and $r > 0$. If $K \subseteq D(a; r)$ is compact, prove that there is some $\rho \in (0, r)$ such that $K \subseteq D(a; \rho)$.

14. Prove Theorem 1.6.17. (*Hint:* To show a Cauchy sequence converges, first show it is bounded, and then turn to the Bolzano–Weierstrass theorem.)

15. Let $\{z_n\}_{n=1}^{\infty}$ be a sequence of complex numbers, and let $z \in \mathbb{C}_{\infty}$. Prove that $z_n \to z$ as $n \to \infty$ if and only if $Z_n \to Z$ on the Riemann sphere S, where Z_n, Z correspond to z_n, z under the stereographic projection.

1.7 Complex Series

A series of complex numbers can intuitively be thought of as an infinite sum, but, in actuality, is a special kind of sequence. Our approach to studying function theory will rest squarely on series.

1.7.1 Definition. Let $m \in \mathbb{Z}$ and $\{z_n\}_{n=m}^{\infty}$ be a sequence of complex numbers. We define the *partial sums* of $\{z_n\}$ to be the complex numbers

$$s_N = \sum_{n=m}^{N} z_n = z_m + \cdots + z_N, \tag{1.7.1}$$

for each $N \in \mathbb{Z}$ such that $N \geq m$.

The partial sum sequence $\{s_N\}_{N=m}^{\infty}$ is well defined for any complex sequence.

1.7.2 Definition. If the sequence of partial sums in Definition 1.7.1 converges to some $s \in \mathbb{C}$, then we say that the *infinite series* $\sum_{n=m}^{\infty} z_n$ with *terms* $\{z_n\}_{n=m}^{\infty}$ *converges* to s. This is denoted by

$$s = \sum_{n=m}^{\infty} z_n = z_m + z_{m+1} + \cdots, \tag{1.7.2}$$

and we accordingly also refer to s as the *sum* of the series. Any series that fails to converge is said to *diverge*. If $s_N \to \infty$ as $N \to \infty$, then we say that the series *diverges to* ∞. This special type of divergence is written

$$\sum_{n=m}^{\infty} z_n = \infty.$$

Many series we consider will have beginning index $m = 1$ or $m = 0$. As with sequences, series can be reindexed, so there is no loss of generality in proving results for these specific cases.

The first theorem is a direct result of Theorem 1.6.4.

1.7.3 Theorem. *A convergent series has a unique sum.*

The next theorem establishes some rules concerning the algebra of series. The proof follows from the properties of the sequence of partial sums and Theorem 1.6.8 and is left as an exercise.

1.7.4 Theorem. *Let $\sum_{n=1}^{\infty} z_n$ and $\sum_{n=1}^{\infty} w_n$ be infinite series of complex numbers converging to a and b, respectively, and let $c \in \mathbb{C}$. Then*

(a) $\sum_{n=1}^{\infty} cz_n = ca$ *and*

(b) $\sum_{n=1}^{\infty}(z_n + w_n) = a + b$.

We refer the reader to Appendix B where the concept of a product of two series is considered.

The central problem when dealing with series is the determination of convergence. Often, we are only concerned with *whether or not* a series converges, not *to what* the series converges. The following is the first and simplest test. Its proof is left as an exercise.

1.7.5 Theorem. *Let $\sum_{n=1}^{\infty} z_n$ be a convergent series of complex numbers. Then $z_n \to 0$ as $n \to \infty$.*

The usefulness of the preceding theorem lies in the contrapositive. Divergence of a series can be detected by observing that its sequence of terms fails to converge to 0. The converse is not addressed; nothing is said about the behavior of the series when its terms do converge to 0.

1.7.6 Example. Consider the series $\sum_{n=0}^{\infty} z^n$, and take note of Exercise 7 in Section 1.6. (We adopt the traditional convention that $z^0 = 1$ for all $z \in \mathbb{C}$.) If $|z| \geq 1$, then $\{z^n\}$ fails to converge to 0, and hence the series is divergent by Theorem 1.7.5.

For $z \in \mathbb{D}$, consider the partial sums $s_N = \sum_{n=0}^{N} z^n$. Observe that

$$(1 - z)s_N = \sum_{n=0}^{N} z^n - \sum_{n=1}^{N+1} z^n = 1 - z^{N+1}.$$

Therefore

$$s_N = \frac{1 - z^{N+1}}{1 - z}.$$

Since $z^{N+1} \to 0$ as $N \to \infty$, it follows that

$$\sum_{n=0}^{\infty} z^n = \frac{1}{1 - z} \tag{1.7.3}$$

for $z \in \mathbb{D}$.

This series is called the *geometric series*. Its elementary nature should not be underestimated; we will find it to be a useful tool on several upcoming occasions.

Detecting convergence of a series is usually a more difficult task than what we faced in Example 1.7.6. Here is a start. Its proof is standard from calculus and is left as an exercise.

1.7.7 Comparison Test. *Suppose that $\{a_n\}_{n=1}^{\infty}$ and $\{b_n\}_{n=1}^{\infty}$ are sequences of nonnegative real numbers such that $a_n \le b_n$ for all $n \in \mathbb{N}$.*

(a) *If $\sum_{n=1}^{\infty} b_n$ converges, then $\sum_{n=1}^{\infty} a_n$ converges and*

$$\sum_{n=1}^{\infty} a_n \le \sum_{n=1}^{\infty} b_n.$$

(b) *If $\sum_{n=1}^{\infty} a_n$ diverges, then $\sum_{n=1}^{\infty} b_n$ diverges.*

The comparison test only applies to series of nonnegative real numbers. This seems to address only a slight number of the series that we are likely to consider in complex analysis! The following concept helps to make a connection.

1.7.8 Definition. The series $\sum_{n=1}^{\infty} z_n$ of complex numbers is said to be *absolutely convergent* if the series $\sum_{n=1}^{\infty} |z_n|$ is convergent.

The phrase "absolutely convergent" seems to imply convergence. That is no accident.

1.7.9 Theorem. *If the series $\sum_{n=1}^{\infty} z_n$ of complex numbers is absolutely convergent, then it is convergent and*

$$\left| \sum_{n-1}^{\infty} z_n \right| \le \sum_{n=1}^{\infty} |z_n|. \tag{1.7.4}$$

Proof. Let $\varepsilon > 0$. Absolute convergence implies that there is $N \in \mathbb{N}$ such that

$$\left| \sum_{n=1}^{N} |z_n| - \sum_{n=1}^{\infty} |z_n| \right| = \sum_{n=N+1}^{\infty} |z_n| < \varepsilon.$$

Let $\{s_n\}$ be the sequence of partial sums of the series $\sum_{k=1}^{\infty} z_k$. Then for any $n, m \geq N$ with $m < n$,

$$|s_n - s_m| = \left| \sum_{k=m+1}^{n} z_k \right| \leq \sum_{k=m+1}^{n} |z_k| < \varepsilon,$$

due (inductively) to the triangle inequality. This shows that $\{s_n\}$ is a Cauchy sequence, and is thus convergent. Hence $\sum_{n=1}^{\infty} z_n$ is convergent.

Now that we know that both sides of the (triangle) inequality

$$\left| \sum_{k=1}^{n} z_k \right| \leq \sum_{k=1}^{n} |z_k|$$

converge as $n \to \infty$, taking this limit and using limit inequalities gives (1.7.4). See Exercises 5 and 9 of Section 1.6 for the properties of sequences used here. $\square$

1.7.10 Example. Consider the complex series

$$\sum_{n=0}^{\infty} \frac{(2i)^n - 1}{3^n + 2} = -\frac{1}{2} - \frac{1 - 2i}{5} - \frac{5}{11} + \cdots.$$

For each n, the triangle inequality gives

$$\left| \frac{(2i)^n - 1}{3^n + 2} \right| \leq \frac{2^n + 1}{3^n + 2} \leq \frac{2^n + 2^n}{3^n} = 2 \left(\frac{2}{3} \right)^n.$$

The series $\sum_{n=0}^{\infty} 2(2/3)^n$ is seen to converge using the geometric series and Theorem 1.7.4. Therefore the given series converges absolutely by the comparison test.

1.7.11 Definition. Any series that is convergent but not absolutely convergent is called *conditionally convergent*.

While distinguishing between those series that converge absolutely and those that converge conditionally is a prevalent theme in calculus, absolute convergence will be our primary need. Many tests for convergence have been purposely postponed until the discussion of power series in Section 2.3. This will be the setting of most interest to us. The following exercises are more focused on developing the algebraic aspects of series, rather than addressing the convergence of specific series.

Summary and Notes for Section 1.7.

An infinite series of complex numbers is nothing more than a special type of sequence, the sequence of partial sums. We casually denote the sum (limit) of the series by the series itself which is terribly convenient and causes no confusion. Absolute convergence allows for comparison to series of nonnegative terms and their convergence tests learned in calculus. We will see that infinite series, in particular geometric series, are of essential importance in the study of function theory.

Infinite series (and sequences) have been considered for millennia, dating back to the ancient Greeks. The concept of convergence was historically quite fluid until the 19th century efforts to add rigor to analysis. Indeed, for centuries, infinite series were treated in a casual manner that seems a bit sloppy by today's standards.

Exercises for Section 1.7.

1. Determine the sum of the series

$$\sum_{n=0}^{\infty} \frac{1}{(n+i)(n+1+i)}.$$

(*Hint*: Use partial fractions and the sequence of partial sums.)

2. For each of the following series, list the first three terms of the series and determine whether or not the series converges. Can you find the sum of some of them?

(a) $\displaystyle\sum_{n=1}^{\infty} \frac{(1-2i)^{n+2}}{(4+i)^n}$

(b) $\displaystyle\sum_{n=1}^{\infty} \frac{(3-i)^n}{(1+2i)^{n+1}}$

(c) $\displaystyle\sum_{n=0}^{\infty} \frac{3-(2i)^n}{4^{2n}}$

(d) $\displaystyle\sum_{n=2}^{\infty} \frac{(3i)^n}{3+4^n}$

3. Prove Theorem 1.7.4.

4. Prove Theorem 1.7.5.

5. Prove Theorem 1.7.7.

6. Suppose that $\sum_{n=1}^{\infty} z_n$ is a convergent series of complex numbers. Prove that both of the real series $\sum_{n=1}^{\infty} \operatorname{Re} z_n$ and $\sum_{n=1}^{\infty} \operatorname{Im} z_n$ converge and

$$\sum_{n=1}^{\infty} \operatorname{Re} z_n = \operatorname{Re} \sum_{n=1}^{\infty} z_n, \qquad \sum_{n=1}^{\infty} \operatorname{Im} z_n = \operatorname{Im} \sum_{n=1}^{\infty} z_n.$$

Is the converse true?

7. Let $\{z_n\}_{n=1}^{\infty}$ and $\{w_n\}_{n=1}^{\infty}$ be sequences of complex numbers.

(a) Prove that if the series $\sum_{n=1}^{\infty} z_n$ and $\sum_{n=1}^{\infty} w_n$ are absolutely convergent, then $\sum_{n=1}^{\infty} (z_n + w_n)$ is absolutely convergent.

(b) If $\sum_{n=1}^{\infty} z_n$ is absolutely convergent and $\sum_{n=1}^{\infty} w_n$ is conditionally convergent, what can be said about the convergence of $\sum_{n=1}^{\infty} (z_n + w_n)$?

(c) If $\sum_{n=1}^{\infty} z_n$ and $\sum_{n=1}^{\infty} w_n$ are conditionally convergent, what can be said about the convergence of $\sum_{n=1}^{\infty} (z_n + w_n)$?

CHAPTER 2

COMPLEX FUNCTIONS AND MAPPINGS

Analysis is the study of functions using limits and tools built from them, and here we consider the familiar concepts of continuity, the derivative, the integral, and power series representations in the setting of the complex plane. The concept of uniform convergence of a sequence of functions is fundamental to our work in later chapters and is addressed in Section 2.2.

Many functions familiar from calculus behave in unexpected ways when extended to the complex field. In particular, we will see connections between exponential and trigonometric functions that lead to the famous Euler's formula and provide many computational advantages, and we will learn of complications with the complex logarithm that are key to complex integration theory. An alternative to the graph of a function will provide geometric intuition for complex functions, and we further explore this by considering the valuable family of linear fractional transformations.

2.1 Continuous Functions

Almost every function we will study is continuous, and so the following definition is a natural starting point.

Complex Analysis: A Modern First Course in Function Theory, First Edition. Jerry R. Muir, Jr.
© 2015 John Wiley & Sons, Inc. Published 2015 by John Wiley & Sons, Inc.

2.1.1 Definition. Let $A \subseteq \mathbb{C}$ and $a \in A$. A function $f \colon A \to \mathbb{C}$ is *continuous* at a provided that for every $\varepsilon > 0$, there exists some $\delta > 0$ such that $|f(z) - f(a)| < \varepsilon$ whenever $z \in A$ and $|z - a| < \delta$. If f is continuous at every point of A, then f is *continuous on A*.

Notice that since $\mathbb{R} \subseteq \mathbb{C}$, the above definition is really just an extension of the definition of continuity of a function defined on a set of real numbers. As is the case with a function of a real variable, continuity can be expressed in terms of the limit of the function at different points in its domain. Consider the following definition.

2.1.2 Definition. Let $A \subseteq \mathbb{C}$, a be a limit point of A, and $f \colon A \to \mathbb{C}$. The *limit* of f as z approaches a is equal to $L \in \mathbb{C}$, denoted

$$\lim_{z \to a} f(z) = L, \tag{2.1.1}$$

if for every $\varepsilon > 0$, there is $\delta > 0$ such that $|f(z) - L| < \varepsilon$ for every $z \in A$ such that $0 < |z - a| < \delta$.

We may occasionally opt to write $f(z) \to L$ as $z \to a$ instead of (2.1.1). Observe that a need not be in A, and even if it is, the value of $f(a)$ has no bearing on the limit of f as z approaches a. We then have the following theorems, whose proofs are left as exercises.

2.1.3 Theorem. *Let $A \subseteq \mathbb{C}$, a be a limit point of A, and $f \colon A \to \mathbb{C}$. If f has a limit as z approaches a, then the limit is unique.*

2.1.4 Theorem. *Let $A \subseteq \mathbb{C}$ and $f \colon A \to \mathbb{C}$. If $a \in A$, then f is continuous at a if and only if either a is not a limit point of A or $\lim_{z \to a} f(z) = f(a)$.*

2.1.5 Example. Consider the steps used in the following limit evaluation.

$$\lim_{z \to i} \frac{z^2 + 1}{z - i} = \lim_{z \to i} \frac{(z + i)(z - i)}{z - i} = \lim_{z \to i} (z + i) = 2i.$$

The first equality is clear. The second follows because the two expressions within the limits are equal for all $z \neq i$, and their value at $z = i$ is irrelevant to the limit. The last equality can be argued directly from Definition 2.1.2 or by applying the limit and continuity rules that follow.

The concept of the limit of a function is useful in the study of derivatives. In fact, it may be of more use to us in that context than it will be in our study of continuity.

There is a helpful way to consider continuity of a function using sequences.

2.1.6 Theorem. *Let $A \subseteq \mathbb{C}$, $f \colon A \to \mathbb{C}$, and $a \in A$. Then f is continuous at a if and only if $f(z_n) \to f(a)$ for every sequence $\{z_n\}_{n=1}^{\infty} \subseteq A$ such that $z_n \to a$.*

Proof. Suppose f is continuous at a, and let $\{z_n\}_{n=1}^{\infty} \subseteq A$ be such that $z_n \to a$. Given $\varepsilon > 0$, there is $\delta > 0$ such that $|f(z) - f(a)| < \varepsilon$ when $z \in A$ satisfies $|z - a| < \delta$. There exists $N \in \mathbb{N}$ such that $|z_n - a| < \delta$ whenever $n \geq N$. But then $|f(z_n) - f(a)| < \varepsilon$ for all $n \geq N$, showing $f(z_n) \to f(a)$.

Inversely, suppose f is discontinuous at a. There exists $\varepsilon > 0$ such that for all $\delta > 0$, there is some $z \in A$ such that $|z - a| < \delta$ yet $|f(z) - f(a)| \geq \varepsilon$. Hence for each $n \in \mathbb{N}$, we may choose $z_n \in A$ such that $|z_n - a| < 1/n$ and $|f(z_n) - f(a)| \geq \varepsilon$. Then $z_n \to a$ by Theorem 1.6.9, but $\{f(z_n)\}$ fails to converge to $f(a)$. $\qquad\square$

2.1.7 Example. Let $z \in \mathbb{C}$ and $\{z_k\}_{k=1}^{\infty} \subseteq \mathbb{C}$ be any sequence such that $z_k \to z$. Then from Theorem 1.6.8, it follows that $z_k^2 \to z^2$. Inductively, we have that $z_k^n \to z^n$ for all nonnegative integers n. If $a_n \in \mathbb{C}$ for each n, then $a_n z_k^n \to a_n z^n$ by the same theorem. But then the addition rule from that theorem gives that

$$\lim_{k \to \infty} \sum_{n=0}^{N} a_n z_k^n = \sum_{n=0}^{N} a_n z^n$$

for all nonnegative integers N. This shows that all polynomials are continuous on $\mathbb{C}$ by Theorem 2.1.6.

The preceding example hints at the following theorem. Its proof is an exercise.

2.1.8 Theorem. *Let $A \subseteq \mathbb{C}$, $f, g \colon A \to \mathbb{C}$ each be continuous at some $a \in A$, and $c \in \mathbb{C}$. Then the following functions defined on A are continuous at a.*

(a) cf,

(b) $f + g$,

(c) fg, *and*

(d) f/g, *provided that $g(z) \neq 0$ for all $z \in A$.*

Here is a topological approach to continuity. The proof is left as an exercise.

2.1.9 Theorem. *Let $A \subseteq \mathbb{C}$, and $f \colon A \to \mathbb{C}$. Then f is continuous on A if and only if for every open set $U \subseteq \mathbb{C}$, there exists an open set $V \subseteq \mathbb{C}$ such that*

$$f^{-1}(U) = A \cap V.$$

In particular, if A is open, this condition reduces to $f^{-1}(U)$ being open whenever $U \subseteq \mathbb{C}$ is open.

Certain topological properties of sets in the plane are inherited by the images of those sets under continuous functions.

2.1.10 Theorem. *Let $K \subseteq \mathbb{C}$ be a compact set, and suppose $f \colon K \to \mathbb{C}$ is continuous. Then $f(K)$ is a compact set.*

Proof. Let $\{w_n\}_{n=1}^{\infty}$ be a sequence in $f(K)$. For each $n \in \mathbb{N}$, there exists $z_n \in K$ such that $f(z_n) = w_n$. The Bolzano–Weierstrass theorem gives that there is a subsequence $\{z_{n_k}\}_{k=1}^{\infty}$ of $\{z_n\}$ that converges to some $a \in K$. By Theorem 2.1.6, $w_{n_k} = f(z_{n_k}) \to f(a) \in f(K)$, and hence $f(K)$ is compact by the Bolzano–Weierstrass theorem. $\qquad\square$

2.1.11 Definition. Let $A \subseteq \mathbb{C}$. A function $f \colon A \to \mathbb{C}$ is *bounded* if there exists $M > 0$ such that $|f(z)| \leq M$ for all $z \in A$.

We will make frequent use of the following immediate consequence of Theorem 2.1.10.

2.1.12 Corollary. *Every continuous function on a compact set is bounded.*

2.1.13 Theorem. *Let $E \subseteq \mathbb{C}$ be a connected set, and suppose that $f \colon E \to \mathbb{C}$ is continuous. Then $f(E)$ is a connected set.*

Proof. Suppose, to prove the contrapositive, that $f(E)$ is disconnected. There are then separated sets $A, B \subseteq \mathbb{C}$ such that $f(E) = A \cup B$. Set $C = f^{-1}(A)$ and $D = f^{-1}(B)$. Then C and D are nonempty and $E = C \cup D$. If $z \in \overline{C}$, there is a sequence $\{z_n\}_{n=1}^{\infty} \subseteq C$ such that $z_n \to z$. (See Exercise 11 in Section 1.6.) Continuity then gives that $f(z_n) \to f(z)$. This shows that $f(z) \in \overline{A}$. But then $f(z) \notin B$, so $z \notin D$. Hence $\overline{C} \cap D = \varnothing$. Similarly, it can be shown that $C \cap \overline{D} = \varnothing$, proving E disconnected. $\qquad\square$

Uniform continuity, a stronger notion of continuity that is occasionally useful, is considered in Appendix B.

Summary and Notes for Section 2.1.

Continuity is a fundamental analytic concept, and its definition for functions of a complex variable is no different than its definition for real functions in calculus. There are three ways to consider continuity: through functional limits and the traditional ε–δ definition, sequentially with the convergent images of convergent sequences, and topologically using inverse images of open sets. These three approaches each have their strengths. Notice that the sequential approach was useful for proving the preservation of compact and connected sets under continuous functions.

It took several refinements before the final ε–δ definition of continuity was formulated. Even Leonhard Euler, in the 18th century, used a shaky definition. Bernard Bolzano and Augustin-Louis Cauchy independently came up with something close in the early 19th century, but Karl Weierstrass is credited with the version we see today. He also clarified the difference between continuity at a point and continuity on a set.

Exercises for Section 2.1.

1. Evaluate the following limits.

    (a) $\displaystyle \lim_{z \to 1+i} (1 - iz^2)$

    (b) $\displaystyle \lim_{z \to i} \frac{z^2 + 1}{z^4 - 1}$

    (c) $\displaystyle \lim_{z \to -2i} \frac{z^2 + 4iz - 4}{z^2 + 4}$

2. Let $A \subseteq \mathbb{C}$, $f: A \to \mathbb{C}$, and $a \in A$ be such that f is continuous at a. Verify the continuity of $g: A \to \mathbb{C}$ at a, where g is each of the following.

(a) $g(z) = \operatorname{Re} f(z)$

(b) $g(z) = \operatorname{Im} f(z)$

(c) $g(z) = |f(z)|$

(d) $g(z) = \overline{f(z)}$

3. Prove Theorem 2.1.3. (*Hint*: Consider the proof of Theorem 1.6.4.)

4. Prove Theorem 2.1.4.

5. Prove Theorem 2.1.8.

6. ▷ Formulate a definition for the expression $\lim_{z \to a} f(z) = \infty$. (*Hint*: Definition 1.6.3 may be inspirational.) Use it to prove the following limits are infinite.

(a) $\lim\limits_{z \to 0} \dfrac{i}{z}$

(b) $\lim\limits_{z \to i} \dfrac{z - 1}{z^2 + 1}$

7. Formulate a definition for the expression $\lim_{z \to \infty} f(z) = L$. Use it to evaluate the following limits.

(a) $\lim\limits_{z \to \infty} \dfrac{z + 1}{z - 1}$

(b) $\lim\limits_{z \to \infty} \dfrac{z^2 + 1}{z^3 - 2}$

8. Suppose $f: B \to \mathbb{C}$ and $g: A \to \mathbb{C}$ are continuous for sets $A, B \subseteq \mathbb{C}$. Prove that if $g(A) \subseteq B$, then $f \circ g: A \to \mathbb{C}$ is continuous. (Recall that $(f \circ g)(z) = f(g(z))$.)

9. State and prove a theorem that characterizes the functional limit $\lim_{z \to a} f(z) = L$ using sequences in a manner similar to the characterization of continuity in Theorem 2.1.6.

10. State and prove algebraic rules for functional limits similar to Theorems 1.6.8 and 2.1.8.

11. Prove Theorem 2.1.9.

12. For each of the following, find an example of a set $A \subseteq \mathbb{C}$ and continuous function $f: A \to \mathbb{C}$ that satisfy the given conditions.

(a) The set A is open, but the set $f(A)$ is not.

(b) The set A is closed, but the set $f(A)$ is not.

(c) The set A is bounded, but the set $f(A)$ is not.

13. Let $A \subseteq \mathbb{C}$ and $f: A \to \mathbb{C}$. Prove that f is continuous on A if and only if for every closed set $E \subseteq \mathbb{C}$, there is a closed set $F \subseteq \mathbb{C}$ such that $f^{-1}(E) = F \cap A$. Explain how when A is closed, this condition reduces to $f^{-1}(E)$ being closed when E is closed.

14. ▷ Let $K \subseteq \mathbb{C}$ be compact and $f: K \to \mathbb{C}$ be continuous. Show that there are $a, b \in K$ such that $|f(a)| \leq |f(z)| \leq |f(b)|$ for all $z \in K$.

15. This exercise requires the Heine–Borel theorem in Exercise 18 of Section 1.4. Give proofs of the following theorems using open covers in lieu of sequences.

(a) Theorem 2.1.10

(b) Theorem B.13

16. We consider continuity of a function of two variables. Let $A, B \subseteq \mathbb{C}$, $\varphi \colon A \times B \to \mathbb{C}$, $a \in A$, and $b \in B$. Then φ is *continuous* at (a, b) provided that for every $\varepsilon > 0$, there exists $\delta > 0$ such that $|\varphi(z, w) - \varphi(a, b)| < \varepsilon$ whenever $z \in A$, $w \in B$, and $\sqrt{|z - a|^2 + |w - b|^2} < \delta$.

(a) Prove that φ is continuous at (a, b) provided that for any $\varepsilon > 0$, there is $\delta > 0$ such that $|\varphi(z, w) - \varphi(a, b)| < \varepsilon$ whenever $z \in A$, $w \in B$, $|z - a| < \delta$, and $|w - b| < \delta$.

(b) Show that φ is continuous at (a, b) if and only if $\varphi(z_n, w_n) \to \varphi(a, b)$ for any pair of sequences $\{z_n\}_{n=1}^{\infty} \subseteq A$ and $\{w_n\}_{n=1}^{\infty} \subseteq B$ such that $z_n \to a$ and $w_n \to b$.

(c) We say that φ is *uniformly continuous* on $A \times B$ if for any $\varepsilon > 0$, there is $\delta > 0$ such that $|\varphi(z, w) - \varphi(\zeta, \eta)| < \varepsilon$ whenever $z, \zeta \in A$ and $w, \eta \in B$ satisfy $|z - \zeta| < \delta$ and $|w - \eta| < \delta$. Show that if A and B are compact and φ is continuous at each point of $A \times B$, then φ is uniformly continuous on $A \times B$. (See Appendix B.)

2.2 Uniform Convergence

We have dealt with the notion of convergence of a sequence of complex numbers. Now, we take this idea a step further and examine the limit of a sequence of functions. Consider the following definition.

2.2.1 Definition. Let $E \subseteq \mathbb{C}$, and define $\mathcal{F}$ to be the set of all functions $f \colon E \to \mathbb{C}$. A *sequence* of functions on E is a function $F \colon \{m, m + 1, \dots\} \to \mathcal{F}$ for some $m \in \mathbb{Z}$.

It is common to write f_n for the function $F(n)$. Note then that f_n is a function from E into $\mathbb{C}$ for each n. Often, we denote a sequence of functions by $\{f_n\}_{n=m}^{\infty}$. As is the case with sequences of complex numbers, a sequence of functions can be reindexed, and hence we commonly begin our sequences with index $m = 1$ or $m = 0$.

Sequences demand a notion of convergence. Unlike with numerical sequences, there are multiple types of convergence for sequences of functions, requiring more a careful treatment. Here is the first and most elementary.

2.2.2 Definition. Let $E \subseteq \mathbb{C}$ and $\{f_n\}_{n=1}^{\infty}$ be a sequence of functions on E. We say that $\{f_n\}$ *converges pointwise* to the function $f \colon E \to \mathbb{C}$ provided that

$$\lim_{n \to \infty} f_n(z) = f(z)$$

for all $z \in E$.

Pointwise convergence is simply the convergence of the sequence of complex numbers $\{f_n(z)\}_{n=1}^{\infty}$ to the number $f(z)$ for each $z \in E$. It is important to point out that if a sequence of functions is said to converge, its convergence is assumed to be pointwise unless otherwise noted. We often denote convergence of functions, as described in the above definition, by $f_n \to f$. These limits are unique by Theorem 1.6.4.

The following form of convergence is much stronger and more useful.

2.2.3 Definition. Let $E \subseteq \mathbb{C}$ and $\{f_n\}_{n=1}^{\infty}$ be a sequence of functions on E. We say that $\{f_n\}$ *converges uniformly* to a function $f: E \to \mathbb{C}$ provided that for all $\varepsilon > 0$, there exists $N \in \mathbb{N}$ such that

$$|f_n(z) - f(z)| < \varepsilon$$

for every $n \geq N$ and every $z \in E$.

Let us consider the difference between uniform and pointwise convergence. If $f_n \to f$ pointwise on a set E, then given $\varepsilon > 0$, for each $z \in E$ there is some $N_z \in \mathbb{N}$ such that $|f_n(z) - f(z)| < \varepsilon$ whenever $n \geq N_z$. The difficulty that can arise is that N_z will likely differ for different values of z. If $f_n \to f$ uniformly on E, then N_z can be chosen to be the same for all $z \in E$. This is a much stronger condition, as the next theorem suggests.

2.2.4 Example. For each $n \in \mathbb{N}$, define $f_n: \mathbb{C} \to \mathbb{C}$ by

$$f_n(z) = \frac{z + in}{n + 1}.$$

For any $z \in \mathbb{C}$,

$$\lim_{n \to \infty} f_n(z) = \lim_{n \to \infty} \frac{z/n + i}{1 + 1/n} = i,$$

and hence $f_n \to f$ pointwise, where $f(z) = i$. For all n, $f_n(-in) - f(-in) = -i$, and hence there is no $n \in \mathbb{N}$ such that $|f_n(z) - f(z)| < 1$ for all $z \in \mathbb{C}$, showing the convergence is not uniform. However, suppose $K \subseteq \mathbb{C}$ is compact, so that $K \subseteq D(0; R)$ for some $R > 0$. Using the triangle inequality, we have

$$|f_n(z) - f(z)| = \left| \frac{z - i}{n + 1} \right| \leq \frac{|z| + 1}{n + 1} < \frac{R + 1}{n}$$

for all $z \in K$. If $\varepsilon > 0$, then letting $N \in \mathbb{N}$ such that $N > (R + 1)/\varepsilon$ gives $|f_n(z) - f(z)| < \varepsilon$ for all $n \geq N$, and hence $f_n \to f$ uniformly on K.

2.2.5 Theorem. *If $E \subseteq \mathbb{C}$ and $\{f_n\}_{n=1}^{\infty}$ is a sequence of continuous functions on E that converges uniformly to the function $f: E \to \mathbb{C}$, then f is continuous.*

Proof. Choose $a \in E$ and $\varepsilon > 0$. Uniform convergence guarantees the existence of an $n \in \mathbb{N}$ such that $|f_n(z) - f(z)| < \varepsilon/3$ for all $z \in E$.

Since f_n is continuous at a, there is $\delta > 0$ such that for all $z \in E$ satisfying $|z - a| < \delta$, $|f_n(z) - f_n(a)| < \varepsilon/3$. Now for all $z \in E$ such that $|z - a| < \delta$,

$$|f(z) - f(a)| \leq |f(z) - f_n(z)| + |f_n(z) - f_n(a)| + |f_n(a) - f(a)|$$

$$< \frac{\varepsilon}{3} + \frac{\varepsilon}{3} + \frac{\varepsilon}{3} = \varepsilon.$$

Therefore f is continuous at a. $\qquad\square$

Certainly, pointwise convergence of continuous functions is not sufficient to guarantee a continuous limit. See the exercises for an example.

Just as the study of sequences of complex numbers led naturally to the study of infinite series, the notion of uniform convergence carries over nicely to series of functions. While the definition here is for series beginning with index $n = 1$, any other starting value may be attained through reindexing.

2.2.6 Definition. Let $E \subseteq \mathbb{C}$ and $\{f_n\}_{n=1}^{\infty}$ be a sequence of functions on E. For each $N \in \mathbb{N}$, define $s_N \colon E \to \mathbb{C}$ by

$$s_N(z) = \sum_{n=1}^{N} f_n(z). \tag{2.2.1}$$

The functions s_N are called the *partial sums* of the sequence $\{f_n\}$. If $\{s_N\}_{N=1}^{\infty}$ converges pointwise to a function $s \colon E \to \mathbb{C}$, then the series $\sum_{n=1}^{\infty} f_n$ with *terms* $\{f_n\}$ *converges pointwise* to s, denoted by

$$s = \sum_{n=1}^{\infty} f_n. \tag{2.2.2}$$

If $\{s_N\}$ converges to s uniformly, then the series *converges uniformly* to s.

As was the case with numerical sequences and series, properties of sequences of functions transfer naturally to series of functions. For instance, if each f_n is continuous, then so is each s_N, and hence if the series converges uniformly, s is continuous.

The following result is the "go-to" test for showing that a series of functions converges uniformly.

2.2.7 Weierstrass M-Test. *Let $E \subseteq \mathbb{C}$ and $\{f_n\}_{n=1}^{\infty}$ be a sequence of functions on E. If for each $n \in \mathbb{N}$, there is $M_n \geq 0$ such that $|f_n(z)| \leq M_n$ for all $z \in E$ and $\sum_{n=1}^{\infty} M_n$ converges, then $\sum_{n=1}^{\infty} f_n$ converges uniformly.*

Proof. For each $z \in E$, $\sum_{n=1}^{\infty} |f_n(z)|$ converges by comparison to $\sum_{n=1}^{\infty} M_n$. (See Theorem 1.7.7.) Therefore $\sum_{n=1}^{\infty} f_n(z)$ is absolutely convergent for each $z \in E$, and hence we may define $f \colon E \to \mathbb{C}$ by

$$f(z) = \sum_{n=1}^{\infty} f_n(z).$$

Let $\{s_N\}$ be the sequence of partial sums of the series. To show that $s_N \to f$ uniformly, let $\varepsilon > 0$. For some $K \in \mathbb{N}$,

$$\sum_{k=K+1}^{\infty} M_k < \varepsilon.$$

Then for all $N \geq K$ and $z \in E$,

$$|s_N(z) - f(z)| = \left| \sum_{k=N+1}^{\infty} f_k(z) \right| \leq \sum_{k=N+1}^{\infty} |f_k(z)| \leq \sum_{k=N+1}^{\infty} M_k < \varepsilon.$$

Note that the comparison test was used again, as was Theorem 1.7.9. We have shown that $s_N \to f$ uniformly, as needed. $\qquad\square$

2.2.8 Example. Consider the series of functions

$$\sum_{n=0}^{\infty} \frac{z^n + 6i}{2^n + 1}, \qquad z \in \mathbb{D}.$$

For each $n \in \mathbb{N}$ and $z \in \mathbb{D}$, we have

$$\left| \frac{z^n + 6i}{2^n + 1} \right| \leq \frac{|z|^n + 6}{2^n + 1} \leq \frac{7}{2^n}.$$

Since the series $\sum_{n=0}^{\infty} 7/2^n$ converges (why?), the given series converges uniformly on $\mathbb{D}$ by the Weierstrass M-test. Now, because each term in the series is a continuous (polynomial) function, we see that the sum of the series is continuous on $\mathbb{D}$ by Theorem 2.2.5.

Summary and Notes for Section 2.2.

Uniform convergence is a concept that may seem subtle at first glance. In both the definitions of pointwise and uniform convergence, N depends on the choice of ε. However, in pointwise convergence N also depends on the initial point chosen, making the convergence weaker.

The major concepts in analysis are defined using some type of limit, and many difficult problems boil down to knowing when we are permitted to interchange limit processes. Theorem 2.2.5 is an example of this. Indeed, using the functional limit expression of continuity, its conclusion could be rephrased: For each $z_0 \in E$ that is a limit point of E,

$$\lim_{z \to z_0} \lim_{n \to \infty} f_n(z) = \lim_{n \to \infty} \lim_{z \to z_0} f_n(z).$$

(Of course, $\lim_{z \to z_0} f_n(z) = f_n(z_0)$ here.)

By missing the subtlety of N depending on the chosen point, Cauchy mistakenly wrote that the limit of a series of continuous functions is continuous. This "result" has been called "Cauchy's wrong theorem." It was known to be incorrect for some time before Weierstrass first introduced uniform convergence as the necessary patch. We will enjoy the benefits of uniform convergence on a number of upcoming occasions.

Exercises for Section 2.2.

1. Use the Weierstrass M-test to show that the following series of functions converge uniformly on the unit disk $\mathbb{D}$.

    (a) $\displaystyle\sum_{n=1}^{\infty} \text{Re}\left[\left(\frac{z+i}{3}\right)^n\right]$

    (b) $\displaystyle\sum_{n=1}^{\infty} \frac{1+z^n}{2^n - z}$

2. For each $n \in \mathbb{N}$, define $f_n : \mathbb{C} \to \mathbb{C}$ by $f_n(z) = z + 1/n$, and let $f(z) = z$ for all $z \in \mathbb{C}$.

    (a) Does $f_n \to f$ uniformly on $\mathbb{C}$? Explain.

    (b) Does $f_n^2 \to f^2$ uniformly on $\mathbb{C}$? Explain.

3. Let $E = \mathbb{C} \setminus \mathbb{N}$. Define $f_n : E \to \mathbb{C}$ for each $n \in \mathbb{N}$ by

    $$f_n(z) = \frac{nz}{n - z}.$$

    Prove that $\{f_n\}_{n=1}^{\infty}$ converges uniformly on any compact subset of E, but $\{f_n\}$ does not converge uniformly on E.

4. Suppose that $\{f_n\}_{n=1}^{\infty}$ and $\{g_n\}_{n=1}^{\infty}$ are sequences of functions defined on a set $E \subseteq \mathbb{C}$. Prove that if $f, g : E \to \mathbb{C}$ exist such that $f_n \to f$ uniformly on E and $g_n \to g$ uniformly on E, then $(f_n + g_n) \to f + g$ uniformly on E.

5. Give an example of a set $E \subseteq \mathbb{C}$ and a sequence $\{f_n\}_{n=1}^{\infty}$ of continuous functions on E that converges to a discontinuous function $f : E \to \mathbb{C}$. (*Hint*: It suffices to consider real functions.)

6. Let $K \subseteq \mathbb{C}$ be compact and $\{f_n\}_{n=1}^{\infty}$ and $\{g_n\}_{n=1}^{\infty}$ be sequences of continuous functions on K converging uniformly to $f, g : K \to \mathbb{C}$, respectively.

    (a) Prove that $f_n g_n \to fg$ uniformly.

    (b) Prove that $f_n/g_n \to f/g$ uniformly if $g(z) \neq 0$ and $g_n(z) \neq 0$ for all $z \in K$ and $n \in \mathbb{N}$.

    (c) Give an example showing that (a) fails if K is not compact.

7. Let $E \subseteq \mathbb{C}$, $\{f_n\}_{n=1}^{\infty}$ be a sequence of functions on E, and $f : E \to \mathbb{C}$. Assume that for each $n \in \mathbb{N}$,
    $$M_n = \sup\{|f_n(z) - f(z)| : z \in E\}$$
    exists. Prove that $f_n \to f$ uniformly in E if and only if $M_n \to 0$.

2.3 Power Series

When analyzing complex functions, we will often utilize power series representations. In this section, we are concerned with power series themselves – specifically, with where and how they converge. At a later time, we will address when a given function is equal to a power series on a given set.

2.3.1 Definition. Let $\{c_n\}_{n=0}^{\infty}$ be a sequence of complex numbers, and let $a \in \mathbb{C}$. The *power series* based at a with coefficients $\{c_n\}$ is the series

$$\sum_{n=0}^{\infty} c_n (z-a)^n. \tag{2.3.1}$$

Notice that the power series defines a function of the complex variable z whose domain naturally consists of those $z \in \mathbb{C}$ for which the power series converges. (Of course, one may choose a different domain in a subset of the natural domain if so desired.)

2.3.2 Example. The geometric series (1.7.3) is a power series based at $a = 0$ with coefficients $c_n = 1$ for all $n = 0, 1, \dots$, converging on the unit disk $\mathbb{D}$ to $1/(1-z)$.

We will now use the geometric series to prove the following theorem, which details the region of convergence of a power series. The limit superior is defined and discussed in Appendix B.

2.3.3 Theorem. *Let $\{c_n\}_{n=0}^{\infty}$ be a sequence of complex numbers, and let $a \in \mathbb{C}$. Define*

$$\alpha = \limsup_{n \to \infty} |c_n|^{1/n},$$

and set $R = 1/\alpha$, with the natural definition of R if $\alpha = 0$ or $\alpha = \infty$.

(a) *If $R > 0$, the power series $\sum_{n=0}^{\infty} c_n (z-a)^n$ converges absolutely for each $z \in D(a; R)$ and converges uniformly on compact subsets of $D(a; R)$.*

(b) *If $R < \infty$, the power series $\sum_{n=0}^{\infty} c_n (z-a)^n$ diverges for all $z \in \mathbb{C} \setminus \overline{D}(a; R)$.*

Notice that no information is given about the convergence of the series for z on the circle $\partial D(a; R)$.

Proof. By definition, we have $\alpha = \lim_{n \to \infty} \alpha_n$ where

$$\alpha_n = \sup\{|c_k|^{1/k} : k \geq n\}, \qquad n \geq 0.$$

For $R > 0$, let $K \subseteq D(a; R)$ be compact. From Exercise 13 in Section 1.6, there exists $r \in (0, R)$ such that $K \subseteq \overline{D}(a; r)$. Clearly, $\alpha < 1/r$, and therefore choose $\varepsilon > 0$ such that $\varepsilon < 1/r - \alpha$. There exists $N \in \mathbb{N}$ such that $\alpha_N \leq \alpha + \varepsilon$. Thus for $n \geq N$, $|c_n|^{1/n} \leq \alpha + \varepsilon$. It follows that for all such n and for all $z \in \overline{D}(a; r)$,

$$|c_n (z-a)^n| \leq [(\alpha + \varepsilon) r]^n.$$

But $(\alpha + \varepsilon) r < 1$, and hence

$$\sum_{n=N}^{\infty} [(\alpha + \varepsilon) r]^n$$

is a convergent geometric series. The Weierstrass M-test (Theorem 2.2.7) implies that

$$\sum_{n=N}^{\infty} c_n(z-a)^n$$

converges uniformly on $\overline{D}(a;r)$, and hence on K. The series also converges absolutely for each $z \in \overline{D}(a;r)$. Part (a) follows.

For $R < \infty$, let $z \in \mathbb{C} \setminus \overline{D}(a;R)$. Then $\alpha_n \geq \alpha > 1/|z-a|$ for all $n \in \mathbb{N}$. Therefore, for every $n \in \mathbb{N}$, there is $k \geq n$ such that $|c_k|^{1/k} > 1/|z-a|$, which is equivalent to $|c_k(z-a)^k| > 1$. We conclude that $\{c_n(z-a)^n\}$ fails to converge to 0, and thus the series diverges by Theorem 1.7.5. This proves (b). $\qquad \square$

2.3.4 Definition. The number R given in Theorem 2.3.3 is called the *radius of convergence* of the power series $\sum_{n=0}^{\infty} c_n(z-a)^n$.

2.3.5 Corollary. *Suppose that f is the limit of a power series*

$$f(z) = \sum_{n=0}^{\infty} c_n(z-a)^n$$

with radius of convergence $R > 0$. Then f is continuous on $D(a;R)$.

Proof. Let $z_0 \in D(a;R)$, and choose $r = (R - |z_0 - a|)/2$ so that $\overline{D}(z_0;r) \subseteq D(a;R)$. Since the series converges uniformly on $\overline{D}(z_0;r)$, f is continuous on $\overline{D}(z_0;r)$ by Theorem 2.2.5. Since z_0 is an interior point of $\overline{D}(z_0;r)$, f is continuous at the point z_0. $\qquad \square$

In many practical circumstances, it is difficult to find $\limsup_{n\to\infty} |c_n|^{1/n}$. It turns out that the ratios of the coefficients of a power series can often be used to find the radius of convergence, resulting in an easier calculation.

2.3.6 Theorem. *Let $\{c_n\}_{n=0}^{\infty}$ be a sequence of nonzero complex numbers. If the sequence $\{|c_{n+1}/c_n|\}_{n=1}^{\infty}$ converges to a finite or infinite value, then*

$$\lim_{n\to\infty} \left| \frac{c_{n+1}}{c_n} \right| = \limsup_{n\to\infty} |c_n|^{1/n}.$$

Proof. Assume that $|c_{n+1}/c_n| \to \alpha$. To prove the result, it suffices to show that the radius of convergence of the power series $\sum_{n=0}^{\infty} c_n z^n$ is $1/\alpha$. We will consider the case $\alpha \in (0,\infty)$ and leave the remaining cases as an exercise.

Choose $z \in D(0;1/\alpha) \setminus \{0\}$. Let $\varepsilon > 0$ such that $\varepsilon < 1/|z| - \alpha$. It follows that $(\alpha + \varepsilon)|z| < 1$. For some $N \in \mathbb{N}$, $|c_{n+1}/c_n| < \alpha + \varepsilon$ for all $n \geq N$. Therefore

$$|c_{n+1}z^{n+1}| < (\alpha + \varepsilon)|z||c_n z^n|$$

for all $n \geq N$. Inductively this gives

$$|c_{N+k}z^{N+k}| \leq |c_N z^N|[(\alpha + \varepsilon)|z|]^k$$

for $k = 0, 1, \ldots$. Therefore, since

$$\sum_{k=0}^{\infty} |c_N z^N| [(\alpha + \varepsilon)|z|]^k$$

converges (it's geometric),

$$\sum_{k=0}^{\infty} |c_{N+k} z^{N+k}|$$

converges. It follows immediately that $\sum_{n=0}^{\infty} c_n z^n$ converges absolutely.

If $|z| > 1/\alpha$, then $1/|z| < \alpha$, and there exists $N \in \mathbb{N}$ such that for all $n \geq N$, $|c_{n+1}/c_n| > 1/|z|$. But then $|c_{n+1} z^{n+1}| > |c_n z^n|$ for all $n \geq N$. This shows $\{c_n z^n\}$ fails to converge to 0, and hence $\sum_{n=0}^{\infty} c_n z^n$ diverges by Theorem 1.7.5.

We have shown that the series $\sum_{n=0}^{\infty} c_n z^n$ converges if $0 < |z| < 1/\alpha$ and diverges if $|z| > 1/\alpha$. It follows that $1/\alpha$ is the radius of convergence of the series by Theorem 2.3.3. $\qquad\square$

2.3.7 Example. Consider the power series

$$\sum_{n=0}^{\infty} \frac{(z - 1 + i)^n}{(2 - i)^n}.$$

Here, the series is based at $a = 1 - i$ and has coefficients $c_n = 1/(2 - i)^n$. We calculate

$$\alpha = \limsup_{n \to \infty} |c_n|^{1/n} = \limsup_{n \to \infty} \frac{1}{|2 - i|} = \frac{1}{\sqrt{5}}.$$

It follows that the series has radius of convergence $R = \sqrt{5}$ and converges absolutely at each point of, and uniformly on compact subsets of, the open disk $D(1 - i; \sqrt{5})$, using Theorem 2.3.3.

Alternatively, we see that the calculation

$$\lim_{n \to \infty} \left| \frac{c_{n+1}}{c_n} \right| = \lim_{n \to \infty} \left| \frac{(2 - i)^n}{(2 - i)^{n+1}} \right| = \lim_{n \to \infty} \frac{1}{|2 - i|} = \frac{1}{\sqrt{5}}$$

gives the same result using Theorem 2.3.6.

Unlike we see here, it is often the case that one of the methods to determine the radius of convergence is significantly easier to use than the other.

Summary and Notes for Section 2.3.

Power series, first encountered in calculus, are now naturally defined for complex numbers. Each series has a radius of convergence $R \in [0, \infty]$, which can be calculated using limits of expressions involving roots, and sometimes ratios, of the series' coefficients. It is important to note that power series converge absolutely at each point of, and uniformly on compact subsets of, an open disk of radius R centered at the base point, when $R > 0$. In particular, uniform convergence on compact sets will have many benefits. Already, it guarantees that a power series sums to a continuous function on its open disk of convergence.

Exercises for Section 2.3.

1. Determine the radius of convergence of each power series, and give the open set of convergence of the series.

   (a) $\displaystyle\sum_{n=0}^{\infty}(1+i)^n z^n$

   (b) $\displaystyle\sum_{n=1}^{\infty}\frac{(-1)^n(z-i)^n}{2in^2}$

   (c) $\displaystyle\sum_{n=0}^{\infty}\frac{(z-4i+2)^n}{n!}$

   (d) $\displaystyle\sum_{n=0}^{\infty}\frac{\sqrt{n}(z+i+2)^n}{(2-i)^n}$

   (e) $\displaystyle\sum_{n=0}^{\infty}(2-(-1)^n)^n(z+i)^n$

   (f) $\displaystyle\sum_{n=0}^{\infty}z^{n^2}$

2. Show how the root and ratio tests, familiar from calculus, for convergence of a series $\sum_{n=1}^{\infty} a_n$ of complex numbers can be concluded from the results of this section.

   (a) *Root test*: Suppose $\lim_{n\to\infty}|a_n|^{1/n} = L$. If $L < 1$, the series converges absolutely. If $L > 1$, the series diverges.

   (b) *Ratio test*: Suppose $\lim_{n\to\infty}|a_{n+1}/a_n| = L$. If $L < 1$, the series converges absolutely. If $L > 1$, the series diverges.

3. Complete the proof of Theorem 2.3.6 in the cases $\alpha = 0$ and $\alpha = \infty$.

4. Suppose that $\sum_{n=0}^{\infty} c_n(z-a)^n$ and $\sum_{n=0}^{\infty} d_n(z-a)^n$ have radii of convergence R and S, respectively. What is the radius of convergence of the series $\sum_{n=0}^{\infty}(c_n+d_n)(z-a)^n$? (*Hint*: It is useful to consider cases.)

5. Let $\{c_n\}_{n=0}^{\infty}$ be a sequence of nonzero complex numbers. Suppose that $\sum_{n=0}^{\infty} c_n z^n$ has radius of convergence R.

   (a) Under what conditions will the radius of convergence of the series

   $$\sum_{n=0}^{\infty}\frac{z^n}{c_n}$$

   be $1/R$?

   (b) Give an example that shows that the series in part (a) does not necessarily have radius of convergence $1/R$.

6. This exercise uses material from Appendix B. Let $\{a_n\}_{n=0}^{\infty}$ and $\{b_n\}_{n=0}^{\infty}$ be sequences of complex numbers with Cauchy product $\{c_n\}_{n=0}^{\infty}$. Show that if the power series

$\sum_{n=0}^{\infty} a_n(z-a)^n$ and $\sum_{n=0}^{\infty} b_n(z-a)^n$ converge on $D(a; R)$ for some $R > 0$, then so does $\sum_{n=0}^{\infty} c_n(z-a)^n$ and that for each $z \in D(a; R)$,

$$\left(\sum_{n=0}^{\infty} a_n(z-a)^n \right) \left(\sum_{n=0}^{\infty} b_n(z-a)^n \right) = \sum_{n=0}^{\infty} c_n(z-a)^n.$$

2.4 Elementary Functions and Euler's Formula

Our newfound ability to define functions of a complex variable using power series provides us with a natural way to extend the exponential and trigonometric functions from $\mathbb{R}$ to $\mathbb{C}$.

2.4.1 Definition. We define the *exponential*, *sine*, and *cosine* functions of a complex variable by the power series

$$\exp z = \sum_{n=0}^{\infty} \frac{z^n}{n!}, \tag{2.4.1}$$

$$\sin z = \sum_{n=0}^{\infty} \frac{(-1)^n z^{2n+1}}{(2n+1)!}, \tag{2.4.2}$$

$$\cos z = \sum_{n=0}^{\infty} \frac{(-1)^n z^{2n}}{(2n)!}. \tag{2.4.3}$$

The remaining four trigonometric functions are defined using sine and cosine, as usual.

The above series coincide with the power series found in calculus for the real-variable versions of the functions, convergent to the functions on all of $\mathbb{R}$. These series thus have an infinite radius of convergence and provide definitions of the functions for all $z \in \mathbb{C}$. By Corollary 2.3.5, these functions are continuous on $\mathbb{C}$.

As is done in calculus, e^z will be written for $\exp z$. The fact that the following rule holds gives some credibility to this notation. Its proof relies on Mertens' theorem given in Appendix B and is left as an exercise.

2.4.2 Theorem. *Let $a, b \in \mathbb{C}$. Then $\exp(a+b) = \exp a \exp b$.*

By adding the respective power series, we obtain the following important formula relating the exponential function to the sine and cosine functions.

2.4.3 Euler's Formula. *For all $z \in \mathbb{C}$,*

$$e^{iz} = \cos z + i \sin z. \tag{2.4.4}$$

If $z = \theta$, where $\theta \in \mathbb{R}$, then Euler's formula becomes

$$e^{i\theta} = \cos \theta + i \sin \theta. \tag{2.4.5}$$

The expression on the right-hand side of the above equation gives the point on the unit circle at the angle θ with respect to the positive real axis. It follows that *every number $z \in \mathbb{C}$ can be written in the form*

$$z = re^{i\theta}, \tag{2.4.6}$$

where $r = |z|$ and θ is the angle between the positive real axis and the ray from the origin through z. See Figure 2.1. Of course, if $z = 0$, then θ may be any real value.

2.4.4 Definition. If $z = x + iy \in \mathbb{C}$ and (r, θ) are polar coordinates of (x, y) in $\mathbb{R}^2$ with $r \geq 0$, then (2.4.6) is called the *polar form* of z.

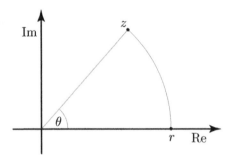

Figure 2.1 $z = re^{i\theta}$

2.4.5 Example. Using the polar form of a complex number is often key to simplifying calculations. For instance, $(\sqrt{3} - i)^6$ could be found in rectangular form by successive multiplications. However, switching to polar form allows the comparatively simple calculation

$$(\sqrt{3} - i)^6 = (2e^{-i\pi/6})^6 = 2^6 e^{-i\pi} = -64.$$

2.4.6 Definition. Let $z \in \mathbb{C} \setminus \{0\}$. The angle θ in (2.4.6) is called the *argument* of z, written

$$\theta = \arg z. \tag{2.4.7}$$

There is some ambiguity here because there are an infinite number of choices for θ. (They differ by $2n\pi$, $n \in \mathbb{Z}$.) Therefore we must realize that *the argument is not a function!* To be rigorous, $\arg z$ is a set of real numbers, and we could technically write $\theta \in \arg z$. (But we won't, out of respect for tradition.) We can partially solve this problem with the following definition.

2.4.7 Definition. The *principal value of the argument* of the nonzero complex number z is the value θ of $\arg z$ in the interval $(-\pi, \pi]$. This is written as

$$\theta = \operatorname{Arg} z. \tag{2.4.8}$$

Now the principal value of the argument is a function of $z \in \mathbb{C} \setminus \{0\}$. Note that it is not continuous. (Why?)

Solving the pair of equations from Euler's formula,

$$e^{iz} = \cos z + i \sin z$$
$$e^{-iz} = \cos z - i \sin z,$$

for $\cos z$ and $\sin z$ gives the identities

$$\cos z = \frac{e^{iz} + e^{-iz}}{2}, \qquad \sin z = \frac{e^{iz} - e^{-iz}}{2i}. \tag{2.4.9}$$

These are useful expressions for the sine and cosine functions, and they immediately reveal something unexpected – there are complex numbers z for which $\sin z$ and $\cos z$ are arbitrarily large. They are not bounded by 1!

By use of the exponential function we can prove the following.

2.4.8 De Moivre's Formula. *For $\theta \in \mathbb{R}$ and $n \in \mathbb{N}$,*

$$(\cos \theta + i \sin \theta)^n = \cos n\theta + i \sin n\theta.$$

Proof. Both sides are equal to $e^{in\theta}$. $\qquad\qquad\square$

Note that if $n = 2$, then

$$\cos 2\theta + i \sin 2\theta = (\cos \theta + i \sin \theta)^2 = \cos^2 \theta - \sin^2 \theta + 2i \sin \theta \cos \theta.$$

Setting real and imaginary parts of the above equation equal gives

$$\cos 2\theta = \cos^2 \theta - \sin^2 \theta$$
$$\sin 2\theta = 2 \sin \theta \cos \theta.$$

These are the double angle identities from trigonometry! The connections between exponential and trigonometric functions are often handy.

Euler's formula reveals even more about the exponential function. Unlike the case with a real variable, *the exponential function is not one-to-one*. In fact, it is periodic with period $2\pi i$. In other words,

$$e^{z+2\pi i} = e^z$$

for all $z \in \mathbb{C}$. Our desire to find an inverse to the exponential function, a logarithm, is impeded by this. We must first restrict the domain of the exponential function. So let us suppose that $\alpha \in \mathbb{R}$ and $B_\alpha = \{z \in \mathbb{C} : \alpha < \text{Im}\, z < \alpha + 2\pi\}$. The exponential function is one-to-one on B_α. (See the exercises.) We use B_α for the following definition.

2.4.9 Definition. The *branch of the logarithm* corresponding to B_α is the inverse of the exponential function with domain restricted to B_α. We denote the logarithm by $\log z$.

As is the case with the argument, there is some (tolerable) imprecision in using $\log z$ to denote any branch of the logarithm. Now if

$$\log z = u + iv$$

is the branch of the logarithm corresponding to the domain restriction B_α of the exponential function, then $\alpha < v < \alpha + 2\pi$. To solve for u and v, we exponentiate to get

$$z = e^{u+iv} = e^u e^{iv}.$$

Since $u, v \in \mathbb{R}$ and $e^u > 0$, this is the polar form of z. Thus $e^u = |z|$ and $v = \arg z$, chosen in the appropriate range. In other words,

$$\log z = \ln |z| + i \arg z, \qquad \alpha < \arg z < \alpha + 2\pi. \tag{2.4.10}$$

It seems convenient to use $\ln x$ to denote the familiar real natural logarithm defined for $x \in (0, \infty)$. Therefore the domain of the branch of the logarithm is

$$D_\alpha = \{z \in \mathbb{C} \setminus \{0\} : \alpha < \arg z < \alpha + 2\pi\}.$$

(Incidentally, choosing B_α so that, in addition, $\operatorname{Im} z \le \alpha + 2\pi$ is also possible, but it would destroy the continuity of the logarithm.) Consider Figure 2.4 in Section 2.5 to see the domains D_α and B_α, respectively.

As we did with the argument, we define the following.

2.4.10 Definition. The *principal branch of the logarithm* is defined by

$$\operatorname{Log} z = \ln |z| + i \operatorname{Arg} z \tag{2.4.11}$$

for nonzero $z \in \mathbb{C}$ such that $-\pi < \arg z < \pi$. The *principal value of the logarithm* is the extension of $\operatorname{Log} z$ to the negative real numbers using the principal value of the argument.

Note the difference between the principal branch and the principal value. The principal branch of the logarithm is a continuous function on its domain. The principal value of the logarithm allows $\operatorname{Log} z$ to be defined for all $z \in \mathbb{C} \setminus \{0\}$, but is discontinuous.

There are more "bizarre" ways to define branches to the logarithm. They are easier to define by restricting the domain of the logarithm instead of the domain of the exponential function. See the exercises.

We may use the logarithm to define complex powers. (Prior to this section, we have only seen integer powers of a nonreal number, which are defined using products and reciprocals.) As one would expect, since different branches of the logarithm can be defined, there must be different branches of a power function.

2.4.11 Definition. Let $a \in \mathbb{C}$ and let $\log$ denote some branch of the logarithm. We define the corresponding branch of the power z^a by

$$z^a = e^{a \log z}. \tag{2.4.12}$$

The *principal branch* (respectively, *principal value*) of z^a is given when the principal branch (respectively, principal value) of the logarithm is used.

As an example, we can calculate the nth roots of 1.

2.4.12 Example. Let $n \in \mathbb{N}$. We wish to find all possible values of $1^{1/n}$. Use the above definition to write

$$1^{1/n} = \exp\left(\frac{1}{n}\log 1\right),$$

for some value of $\log 1$. We can verify that $\log 1 = 2k\pi i$ for $k \in \mathbb{Z}$. Therefore $1^{1/n} = e^{2k\pi i/n}$ for $k \in \mathbb{Z}$. The nature of the polar form gives that $e^{2k\pi i/n} = e^{2j\pi i/n}$ if $j - k$ is a multiple of n. Hence there are n distinct values of $1^{1/n}$, equally spaced about the unit circle with angle $2\pi/n$ between consecutive values. These are called the nth *roots of unity*. Figure 2.2 illustrates the 7th roots of unity.

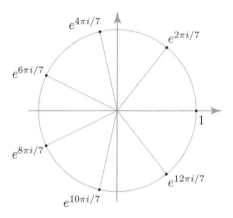

Figure 2.2   The seven 7th roots of unity

It is important to expose some notational inconsistencies. The expression 2^z has multiple values depending on the choice of the value of $\log 2$. Similarly e^z is a multiple-valued expression, as $\log e = 1 + 2n\pi i$ for $n \in \mathbb{Z}$. However, when we see e^z written, we will always assume its principal value, where $\log e = 1$. This is the only value that is consistent with the power series definition of $\exp z$.

2.4.13 Example. We finish off this section with an example of a series manipulation involving a trigonometric (and hence exponential) function. We will consider this series further in Sections 2.7 and 2.9.

Let $f(z) = \cos^2 z$. Then

$$f(z) = \left(\frac{e^{iz} + e^{-iz}}{2}\right)^2$$

$$= \frac{1}{4}(e^{2iz} + 2 + e^{-2iz}) \tag{2.4.13}$$

$$= \frac{1}{2} + \frac{1}{4} \sum_{n=0}^{\infty} \frac{2^n i^n z^n}{n!} + \frac{1}{4} \sum_{n=0}^{\infty} \frac{(-1)^n 2^n i^n z^n}{n!}$$

$$= \frac{1}{2} + \sum_{n=0}^{\infty} \frac{(1 + (-1)^n) 2^{n-2} i^n z^n}{n!}.$$

The series terms with odd index are equal to 0, and hence we let $n = 2k$, $k = 0, 1, \ldots$, to obtain

$$f(z) = \frac{1}{2} + \sum_{k=0}^{\infty} \frac{2^{2k-1} i^{2k} z^{2k}}{(2k)!} = 1 + \sum_{k=1}^{\infty} \frac{(-1)^k 2^{2k-1} z^{2k}}{(2k)!}.$$

The series has radius of convergence ∞ because it is built from the exponential series.

In the case that $z \in \mathbb{R}$, the above is a real series that could be obtained in calculus, although not using the same method with complex exponential functions. One could observe that the familiar trigonometric identity

$$\cos^2 z = \frac{1 + \cos 2z}{2} \tag{2.4.14}$$

holds for $z \in \mathbb{C}$ by equating the right-hand side of the identity to (2.4.13) and then substitute $2z$ into the cosine series as an alternative method for finding the expansion of f.

Summary and Notes for Section 2.4.

It is natural to extend algebraic functions from $\mathbb{R}$ to $\mathbb{C}$ using algebraic operations. Power series allow us to define the exponential and trigonometric functions on $\mathbb{C}$ in the same way. Euler's formula immediately follows, exposing a surprising relationship between exponential and trigonometric functions. (Other texts may *define* $e^{i\theta}$ to be $\cos\theta + i\sin\theta$, but we are left to wonder why.) This results in the marvelous equation

$$e^{i\pi} + 1 = 0,$$

which relates what are considered to be the five most important numbers in mathematics. (The log of) Euler's formula was first discovered by Roger Cotes in the early 1700s.

The relationship between exponential and trigonometric functions allows for easy proofs of many trigonometric identities. This is the first example we have seen of obtaining real-variable results by moving into the complex plane. We now have the polar form of a complex number and its geometric interpretation, which was first noted by Caspar Wessel in 1797.

The complex logarithm, with its branches, is a much more complicated entity than the real log. While this initially takes some getting used to, the multivalued nature of the logarithm, and in particular the argument, helps give complex contour integration theory its flavor and leads to a number of profound results, as we shall see.

Exercises for Section 2.4.

1. Use the polar form of a complex number to write the following in rectangular form.

   (a) $(1+i)^{10}$

   (b) $(1+i\sqrt{3})^5(2\sqrt{3}+2i)^3$

2. Find all values of the following logarithms. Which value is the principal value?

   (a) $\log 2e^{i\pi/3}$

   (b) $\log(1+i)$

   (c) $\log 2$

   (d) $\log(-1)$

   (e) $\log(\sqrt{3}-i)$

3. Find all values of the following complex powers. Which value is the principal value?

   (a) $(-1)^{1/4}$

   (b) i^i

   (c) 1^i

   (d) $(1+i)^{\pi}$

4. Evaluate each of the following and write in rectangular form.

   (a) $\sin i$

   (b) $\cos(1+i)$

   (c) $\tan\left(\dfrac{\pi+i}{2}\right)$

5. Use the polar form of a complex number to give a geometric interpretation of complex multiplication.

6. Use de Moivre's formula to derive the triple angle identities. That is, calculate identities for $\cos 3\theta$ and $\sin 3\theta$.

7. Use the method of Example 2.4.13 to find a power series expansion for $f(z) = \sin^2 z$.

8. Verify that the exponential function is one-to-one over the set $B_\alpha = \{z \in \mathbb{C} : \alpha < \operatorname{Im} z < \alpha + 2\pi\}$, where $\alpha \in \mathbb{R}$ is arbitrary.

9. Prove Theorem 2.4.3.

10. Let $n \geq 2$ be an integer and let $z \neq 1$ be an nth root of unity. Show that

$$\sum_{k=0}^{n-1} z^k = 0.$$

11. Prove the identity given in Theorem 2.4.2. Use Mertens' theorem in Appendix B along with the *binomial formula*: For $a, b \in \mathbb{C}$ and $n \in \mathbb{N}$,

$$(a+b)^n = \sum_{k=0}^{n} \binom{n}{k} a^k b^{n-k}, \qquad \binom{n}{k} = \frac{n!}{k!(n-k)!}.$$

12. A familiar identity involving the logarithm could be written $\log zw = \log z + \log w$ for $z, w \in \mathbb{C} \setminus \{0\}$.

    (a) Prove that the identity holds if we are allowed to switch branches of the logarithm.

    (b) What are conditions on z and w that allow for $\text{Log } zw = \text{Log } z + \text{Log } w$ to hold?

13. Show that the function $\text{Arg} \colon \mathbb{C} \setminus \{0\} \to \mathbb{R}$ is discontinuous. What are its points of discontinuity?

14. Find all solutions to the equation $\cos z = 10$.

15. Prove the Pythagorean identity $\sin^2 z + \cos^2 z = 1$ for all $z \in \mathbb{C}$.

16. Let $z \in \mathbb{C} \setminus \{0\}$ and $r = m/n \in \mathbb{Q}$ be reduced. Show that z^r has exactly n values and that these values are evenly spaced on a circle centered at 0. (In particular, z^r is single valued when $r \in \mathbb{Z}$.)

17. Let $x < 0$. If $r = m/n \in \mathbb{Q}$ is reduced, we know from elementary algebra that x^r has a real value if n is odd and does not have a real value if n is even. For exactly which $a \in \mathbb{R}$ does there exist a real value of x^a?

18. Find all $z \in \mathbb{C}$ such that
$$\overline{e^{iz}} = e^{i\overline{z}}.$$

19. Find the domain of the function
$$f(z) = \text{Log}(e^z + 1).$$

20. Let $\alpha \in \mathbb{R}$, and let $\log z$ denote the branch of the logarithm such that $\alpha < \arg z < \alpha + 2\pi$. Let $z^{1/2}$ denote the branch of the square root that is consistent with this branch of the logarithm.

    (a) For what values of z is the expression $\log z^{1/2}$ defined? (Be sure to consider the domains of $\log z$ and $z^{1/2}$ separately.)

    (b) Determine the domain of
$$f(z) = \tan\left[\frac{1}{i} \log \left(\frac{1 + iz}{1 - iz}\right)^{1/2}\right].$$

    (From part (a), it should be clear that your answer must depend upon α.)

    (c) Show that for all z in the domain of f, $f(z) = z$.

21. Let $S = \{te^{it} : t \geq 0\}$. Discuss how the logarithm can be defined on $\mathbb{C} \setminus S$. Specifically, suppose that $\log 1 = 0$ in your definition. What are the values of $\log \pi$ and $\log 3\pi$?

2.5   Continuous Functions as Mappings

We know that the function $f \colon \mathbb{C} \to \mathbb{C}$ given by $f(z) = z^2$ is continuous from Example 2.1.7. If we replaced $\mathbb{C}$ by $\mathbb{R}$ and asked for a graphical interpretation of f, then the response would most likely be the graph of the parabola $y = x^2$ in the plane. But as a function of a complex variable, what graphical interpretation is available for f?

This is an example of a larger problem. If we were to try to graph a complex-valued function of a complex variable, we would have to draw the graph in four-dimensional space! Obviously, we should seek a more practical alternative. The best way to attempt to visualize a complex function geometrically is to isolate subsets of its domain and examine their images under the function. In other words, if A is a (smartly chosen) subset of the domain of a function f, how does the set $f(A)$ look?

2.5.1 Example. Let $f(z) = z^2$ for all $z \in \mathbb{C}$. We will determine the image of a vertical line in the plane under f. The simplest line to consider first is the imaginary axis. If z lies on that line, then $z = it$ for some $t \in \mathbb{R}$, and hence $f(z) = -t^2$. Therefore the image of the imaginary axis under f is the nonpositive real axis. Furthermore, we see that f is two-to-one at all points on the line but 0.

Now consider the line L described by the equation $\operatorname{Re} z = a$, where $a \neq 0$. Then $z \in L$ if and only if $z = a + it$ for some $t \in \mathbb{R}$. In this case,

$$f(z) = (a^2 - t^2) + 2iat.$$

Therefore $w = u + iv \in f(L)$ if and only if for some $t \in \mathbb{R}$, the following system of parametric equations is satisfied.

$$u = a^2 - t^2$$
$$v = 2at.$$

If we solve this system to eliminate the parameter t, we obtain

$$u = a^2 - \frac{v^2}{4a^2}.$$

This is a parabola symmetric in the u-axis opening to the left with u-intercept a^2 and v-intercepts $\pm 2a^2$. We see that f is one-to-one on L, but the image is the same if a is replaced by $-a$. See Figure 2.3.

2.5.2 Example. Let us study the exponential function $f(z) = e^z$ as a mapping by determining the image of horizontal lines under f. Consider the line $\operatorname{Im} z = b$. For z on this line, $z = t + ib$ for some $t \in \mathbb{R}$, and therefore

$$f(z) = e^t e^{ib}.$$

This is the polar form of a complex number with argument b and modulus e^t. Letting t vary, we see that $w = f(z)$ is the ray $\arg w = b$ in $\mathbb{C}$, where $w \to 0$ as $t \to -\infty$ and $w \to \infty$ as $t \to \infty$.

2.5.3 Example. Since the logarithm is the inverse of the exponential function, we consider reversing the process done in Example 2.5.2. Let $\alpha \in \mathbb{R}$ and $\Omega = \{z \in \mathbb{C} : \alpha < \arg z < \alpha + 2\pi\}$. (Since $z = 0$ has no argument, $0 \notin \Omega$.) Now let $f \colon \Omega \to \mathbb{C}$ be the branch of the logarithm given by

$$f(z) = \log z, \qquad \alpha < \arg z < \alpha + 2\pi.$$

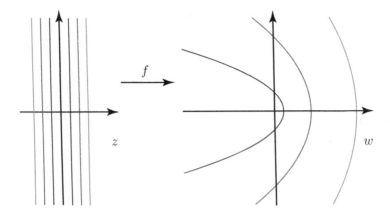

Figure 2.3   Mapping of vertical lines under $w = f(z) = z^2$

If we write $z = re^{i\theta}$ with $\alpha < \theta < \alpha + 2\pi$, then

$$f(z) = \ln r + i\theta.$$

For a fixed value of θ, $\{re^{i\theta} : r > 0\}$ is a ray emanating from the origin, and the image of that ray under $w = f(z)$ is the horizontal line $\operatorname{Im} w = \theta$, because $\ln r$ ranges through all of $\mathbb{R}$. Allowing for the entire range of θ-values shows that the image of Ω under the branch of the logarithm is the horizontal *infinite strip*

$$\{w \in \mathbb{C} : \alpha < \operatorname{Im} w < \alpha + 2\pi\}.$$

See Figure 2.4, and note that we may also observe $f^{-1}(w) = e^w$ mapping horizontal lines to rays, as seen in Example 2.5.2.

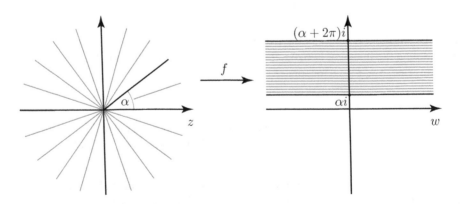

Figure 2.4   The domain and image of the branch of $w = f(z) = \log z$ given in Example 2.5.3

Summary and Notes for Section 2.5.

In calculus, the graph of a function is central to understanding the function's behavior. In the complex setting, we are forced to find a different way to geometrically analyze functions. Here, we see that examining the images of carefully selected subsets of the domain of a complex function gives intuition about the function's nature.

A more complicated problem is finding a function that nicely maps one given set onto another. This will be considered in the next section.

Exercises for Section 2.5.

1. Find the images of horizontal lines in $\mathbb{C}$ under the function $f(z) = z^2$. Include a plot similar to Figure 2.3.

2. Describe the images of vertical lines in $\mathbb{C}$ under the function $f(z) = e^z$. Include a plot similar to Figure 2.3.

3. Find a polar equation for the image of the line $\{z \in \mathbb{C} : \operatorname{Re} z = \operatorname{Im} z\}$ under the function $f(z) = e^z$, and sketch the curve.

4. Let T be the closed triangle with vertices 0, 2, and $1 + i$. Describe and plot the image of T under the function $f(z) = z^2$.

5. Geometrically analyze the mapping $f(z) = 1/z$ on $\mathbb{C} \setminus \{0\}$ by considering the images of rays, circles centered at 0, and horizontal and vertical lines. Include some plots.

6. ▷ Let $p > 0$ and $f(z) = z^p$, using the principal branch of the power. Describe the image of rays from the origin under f. If $p \in (0, 2]$ and $H = \{z \in \mathbb{C} : \operatorname{Im} z > 0\}$ is the upper half-plane, what is $f(H)$? Include a plot similar to Figure 2.4.

7. Describe the images of vertical lines in the right half-plane $\{z \in \mathbb{C} : \operatorname{Re} z > 0\}$ under the function $f(z) = \operatorname{Log} z$. Include a plot similar to Figure 2.3.

8. Let $\Omega = \{z \in \mathbb{C} : -\pi/2 < \operatorname{Re} z < \pi/2\}$. Answer the following questions concerning $\sin z$, for $z \in \Omega$. The following identity is useful, where $z = x + iy$.

$$\sin z = \sin x \cosh y + i \cos x \sinh y$$

   (a) Let $a \in (0, \pi/2)$. Show that the image of the line $\{z \in \mathbb{C} : \operatorname{Re} z = a\}$ under $\sin$ is the right half of a hyperbola with foci ± 1 and that the image of the line $\{z \in \mathbb{C} : \operatorname{Re} z = -a\}$ is the left half of the hyperbola.

   (b) Let $b > 0$. Show that the image of the segment $\{z \in \Omega : \operatorname{Im} z = b\}$ under $\sin$ is the top half of an ellipse with foci ± 1 and that the image of $\{z \in \Omega : \operatorname{Im} z = -b\}$ is the bottom half of the ellipse.

   (c) Determine the image under $\sin$ of the line $\{z \in \mathbb{C} : \operatorname{Re} z = 0\}$ and the segment $\{z \in \Omega : \operatorname{Im} z = 0\}$.

   (d) Verify that $\sin$ is one-to-one in Ω and find the set $\sin(\Omega)$.

   (e) Identify a set $E \subseteq \partial\Omega$ such that $\sin$ is a one-to-one mapping of $\Omega \cup E$ onto $\mathbb{C}$.

2.6 Linear Fractional Transformations

In the last section, we studied certain complex mappings from a geometric viewpoint by considering their action on particular subsets of the plane. In this section, we turn things around by introducing a class of mappings that interact so naturally with the geometry of $\mathbb{C}$ that we will be able to find complex mappings that send preselected geometric regions to others.

Let $a, b, c, d \in \mathbb{C}$ such that at least one of c or d is nonzero, and consider the function defined by

$$T(z) = \frac{az + b}{cz + d}. \tag{2.6.1}$$

Clearly, T is defined for all $z \in \mathbb{C}$ such that $cz + d \neq 0$. It is left as an exercise to show that T is one-to-one on the set of such z if and only if $ad - bc \neq 0$. In this case, we have the following definition.

2.6.1 Definition. A function T defined as in (2.6.1) with $ad - bc \neq 0$ is called a *linear fractional transformation* or *Möbius transformation*.

Linear fractional transformations are best considered as functions of the extended complex plane $\mathbb{C}_\infty$. If $c = 0$, then T is defined for all $z \in \mathbb{C}$. Moreover, $a \neq 0$, and so $T(z) \to \infty$ as $z \to \infty$, and we consequently define $T(\infty) = \infty$. In the case $c \neq 0$, $T(z) \to \infty$ when $z \to -d/c$ and $T(z) \to a/c$ as $z \to \infty$. We then define $T(-d/c) = \infty$ and $T(\infty) = a/c$. With these definitions, we see that T is a continuous function from $\mathbb{C}_\infty$ to $\mathbb{C}_\infty$.

We now begin a thorough study of these transformations and their relationship to the geometry of the plane.

2.6.2 Theorem. *If T and S are linear fractional transformations, then so is $S \circ T$. Furthermore, T is a bijection of $\mathbb{C}_\infty$ onto $\mathbb{C}_\infty$ and T^{-1} is a linear fractional transformation.*

Proof. Direct calculation shows that $S \circ T$ is a linear fractional transformation. If T has the form (2.6.1), then one can directly verify that the linear fractional transformation

$$R(z) = \frac{dz - b}{-cz + a}$$

is the inverse of T. □

A reader familiar with algebra may note that the set of all linear fractional transformations is a *group* under the operation of function composition.

2.6.3 Definition. Let $E \subseteq \mathbb{C}_\infty$ and $f \colon E \to \mathbb{C}_\infty$. A point $z \in E$ such that $f(z) = z$ is called a *fixed point* of f.

2.6.4 Lemma. *If T is a linear fractional transformation with three distinct fixed points in $\mathbb{C}_\infty$, then T is the identity.*

Proof. Let T have the form (2.6.1). If $c \neq 0$, then ∞ is not a fixed point of T, and the equation $T(z) = z$ reduces to the quadratic equation

$$cz^2 + (d - a)z - b = 0, \tag{2.6.2}$$

which has at most two solutions. If $c = 0$, then ∞ is a fixed point of T. In this case, (2.6.2) is linear and can have at most one solution unless T is the identity. $\qquad\square$

2.6.5 Corollary. *If T and S are two linear fractional transformations that agree at three distinct points of $\mathbb{C}_\infty$, then $T = S$.*

Proof. If T and S agree at three distinct points of $\mathbb{C}_\infty$, then $T \circ S^{-1}$ is fixed at each of these points. Thus $T \circ S^{-1}$ is the identity, proving $T = S$. $\qquad\square$

It is now apparent that at most one linear fractional transformation can have pre-described behavior at three distinct points in $\mathbb{C}_\infty$. The natural question then is, if we specify the images of three points of $\mathbb{C}_\infty$, must there be a linear fractional transformation that satisfies this? Sure enough.

2.6.6 Theorem. *Let $\alpha, \beta, \gamma \in \mathbb{C}_\infty$ be distinct and $\lambda, \mu, \nu \in \mathbb{C}_\infty$ be distinct. Then there is a unique linear fractional transformation T such that*

$$T(\alpha) = \lambda, \qquad T(\beta) = \mu, \qquad T(\gamma) = \nu.$$

Proof. Once the existence is verified, the uniqueness follows from Corollary 2.6.5. Further, using compositions, it suffices to prove the theorem in the specific case $\lambda = 1$, $\mu = 0$, and $\nu = \infty$. We will specify $a, b, c, d \in \mathbb{C}$ so that the linear fractional transformation T of the form (2.6.1) satisfies the desired properties.

Initially, suppose none of α, β, or γ equal ∞. Then a, b, c, and d should satisfy

$$b = -a\beta, \qquad d = -c\gamma, \qquad a(\alpha - \beta) = c(\alpha - \gamma).$$

Solutions to this clearly exist, for instance

$$a = 1, \qquad b = -\beta, \qquad c = \frac{\alpha - \beta}{\alpha - \gamma}, \qquad d = -\gamma\frac{\alpha - \beta}{\alpha - \gamma}.$$

If $\alpha = \infty$, we follow the above reasoning to see that

$$a = c = 1, \qquad b = -\beta, \qquad d = -\gamma$$

is sufficient. The cases $\beta = \infty$ and $\gamma = \infty$ follow similarly. $\qquad\square$

The preceding theorem motivates the following definition.

2.6.7 Definition. Let $\alpha, \beta, \gamma \in \mathbb{C}_\infty$ be distinct, and let $z \in \mathbb{C}_\infty$. The *cross ratio* of z, α, β, and γ, denoted by $\langle z, \alpha, \beta, \gamma \rangle$, is the value $T(z) \in \mathbb{C}_\infty$, where T is the unique linear fractional transformation such that $T(\alpha) = 1, T(\beta) = 0$, and $T(\gamma) = \infty$.

2.6.8 Example. Let us calculate the cross ratio $\langle 2, i, \infty, -1 \rangle$. Because $T(z) = \langle z, i, \infty, -1 \rangle$ defines a linear fractional transformation, we assume that T has the form (2.6.1). Now $T(\infty) = 0$ implies that $a = 0$. Also, $T(-1) = \infty$ gives that $d = c$. Lastly, from $T(i) = 1$ we see that $b = c(1 + i)$. Therefore

$$T(z) = \frac{c(1 + i)}{cz + c} = \frac{1 + i}{z + 1}.$$

We now have

$$\langle 2, i, \infty, -1 \rangle = T(2) = \frac{1+i}{3}.$$

The following theorem gives that the cross ratio is invariant under linear fractional transformations.

2.6.9 Theorem. *Let $\alpha, \beta, \gamma \in \mathbb{C}_\infty$ be distinct. Then for any $z \in \mathbb{C}_\infty$ and linear fractional transformation T,*

$$\langle T(z), T(\alpha), T(\beta), T(\gamma) \rangle = \langle z, \alpha, \beta, \gamma \rangle.$$

Proof. Define the linear fractional transformation S by

$$S(z) = \langle z, T(\alpha), T(\beta), T(\gamma) \rangle.$$

Observe that $(S \circ T)(\alpha) = 1$, $(S \circ T)(\beta) = 0$, and $(S \circ T)(\gamma) = \infty$. Therefore uniqueness gives

$$(S \circ T)(z) = \langle z, \alpha, \beta, \gamma \rangle.$$

Substituting $T(z)$ for z in the definition of S gives the equality. $\square$

Let us pause for a moment to consider circles and lines in the plane. From Exercise 5 in Section 1.5, both lines and circles in $\mathbb{C}$ correspond to circles in $\mathbb{C}_\infty$ (realized as the Riemann sphere), and vice versa. (Lines correspond to circles passing through ∞.) For this reason, we refer to *circles in $\mathbb{C}_\infty$* to mean either lines or circles in $\mathbb{C}$.

From analytic geometry, we know that the solution set in $\mathbb{R}^2$ to the equation

$$A(x^2 + y^2) + Bx + Cy + D = 0 \tag{2.6.3}$$

for $A, B, C, D \in \mathbb{R}$ is a circle, a line, a point, empty, or all of $\mathbb{R}^2$. We leave it as an exercise to determine the appropriate conditions on the constants so that the solution set to the equation is a circle or a line.

The following lemma addresses how particular types of linear fractional transformations behave on circles in $\mathbb{C}_\infty$. The theorem that follows seals the deal for them all.

2.6.10 Lemma. *Let $a, b \in \mathbb{C}$ with $b \neq 0$, and consider the linear fractional transformations*

$$R(z) = z + a, \qquad S(z) = bz, \qquad T(z) = \frac{1}{z}.$$

The image of a circle in $\mathbb{C}_\infty$ under each of R, S, and T is a circle in $\mathbb{C}_\infty$.

The linear fractional transformation R is called a *translation*. If $b = re^{i\theta}$ for $r > 0$, then S is the composition of a *rotation* by angle θ and a *dilation* by factor r. The transformation T is called *inversion*.

Proof. The most complicated transformation to deal with is T. We shall leave the proofs for R and S as exercises. Let E be a circle in $\mathbb{C}_\infty$ that is described in the xy-plane by (2.6.3). If we write $u + iv = T(x + iy)$, then we have

$$u = \frac{x}{x^2 + y^2}, \qquad -v = \frac{y}{x^2 + y^2}, \qquad u^2 + v^2 = \frac{1}{x^2 + y^2}.$$

Dividing (2.6.3) by $x^2 + y^2$ yields

$$D(u^2 + v^2) + Bu - Cv + A = 0$$

as long as $x^2 + y^2 \neq 0$. Since T is a bijection of $\mathbb{C}_\infty$ onto $\mathbb{C}_\infty$, it must be that $T(E)$ is a circle or a line in $\mathbb{C}$. $\qquad\square$

2.6.11 Theorem. *The image of a circle in $\mathbb{C}_\infty$ under a linear fractional transformation is a circle in $\mathbb{C}_\infty$.*

Proof. Suppose that T has the form (2.6.1) for some $a, b, c, d \in \mathbb{C}$. If $c \neq 0$, then $T = T_1 \circ T_2 \circ T_3 \circ T_4$, where

$$T_1(z) = z + \frac{a}{c}, \qquad T_2(z) = \frac{bc - ad}{c^2} z, \qquad T_3(z) = \frac{1}{z}, \qquad T_4(z) = z + \frac{d}{c}.$$

If $c = 0$, then $d \neq 0$ and $T = S_1 \circ S_2$, where

$$S_1(z) = z + \frac{b}{d}, \qquad S_2(z) = \frac{a}{d} z.$$

In either case, the linear fractional transformations composed to get T are of the types in Lemma 2.6.10, and hence map circles in $\mathbb{C}_\infty$ to circles in $\mathbb{C}_\infty$. $\qquad\square$

2.6.12 Corollary. *Let $C \subseteq \mathbb{C}_\infty$ be a circle, $\alpha, \beta, \gamma \in C$ be distinct, and $z \in \mathbb{C}_\infty$. Then $\langle z, \alpha, \beta, \gamma \rangle \in \mathbb{R} \cup \{\infty\}$ if and only if $z \in C$.*

Proof. The linear fractional transformation $T(z) = \langle z, \alpha, \beta, \gamma \rangle$ maps C onto the circle in $\mathbb{C}_\infty$ containing $1, 0$, and ∞. That is, $T(C) = \mathbb{R} \cup \{\infty\}$. $\qquad\square$

The next theorem continues the theme of Theorem 2.6.11.

2.6.13 Theorem. *Let $C_1, C_2 \subseteq \mathbb{C}_\infty$ be circles. Then there is a linear fractional transformation T such that $T(C_1) = C_2$.*

Proof. Let $\alpha, \beta, \gamma \in C_1$ be distinct and $\lambda, \mu, \nu \in C_2$ be distinct. By Theorem 2.6.6, there is a linear fractional transformation T such that $T(\alpha) = \lambda$, $T(\beta) = \mu$, and $T(\gamma) = \nu$. Since three distinct points on $\mathbb{C}_\infty$ determine a unique circle on $\mathbb{C}_\infty$, it must be that $T(C_1) = C_2$ by Theorem 2.6.11. $\qquad\square$

Of course, if we specify the image of three points on one circle, then the linear fractional transformation in the above theorem is unique by Theorem 2.6.6.

There is another useful concept that will be helpful in analyzing and constructing these transformations.

2.6.14 Definition. Let C be a circle in $\mathbb{C}_\infty$. The points $z, z^* \in \mathbb{C}_\infty$ are said to be *symmetric* in C provided that

$$\langle z^*, \alpha, \beta, \gamma \rangle = \overline{\langle z, \alpha, \beta, \gamma \rangle} \tag{2.6.4}$$

for distinct $\alpha, \beta, \gamma \in C$.

In the above definition, we use the shorthand $\overline{\infty} = \infty$. Corollary 2.6.12 shows that if $z \in C$, then $z^* = z$, regardless of the choice of $\alpha, \beta, \gamma \in C$. However, it is not clear, in general, that the definition of symmetric points is independent of the points α, β, and γ chosen on the circle. We now fix that.

2.6.15 Theorem. *Let C be a circle in $\mathbb{C}_\infty$, and suppose that (2.6.4) is satisfied for $z, z^* \in \mathbb{C}_\infty$ and distinct points $\alpha, \beta, \gamma \in C$. Then it is also satisfied with distinct $\alpha_0, \beta_0, \gamma_0 \in C$ in place of α, β, and γ, respectively.*

Proof. Define the linear fractional transformations T and S by

$$T(z) = \langle z, \alpha, \beta, \gamma \rangle, \qquad S(z) = \langle z, \alpha_0, \beta_0, \gamma_0 \rangle.$$

By Corollary 2.6.12, we see that $S \circ T^{-1}$ maps $\mathbb{R} \cup \{\infty\}$ onto $\mathbb{R} \cup \{\infty\}$. It is evident from the calculations in the proof of Theorem 2.6.6 that $S \circ T^{-1}$ may be written in the form (2.6.1) with $a, b, c, d \in \mathbb{R}$. We need only consider $z \in \mathbb{C}_\infty \setminus C$, in which case $T(z)$ and $T(z^*)$ are finite, and we then have

$$
\begin{aligned}
S(z^*) &= (S \circ T^{-1})(T(z^*)) \\
&= \frac{aT(z^*) + b}{cT(z^*) + d} \\
&= \frac{a\overline{T(z)} + b}{c\overline{T(z)} + d} \\
&= \overline{(S \circ T^{-1})(T(z))} \\
&= \overline{S(z)}.
\end{aligned}
$$

This is the desired result. $\qquad\square$

The principal use of symmetric points is outlined in the following theorem.

2.6.16 Theorem. *Let C be a circle in $\mathbb{C}_\infty$. If T is a linear fractional transformation and $z, z^* \in \mathbb{C}_\infty$ are symmetric in C, then $T(z)$ and $T(z^*)$ are symmetric in $T(C)$.*

Proof. Suppose that $\alpha, \beta, \gamma \in C$ are distinct. Then $T(\alpha)$, $T(\beta)$, and $T(\gamma)$ are distinct points on $T(C)$. We use the definition of symmetric points twice and Theorem 2.6.9 to calculate

$$
\begin{aligned}
\langle T(z^*), T(\alpha), T(\beta), T(\gamma) \rangle &= \langle z^*, \alpha, \beta, \gamma \rangle \\
&= \overline{\langle z, \alpha, \beta, \gamma \rangle} \\
&= \overline{\langle T(z), T(\alpha), T(\beta), T(\gamma) \rangle},
\end{aligned}
$$

as required. $\qquad\square$

Knowing that linear fractional transformations preserve symmetry will be of great use once we have a geometric sense of symmetry. We deal with lines and circles separately.

2.6.17 Theorem. *If L is a line and $z, z^* \in \mathbb{C}$ are symmetric in L, then z^* is the reflection of z in L.*

Proof. Note that L is a circle in $\mathbb{C}_\infty$ containing ∞. Let α and β be distinct points in $L \cap \mathbb{C}$, and define the linear fractional transformation T by

$$T(z) = \langle z, \alpha, \beta, \infty \rangle = \frac{z - \beta}{\alpha - \beta}, \qquad z \in \mathbb{C}.$$

For $z \in \mathbb{C}$, the symmetric property gives

$$\frac{z^* - \beta}{\alpha - \beta} = \frac{\overline{z} - \overline{\beta}}{\overline{\alpha} - \overline{\beta}}. \tag{2.6.5}$$

If we take the modulus of both sides of (2.6.5), we can compare numerators to see

$$|z^* - \beta| = |\overline{z} - \overline{\beta}| = |z - \beta|.$$

Since β was any arbitrary finite point on L, this shows us that z and z^* are equidistant from all points of L. Now $T(L) = \mathbb{R} \cup \{\infty\}$. Hence if $z \notin L$, then (2.6.5) shows that $T(z^*) = \overline{T(z)} \neq T(z)$ proving $z \neq z^*$. The only other possibility is the result of the theorem. $\qquad\square$

To deal with symmetry in circles, we require a lemma about conjugation in the cross ratio.

2.6.18 Lemma. *Let $\alpha, \beta, \gamma \in \mathbb{C}_\infty$ be distinct. For any $z \in \mathbb{C}_\infty$,*

$$\overline{\langle z, \alpha, \beta, \gamma \rangle} = \langle \overline{z}, \overline{\alpha}, \overline{\beta}, \overline{\gamma} \rangle.$$

Proof. If T is the linear fractional transformation defined by

$$T(z) = \langle z, \alpha, \beta, \gamma \rangle,$$

then T can be written in the form (2.6.1) for $z \in \mathbb{C}$, where a, b, c, and d are found as in the proof of Theorem 2.6.6. Since a, b, c, and d are rational expressions involving α, β, and γ, conjugation of α, β, and γ results in conjugation of a, b, c, and d. This results in

$$\langle z, \overline{\alpha}, \overline{\beta}, \overline{\gamma} \rangle = \frac{\overline{a} z + \overline{b}}{\overline{c} z + \overline{d}}.$$

Substitution of $\overline{z}$ for z gives the desired equality for $z \in \mathbb{C}$. The equality follows for $z = \infty$ by continuity. $\qquad\square$

2.6.19 Theorem. *Let $C = \partial D(a; R)$ for some $a \in \mathbb{C}$ and $R > 0$. If $z, z^* \in \mathbb{C}_\infty$ are symmetric in C, then either $\{z, z^*\} = \{a, \infty\}$ or there exist $r_1, r_2 > 0$ and $\theta \in \mathbb{R}$ such that $z = a + r_1 e^{i\theta}$, $z^* = a + r_2 e^{i\theta}$, and $r_1 r_2 = R^2$.*

Geometrically, this theorem tells us that points symmetric in C lie on the same ray from a and the product of their distances from a is R^2. Thus R is the *geometric mean* of the distances from the symmetric points to a. See Figure 2.5.

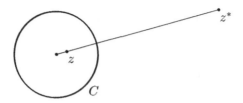

Figure 2.5 z and z^* are symmetric in C

Proof. Suppose $z \in \mathbb{C} \setminus \{a\}$, and let $\alpha, \beta, \gamma \in C$ be distinct. Let $T(\zeta) = \zeta - \bar{a}$. Then T is a linear fractional transformation, and thus by Lemma 2.6.18 and Theorem 2.6.9 using T, we have

$$\langle z^*, \alpha, \beta, \gamma \rangle = \langle \bar{z}, \bar{\alpha}, \bar{\beta}, \bar{\gamma} \rangle = \langle \bar{z} - \bar{a}, \bar{\alpha} - \bar{a}, \bar{\beta} - \bar{a}, \bar{\gamma} - \bar{a} \rangle.$$

Notice that if $\zeta \in C$, then $(\zeta - a)(\bar{\zeta} - \bar{a}) = R^2$. This applies for $\zeta = \alpha, \beta, \gamma$, and hence

$$\langle z^*, \alpha, \beta, \gamma \rangle = \left\langle \bar{z} - \bar{a}, \frac{R^2}{\alpha - a}, \frac{R^2}{\beta - a}, \frac{R^2}{\gamma - a} \right\rangle.$$

Let S be the linear fractional transformation $S(\zeta) = R^2/\zeta$. Then applying S using Theorem 2.6.9, we have

$$\langle z^*, \alpha, \beta, \gamma \rangle = \left\langle \frac{R^2}{\bar{z} - \bar{a}}, \alpha - a, \beta - a, \gamma - a \right\rangle.$$

Lastly, we apply Theorem 2.6.9 with the linear fractional transformation $Q(\zeta) = \zeta + a$ to see that

$$\langle z^*, \alpha, \beta, \gamma \rangle = \left\langle a + \frac{R^2}{\bar{z} - \bar{a}}, \alpha, \beta, \gamma \right\rangle.$$

Hence

$$z^* = a + \frac{R^2}{\bar{z} - \bar{a}}.$$

This gives $(z^* - a)(\bar{z} - \bar{a}) = R^2$. Use polar form to write $z - a = r_1 e^{i\theta_1}$ and $z^* - a = r_2 e^{i\theta_2}$ with $r_1, r_2 > 0$. Since $R^2 = r_1 r_2 e^{i(\theta_2 - \theta_1)}$, we may assume $\theta_1 = \theta_2$. This is the desired form.

The only points in $\mathbb{C}_\infty$ excluded from the previous case are a and ∞. Since they cannot be symmetric to themselves, they must be symmetric to each other.   □

If we fix distinct $z, z^* \in \mathbb{C}_\infty$, it is left as an exercise to show that the family of all circles in $\mathbb{C}_\infty$ in which z and z^* are symmetric is pairwise disjoint. (See Figure 2.6.) *Therefore if we wish to use a linear fractional transformation to map one circle onto another, it suffices to ensure that a pair of points symmetric in the first circle maps to a pair symmetric in the second and that one boundary point of the first circle maps to a boundary point of the second.*

We now put this principle to work in the following examples.

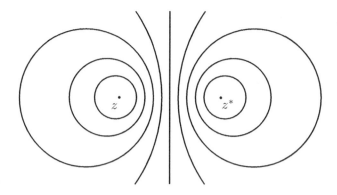

Figure 2.6   Circles in $\mathbb{C}_\infty$ in which z and z^* are symmetric

2.6.20 Example. We will find a linear fractional transformation that maps the unit disk $\mathbb{D}$ onto the right half-plane $H = \{w \in \mathbb{C} : \operatorname{Re} w > 0\}$ and sends 0 to 1. Write T as in (2.6.1). Then $T(0) = 1$ implies $b = d$. Since ∞ is symmetric to 0 in the circle $\partial \mathbb{D}$ and -1 is symmetric to 1 in the line $\{w \in \mathbb{C} : \operatorname{Re} w = 0\} \cup \{\infty\}$, we require $T(\infty) = -1$, which gives $a = -c$. It remains to send a point on the circle to a point on the line. Letting $T(1) = \infty$ gives $c = -d$. We let $d = 1$ to obtain

$$T(z) = \frac{1 + z}{1 - z}.$$

Now we find $T(C)$, where $C = \partial D(0; 2)$. Since $T(1) = \infty$ and $1 \notin C$, we see that $T(C)$ is a circle. Since 4 is symmetric to 1 in C, $T(4) = -5/3$ is symmetric to ∞ in $T(C)$, meaning that it is the center of $T(C)$. Since $T(-2) = -1/3 \in T(C)$, we see that $T(C)$ has radius $4/3$. Thus $T(C) = \partial D(-5/3; 4/3)$.

2.6.21 Example. Let us determine all of the linear fractional transformations that map $\mathbb{D}$ onto $\mathbb{D}$. Assume that T has the form (2.6.1), and choose $\alpha \in \mathbb{D} \setminus \{0\}$ such that $T(\alpha) = 0$. Observe that the point symmetric to α in $\partial \mathbb{D}$ is $1/\overline{\alpha}$. Now $T(\alpha) = 0$ implies $a\alpha + b = 0$, and $T(1/\overline{\alpha}) = \infty$ implies $c/\overline{\alpha} + d = 0$. We solve to see that $b = -a\alpha$ and $c = -\overline{\alpha}d$ and write

$$T(z) = \frac{az - a\alpha}{-\overline{\alpha}dz + d} = \frac{a}{d} \frac{z - \alpha}{1 - \overline{\alpha}z}.$$

Define

$$\varphi_\alpha(z) = \frac{z - \alpha}{1 - \overline{\alpha}z}$$

for any $\alpha \in \mathbb{D}$. Then $\varphi_\alpha(1) = (1 - \alpha)/(1 - \overline{\alpha})$, showing $|\varphi_\alpha(1)| = 1$. By our symmetric point argument, $\varphi_\alpha(\mathbb{D}) = \mathbb{D}$. Therefore $|a/d| = 1$. This shows that $T = \gamma \varphi_\alpha$, where $\gamma \in \partial \mathbb{D}$. Clearly, T must have the same form if $\alpha = 0$. Look ahead to Figure 3.4 for an illustration.

2.6.22 Example. The function $k\colon \mathbb{D} \to \mathbb{C}$ defined by

$$k(z) = \frac{z}{(1-z)^2}$$

is known as the Koebe function and has important extremal properties in geometric function theory. Note that

$$k(z) = \frac{1}{4}\left[\left(\frac{1+z}{1-z}\right)^2 - 1\right].$$

Now the linear fractional transformation T in Example 2.6.20 maps $\mathbb{D}$ onto the right half-plane. Squaring maps the half-plane onto the slit domain $\mathbb{C} \setminus (-\infty, 0]$. (See Exercise 6 in Section 2.5.) Subtracting 1 and multiplying by $1/4$ results in the slit domain $k(\mathbb{D}) = \mathbb{C} \setminus (-\infty, -1/4]$.

Summary and Notes for Section 2.6.

The concept of linear fractional transformations is at once simple and deep. There is no doubt that they are uncomplicated algebraically, but their connection with planar and spherical geometry is elegant and powerful. While these transformations could be defined on $\mathbb{R}^2$ instead of $\mathbb{C}$, it is the combination of complex algebra and geometry that gives this topic its true beauty. In addition to Möbius transformations, they may be called fractional linear transformations, bilinear transformations, or, in older works, simply linear transformations.

Since circles in $\mathbb{C}_\infty$ correspond to circles and lines in $\mathbb{C}$, $\mathbb{C}_\infty$ is the natural domain for these functions. They are one-to-one, onto, and send circles to circles. They can be freely defined on three distinct points, yielding a unique transformation. Symmetric points are also preserved under these transformations, giving a useful way to construct transformations that map one given circle in $\mathbb{C}_\infty$ onto another.

Using the geometric mapping properties of linear fractional transformations together with the mapping properties of functions seen in the previous section, we can now begin to find functions that map one given domain onto another. This is useful for providing changes of variables in two-dimensional problems and gives an application of complex function theory to other areas of analysis.

Exercises for Section 2.6.

1. Write the following linear fractional transformations in the form (2.6.1).

    (a) $T(z) = \langle z, 0, 2, i \rangle$
    (b) $T(z) = \langle z, \infty, 0, 1 \rangle$
    (c) $T(z) = \langle z, -i, \infty, 1 \rangle$

2. Find a linear fractional transformation that maps the half-plane lying below the line $\operatorname{Re} z + \operatorname{Im} z = -1$ onto the disk centered at i of radius 1.

3. Find a linear fractional transformation that maps $\mathbb{C}_\infty \setminus \overline{D}(2i; 1)$ onto $\{w \in \mathbb{C} : \operatorname{Re} w > 3\}$.

4. For the following linear fractional transformations T, find $T(\mathbb{D})$.

   (a) $T(z) = \dfrac{2 + iz}{z + i}$

   (b) $T(z) = \dfrac{i - z}{1 - 2z}$

5. Let $H = \{z \in \mathbb{C} : \operatorname{Im} z > 0\}$ be the upper half-plane. Find $T(H)$ for the linear fractional transformations T given in Exercise 4.

6. Let $\Omega = \mathbb{C} \setminus (\overline{D}(-1; 1) \cup \overline{D}(1; 1))$. Find a linear fractional transformation that maps Ω onto the strip $\{w \in \mathbb{C} : -1 < \operatorname{Im} w < 1\}$.

7. Find a linear fractional transformation that maps the circles $\partial D(0; 4)$ and $\partial D(1; 1)$ onto concentric circles centered at 0.

8. Find all linear fractional transformations that map the open unit disk $\mathbb{D}$ onto the right half-plane $\{w \in \mathbb{C} : \operatorname{Re} w > 0\}$ while mapping 0 to 1.

9. Find all linear fractional transformations that map the upper half-plane $\{z \in \mathbb{C} : \operatorname{Im} z > 0\}$ onto itself.

10. If T is a linear fractional function written as in (2.6.1), verify that T is one-to-one if and only if $ad - bc \neq 0$.

11. Show that $S \circ T$ is a linear fractional transformation if S and T are.

12. Let $\Omega = D(0; 2) \setminus \overline{D}(1; 1)$.

    (a) Find a linear fractional transformation that maps Ω onto the strip $\{w \in \mathbb{C} : 0 < \operatorname{Im} w < \pi\}$.

    (b) Find a continuous one-to-one function that maps Ω onto the upper half-plane $\{w \in \mathbb{C} : \operatorname{Im} w > 0\}$.

    (c) Find a continuous one-to-one function that maps Ω onto $\mathbb{D}$.

13. Let $\Omega = D(-1; 2) \cap D(1; 2)$. Using the properties of linear fractional transformations, find a continuous one-to-one function that maps Ω onto the upper half-plane $\{w \in \mathbb{C} : \operatorname{Im} w > 0\}$.

14. Under what conditions on A, B, C, and D will the solution set to (2.6.3) be a line? a circle?

15. Complete the proof of Lemma 2.6.10 by addressing the transformations R and S.

16. Here we apply geometry to study symmetry in circles. Let $a \in \mathbb{C}$, $R > 0$, and $C = \partial D(a; R)$. Suppose z, z^* are symmetric in C with $z \in D(a; R)$ and $z^* \in \mathbb{C} \setminus \overline{D}(a; R)$. (See Figure 2.5.) Let L be the line through z perpendicular to the line through a, z, and z^*, and choose $w \in L \cap C$. Prove that the line through w and z^* is tangent to C. (*Hint*: Try using similar triangles.)

17. For a linear fractional transformation T written as in (2.6.1), define the corresponding 2×2 matrix

$$\tilde{T} = \begin{bmatrix} a & b \\ c & d \end{bmatrix}.$$

    (a) Show that $\tilde{T}$ is nonsingular (invertible).

    (b) Show that the product of the matrices corresponding to linear fractional transformations corresponds to the composition of the transformations.

(c) Is the matrix corresponding to a given linear fractional transformation T unique? If not, find all matrices that correspond to a fixed T.

18. Let $z, z^* \in \mathbb{C}_\infty$ be distinct. Show that the set of all circles in $\mathbb{C}_\infty$ in which z and z^* are symmetric is a pairwise disjoint family.

2.7 Derivatives

We now return to our study of the calculus of complex functions by considering the derivative with respect to a complex variable.

2.7.1 Definition. Let $\Omega \subseteq \mathbb{C}$ be open and $a \in \Omega$. If $f : \Omega \to \mathbb{C}$, we say that f is *differentiable* at a provided that the limit

$$\lim_{z \to a} \frac{f(z) - f(a)}{z - a} = \lim_{h \to 0} \frac{f(a + h) - f(a)}{h} \tag{2.7.1}$$

exists. (We see that these limits are the same through the relationship $z = a + h$.) The value of the limit is known as the *derivative* of f at a and is denoted $f'(a)$. Furthermore, f is *differentiable* on Ω provided that f is differentiable at each point in Ω, in which case $f' : \Omega \to \mathbb{C}$ is well defined.

As usual, we cannot give a definition without making a couple of remarks. First, we require an open set for our definition because there will then be some open disk $D(a; r) \subseteq \Omega$, and therefore the limit will come from all directions. Second, Leibniz notation is also commonly used to represent derivatives:

$$\frac{d}{dz} f(z) = \frac{df}{dz}(z) = f'(z).$$

The following connection is immediate.

2.7.2 Theorem. *Let $\Omega \subseteq \mathbb{C}$ be open and $f : \Omega \to \mathbb{C}$. If $a \in \Omega$ and f is differentiable at a, then f is continuous at a.*

Proof. The equality

$$0 = f'(a) \lim_{z \to a}(z - a) = \lim_{z \to a} \frac{f(z) - f(a)}{z - a} \lim_{z \to a}(z - a) = \lim_{z \to a}(f(z) - f(a))$$

shows that $f(z) \to f(a)$ as $z \to a$. Thus f is continuous at a. $\qquad\square$

The well-known rules for derivatives (linearity, the product and quotient rules, the chain rule, and the power rule for positive integer powers) from calculus follow here with the same proofs using the definition. (See the exercises.) Consider the following example of a function that one may assume should be differentiable, but is not.

2.7.3 Example. Define $f : \mathbb{C} \to \mathbb{C}$ by $f(z) = \overline{z}$. For any $a \in \mathbb{C}$ and $h \in \mathbb{C} \setminus \{0\}$,

$$\frac{f(a + h) - f(a)}{h} = \frac{\overline{a + h} - \overline{a}}{h} = \frac{\overline{h}}{h}.$$

If h is real, $\bar{h}/h = 1$, and if h is imaginary $\bar{h}/h = -1$. For all $\delta > 0$, $D(0;\delta) \setminus \{0\}$ contains both real and imaginary numbers, and hence the derivative limit fails to converge. Thus f is not differentiable at any $a \in \mathbb{C}$.

The next natural question to consider is whether the derivatives of the elementary functions (trigonometric functions, exponential functions, etc.) behave as they do in the real-variable case. Since these functions are defined using power series, we turn first to the following. Because of the complicated proof, we begin by considering a series based at 0. The general result follows by the chain rule and is addressed in the subsequent corollary.

2.7.4 Theorem. *If f is given by a power series*

$$f(z) = \sum_{n=0}^{\infty} c_n z^n$$

with radius of convergence R, then f is differentiable on $D(0;R)$ and

$$f'(z) = \sum_{n=1}^{\infty} n c_n z^{n-1} \tag{2.7.2}$$

for all $z \in D(0;R)$. The differentiated series also has radius of convergence R.

Clearly, $R = \infty$ is allowable and dealt with in the natural way.

Proof. We first show that the series on the right-hand side of (2.7.2) has radius of convergence R. This series converges for a given $z \in \mathbb{C}$ if and only if $\sum_{n=0}^{\infty} n c_n z^n$ does, and so we will show that the latter has radius of convergence R. Hence we consider for each n,

$$\alpha_n = \sup\{|c_k|^{1/k} : k \geq n\}, \qquad \beta_n = \sup\{k^{1/k}|c_k|^{1/k} : k \geq n\}.$$

Straightforward calculus computations give that the function $u(x) = (\ln x)/x$, $x \geq e$, decreases to a limit of 0 as $x \to \infty$. Therefore $x^{1/x} = e^{u(x)}$ decreases to a limit of 1. For any $n \geq 3$ and $k \geq n$, we see that

$$|c_k|^{1/k} \leq k^{1/k}|c_k|^{1/k} \leq n^{1/n}|c_k|^{1/k}.$$

Taking suprema yields $\alpha_n \leq \beta_n \leq n^{1/n}\alpha_n$ for all $n \geq 3$. (See Theorem B.6.) Both $\{\alpha_n\}$ and $\{\beta_n\}$ have limits in $[0,\infty]$, and we now see that they are equal. Thus their corresponding series have the same radius of convergence.

Now that we know the series (2.7.2) converges on $D(0;R)$, set $g(z)$ to be its sum. We must show that $g = f'$. Let $\varepsilon > 0$, and fix $a \in D(0;R)$. For $N \in \mathbb{N}$, let s_N be the Nth partial sum of f. Then s'_N is a partial sum of g. If $r_N = f - s_N$ is the Nth remainder, then

$$\frac{f(z) - f(a)}{z - a} - g(a)$$

$$= \frac{r_N(z) - r_N(a)}{z - a} + \left(\frac{s_N(z) - s_N(a)}{z - a} - s'_N(a) \right) + (s'_N(a) - g(a))$$

$$\tag{2.7.3}$$

for every $N \in \mathbb{N}$ and $z \in D(0; R) \setminus \{a\}$. To complete the proof, we will show that there exist $N \in \mathbb{N}$ and $\delta > 0$ such that the modulus of right-hand side of (2.7.3) is less than ε for all $z \in D(0; R)$ such that $0 < |z - a| < \delta$.

Since $s'_N(a) \to g(a)$, there exists $N_1 \in \mathbb{N}$ such that $|s'_N(a) - g(a)| < \varepsilon/3$ whenever $N \geq N_1$.

For every $n \in \mathbb{N}$,

$$z^n - a^n = (z - a) \sum_{k=1}^{n} z^{n-k} a^{k-1}. \tag{2.7.4}$$

Choose $r \in (|a|, R)$. Then for all $z \in D(0; r) \setminus \{a\}$,

$$\left| \frac{z^n - a^n}{z - a} \right| \leq \sum_{k=1}^{n} |z|^{n-k} |a|^{k-1} < \sum_{k=1}^{n} r^{n-1} = nr^{n-1}$$

because $|z| < r$ and $|a| < r$. Since the series (2.7.2) converges absolutely for $z = r$ by Theorem 2.3.3, there is $N_2 \in \mathbb{N}$ such that

$$\sum_{n=N_2+1}^{\infty} n|c_n| r^{n-1} < \frac{\varepsilon}{3}.$$

Let $N = \max\{N_1, N_2\}$. Then for all $z \in D(0; r) \setminus \{a\}$,

$$\left| \frac{r_N(z) - r_N(a)}{z - a} \right| = \left| \frac{1}{z - a} \sum_{n=N+1}^{\infty} c_n (z^n - a^n) \right|$$

$$\leq \sum_{n=N+1}^{\infty} \left| c_n \frac{z^n - a^n}{z - a} \right|$$

$$\leq \sum_{n=N+1}^{\infty} n|c_n| r^{n-1}$$

$$< \frac{\varepsilon}{3}.$$

Now the definition of the derivative gives that

$$\lim_{z \to a} \frac{s_N(z) - s_N(a)}{z - a} = s'_N(a).$$

Therefore there must be some $\delta > 0$ (we assume $\delta < r - |a|$ since there is no limit to how small δ can be) such that whenever $0 < |z - a| < \delta$,

$$\left| \frac{s_N(z) - s_N(a)}{z - a} - s'_N(a) \right| < \frac{\varepsilon}{3}.$$

We have shown that for our fixed N and all z such that $0 < |z - a| < \delta$, the modulus of each component of the right-hand side of (2.7.3) is bounded by $\varepsilon/3$. The result then follows by the triangle inequality. $\qquad \square$

The following corollary follows inductively.

2.7.5 Corollary. *Let $f: D(a; R) \to \mathbb{C}$ be given by the power series*

$$f(z) = \sum_{n=0}^{\infty} c_n (z - a)^n$$

with radius of convergence $R > 0$. Then f is infinitely differentiable and the derivatives of f are found by repeatedly differentiating the power series term by term. Furthermore, the resulting series have radius of convergence R. In addition, f has infinitely many antiderivatives found by repeatedly antidifferentiating the series term by term. These antiderivatives also have radius of convergence R.

Using their power series definitions, we now happily note that the familiar derivative formulas

$$\frac{d}{dz} e^z = e^z, \qquad \frac{d}{dz} \sin z = \cos z, \qquad \frac{d}{dz} \cos z = -\sin z$$

hold for the complex-variable versions of the exponential and trigonometric functions.

2.7.6 Example. The series

$$\cos^2 z = 1 + \sum_{k=1}^{\infty} \frac{(-1)^k 2^{2k-1} z^{2k}}{(2k)!}, \qquad z \in \mathbb{C},$$

was found in Example 2.4.13. If we differentiate both sides, we find

$$-2 \sin z \cos z = \sum_{k=1}^{\infty} \frac{(-1)^k 2^{2k-1} z^{2k-1}}{(2k-1)!} = \sum_{k=0}^{\infty} \frac{(-1)^{k+1} 2^{2k+1} z^{2k+1}}{(2k+1)!}, \qquad z \in \mathbb{C}.$$

Substitution into the sine series shows that this is equal to $-\sin 2z$, as expected.

Naturally, the next function to consider is a branch of the logarithm. The following, more general, theorem is useful.

2.7.7 Theorem. *Let $\Omega \subseteq \mathbb{C}$ be open and $f: \Omega \to \mathbb{C}$ be differentiable. If $f(\Omega)$ is open and $g = f^{-1}: f(\Omega) \to \Omega$ exists and is continuous, then g is differentiable and*

$$g'(f(a)) = \frac{1}{f'(a)}$$

for any $a \in \Omega$ such that $f'(a) \neq 0$.

Proof. We know that

$$\frac{1}{f'(a)} = \lim_{z \to a} \frac{z - a}{f(z) - f(a)},$$

since f is one-to-one. As $w \to f(a)$, $g(w) \to a$ due to the continuity of g. Therefore

$$g'(f(a)) = \lim_{w \to f(a)} \frac{g(w) - g(f(a))}{w - f(a)} = \lim_{w \to f(a)} \frac{g(w) - a}{f(g(w)) - f(a)} = \frac{1}{f'(a)},$$

as desired. $\square$

2.7.8 Example. Suppose $\Omega = \{z \in \mathbb{C} : \alpha < \operatorname{Im} z < \alpha + 2\pi\}$ for some $\alpha \in \mathbb{R}$, and let $f \colon \Omega \to \mathbb{C}$ be given by $f(z) = e^z$. Then $f(\Omega) = \{w \in \mathbb{C} : \alpha < \arg w < \alpha + 2\pi\}$ and $g = f^{-1}$ is the corresponding branch of the logarithm. Using Theorem 2.7.7, we see that g is differentiable on $f(\Omega)$. For $w \in f(\Omega)$, let $z = g(w) \in \Omega$. Then

$$g'(w) = \frac{1}{f'(z)} = \frac{1}{e^z} = \frac{1}{w}.$$

This is the derivative of the logarithm, just as in calculus. Notice that the formula for the derivative does not depend upon the choice of the branch. We will find the concept of branches of the logarithm being antiderivatives of $1/w$ central to the complex integral calculus.

Summary and Notes for Section 2.7.

On the surface, the derivative of a complex function looks the same as the derivative of a real function. Indeed, all of the common differentiation rules from calculus hold, including that we can differentiate a power series term by term without changing the radius of convergence.

From one point of view, the complex derivative is significantly different than the real. The limit in the definition of the real derivative only needs to exist from two directions, while the limit for the complex derivative must exist from infinitely many directions. This makes it "harder" for a complex function to have a derivative, and hence being differentiable is more special for a complex function. In fact, this difference holds the key to why complex analysis is so different from real analysis, as we will start to see in Chapter 3.

We also note that, since a function of a complex variable can be thought of as a function of two real variables, one could consider its differentiability in the sense studied in multivariable calculus. This is a different notion of differentiability, and its relationship to complex differentiability is considered in Section 3.6.

Exercises for Section 2.7.

1. Determine where, if anywhere, the following functions $f \colon \mathbb{C} \to \mathbb{C}$ are differentiable. (There are not many places.)

   (a) $f(z) = \operatorname{Re} z$

   (b) $f(z) = \operatorname{Im} z$

   (c) $f(z) = |z|^2$

2. What is the domain of $f(z) = \tan z$? Calculate its derivative on its domain.

3. Use the geometric series and Corollary 2.7.5 to find a power series that converges to $\log(z + 1)$ for all $z \in \mathbb{D}$ for each branch of the logarithm defined in the appropriate region.

4. Find a power series that converges to $f(z) = 1/z^3$ in a disk centered at 1 using the geometric series and the methods in this section. What is the radius of convergence of the series? (*Hint*: $z = 1 - (1 - z)$.)

5. Choose a branch of the logarithm, and define $f(z) = z^a$, where $a \in \mathbb{C}$ and f corresponds to the chosen branch. What can be said about f'?

6. ▷ Let $\Omega \subseteq \mathbb{C}$ be open, and suppose $f \colon \Omega \to \mathbb{C}$ is differentiable. Let $\Omega' = \{\overline{z} : z \in \Omega\}$, and define $g \colon \Omega' \to \mathbb{C}$ by $g(z) = \overline{f(\overline{z})}$. Prove that g is differentiable.

7. Determine a power series that converges to the Koebe function given in Example 2.6.22.

8. Let $\Omega \subseteq \mathbb{C}$ be open and $f, g \colon \Omega \to \mathbb{C}$. If f and g are differentiable at the point $a \in \Omega$, prove the following familiar rules from calculus.

   (a) If $c \in \mathbb{C}$, then the function cf is differentiable at a and $(cf)'(a) = cf'(a)$.

   (b) The function $f + g$ is differentiable at a and $(f + g)'(a) = f'(a) + g'(a)$.

   (c) The function fg is differentiable at a and

   $$(fg)'(a) = f(a)g'(a) + f'(a)g(a).$$

   (d) If $g(z) \neq 0$ in an open set containing a, then f/g is differentiable at a and

   $$\left(\frac{f}{g}\right)'(a) = \frac{f'(a)g(a) - f(a)g'(a)}{[g(a)]^2}.$$

9. Let $n \in \mathbb{N}$, and define $f \colon \mathbb{C} \to \mathbb{C}$ by $f(z) = z^n$.

   (a) Prove that the power rule $f'(z) = nz^{n-1}$ holds for all $z \in \mathbb{C}$. (*Hint*: Use (2.7.4).)

   (b) Explain why this proof is needed for the development of this section. That is, why can we not just conclude it from Theorem 2.7.4?

10. Let $\Omega \subseteq \mathbb{C}$ be open and $a \in \Omega$. If $f \colon \Omega \to \mathbb{C}$, prove that f is differentiable at a with derivative $f'(a)$ if and only if

    $$f'(a) = \lim_{n \to \infty} \frac{f(z_n) - f(a)}{z_n - a}$$

    for every sequence $\{z_n\}_{n=1}^{\infty} \subseteq \Omega \setminus \{a\}$ such that $z_n \to a$.

11. Let $\Omega, U \subseteq \mathbb{C}$ be open sets and suppose $f \colon \Omega \to U$ and $g \colon U \to \mathbb{C}$. If f is differentiable at the point $a \in \Omega$ and g is differentiable at $f(a)$, then prove that $g \circ f$ is differentiable at a and $(g \circ f)'(a) = g'(f(a))f'(a)$. This is the familiar chain rule from calculus. (*Hint*: Define a function h, continuous at $f(a)$, satisfying

    $$h(w)(w - f(a)) = g(w) - g(f(a))$$

    for all $w \in U$.)

12. Find a differentiable function $f \colon \mathbb{C} \to \mathbb{C}$ and distinct $a, b \in \mathbb{C}$ such that there is no c between a and b (on the segment connecting a and b) such that

    $$f'(c) = \frac{f(b) - f(a)}{b - a}.$$

This shows the mean value theorem from calculus does not have a strict analog for the complex derivative. Why is $f(z) = z^2$ necessarily unsuitable?

13. Let $a, b, c \in \mathbb{C}$, and define $f \colon \mathbb{C} \to \mathbb{C}$ by

$$f(z) = az^2 + b|z|^2 + c\bar{z}^2.$$

Prove that f is differentiable at $z \in \mathbb{C}$ if and only if $bz + 2c\bar{z} = 0$.

2.8   The Calculus of Real-Variable Functions

With knowledge of the complex derivative in hand and with an eye toward a complex integral, it is now convenient and appropriate to spend a moment on the calculus of complex-valued functions of a real variable, as they will be an important tool in our upcoming work. We shall see that there is essentially no effect on the rules of calculus by allowing for a complex range, reinforcing the notion that it is the complex variable that is vital to the identity of complex analysis.

2.8.1 Definition. Let $a < b$ and $f \colon (a, b) \to \mathbb{C}$. We say that f is *differentiable* at $t \in (a, b)$ if the limit

$$\lim_{s \to t} \frac{f(s) - f(t)}{s - t} = \lim_{h \to 0} \frac{f(t + h) - f(t)}{h} \tag{2.8.1}$$

exists. (The limits are equal using the substitution $s = t + h$.) In this case, the value of the limit is called the *derivative* of f at t and is denoted $f'(t)$. Furthermore, f is *differentiable* on (a, b) provided that f is differentiable at each $t \in (a, b)$, in which case $f' \colon (a, b) \to \mathbb{C}$ is well defined.

The above definition is precisely the definition of the derivative of a function of a real variable found in calculus with range in $\mathbb{R}$ replaced by range in $\mathbb{C}$. It is important to note that this derivative is significantly different from the complex derivative considered in the previous section because in this limit, h only approaches 0 from two directions. Although "prime" notation is used in both situations, the clear context will not cause confusion. Leibniz notation (d/dt) is commonly used as well.

If $f = u + iv$ for $u, v \colon (a, b) \to \mathbb{R}$, then the properties of limits imply that f is differentiable at some $t \in (a, b)$ if and only if u and v are each differentiable at t. In this case, $f'(t) = u'(t) + iv'(t)$.

Through either the definition or the separation into real and imaginary parts, the basic algebraic derivative rules (sum, product, and quotient) quickly follow. The chain rule requires a little more care due to the possible complex range of the functions. If either

- $f \colon (a, b) \to (c, d)$ is a real-valued differentiable function and $g \colon (c, d) \to \mathbb{C}$ is differentiable in the sense of Definition 2.8.1 or

- f is as in Definition 2.8.1 and g is a differentiable function on an open subset of $\mathbb{C}$ containing $f((a, b))$ (in the sense of Section 2.7), then $g \circ f$ is differentiable on (a, b) in the sense of Definition 2.8.1 and $(g \circ f)'(t) = g'(f(t))f'(t)$, where the "primes" refer to derivatives with respect to real or complex variables as appropriate. See the exercises.

In our upcoming work, we will need flexibility when dealing with continuity and differentiability of real-variable functions.

2.8.2 Definition. Let $[a, b] \subseteq \mathbb{R}$. A complex-valued function f is *piecewise contin-uous* on $[a, b]$ if there is a partition of $[a, b]$,

$$a = t_0 < t_1 < \cdots < t_n = b, \tag{2.8.2}$$

such that for each $k = 1, \ldots, n$, f is defined and continuous on (t_{k-1}, t_k) and the one-sided limits

$$\lim_{t \to t_{k-1}^+} f(t), \qquad \lim_{t \to t_k^-} f(t)$$

exist. (Note that f may not be defined at the points $t_0, \ldots, t_n$.) Furthermore, f is *piecewise continuously differentiable* on $[a, b]$ if f' is piecewise continuous on $[a, b]$.

Piecewise continuous differentiability implies piecewise continuity, an intuitive result whose details are left to an exercise.

2.8.3 Example. Figure 2.7 contains the graph of a real-valued piecewise continu-ously differentiable function on an interval $[a, b]$ with $n = 4$ in (2.8.2). Note that the function is undefined at t_1 and t_4, is defined but discontinuous at t_2, and is continu-ous but not differentiable at t_3.

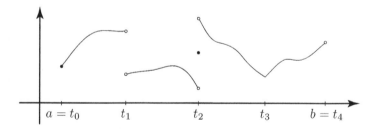

Figure 2.7   A real-valued piecewise continuously differentiable function; see Example 2.8.3

Because the derivative of a complex-valued function of a real variable can be considered by decomposing into real and imaginary parts and using calculus, we can naturally define the integral in a consistent manner.

2.8.4 Definition. Let $[a, b] \subseteq \mathbb{R}$ and $f \colon [a, b] \to \mathbb{C}$ be continuous. Writing $f = u + iv$ for $u, v \colon [a, b] \to \mathbb{R}$, we define the *integral* of f by

$$\int_a^b f(t)\, dt = \int_a^b u(t)\, dt + i \int_a^b v(t)\, dt. \tag{2.8.3}$$

If f is a complex-valued piecewise continuous function on $[a, b]$, then we use the partition (2.8.2) to define

$$\int_a^b f(t)\, dt = \sum_{k=1}^n \int_{t_{k-1}}^{t_k} f(t)\, dt, \tag{2.8.4}$$

where in each integral within the summation, the function f can be extended to be continuous on $[t_{k-1}, t_k]$.

For the sake of clarity, we will unapologetically avoid writing the decomposition (2.8.4) although it is always in the background of calculations. Those who prefer their integrals defined as limits of Riemann sums may note that the separation into real and imaginary parts also works easily in that setting and yields the above.

The following form of the fundamental theorem of calculus is an upgrade from the version seen in calculus. Its proof is left as an exercise.

2.8.5 Fundamental Theorem of Calculus for Functions of a Real Variable. *Let $[a, b] \subseteq \mathbb{R}$ and f be a complex-valued piecewise continuous function on $[a, b]$. If $g: [a, b] \to \mathbb{C}$ is defined by*

$$g(t) = \int_a^t f(s)\, ds,$$

then g is continuous and $g'(t) = f(t)$ for all t at which f is continuous. If the function $F: [a, b] \to \mathbb{C}$ is continuous and satisfies $F'(t) = f(t)$ for all $t \in (a, b)$ at which f is continuous, then

$$\int_a^b f(t)\, dt = F(t)]_a^b = F(b) - F(a).$$

2.8.6 Example. Let $n \in \mathbb{Z}$, and define $f: \mathbb{R} \to \mathbb{C}$ by $f(t) = e^{int}$. As expected, $f'(t) = ine^{int}$. To reason this out, we realize that $f = g \circ h$, where $g(z) = e^z$ (a function of a complex variable) and $h(t) = int$ (a function of a real variable), and apply the chain rule as described above. Alternatively, we can separate into real and imaginary parts, $f(t) = \cos nt + i \sin nt$, and calculate

$$f'(t) = -n \sin nt + in \cos nt = in(\cos nt + i \sin nt) = ine^{int}.$$

When $n \neq 0$, this then allows the calculation

$$\int_0^{2\pi} e^{int}\, dt = \left. \frac{e^{int}}{in} \right]_0^{2\pi} = \frac{e^{2in\pi} - e^0}{in} = 0,$$

using the fundamental theorem of calculus. When $n = 0$, this is the integral of a constant, and so we have, for all $n \in \mathbb{Z}$,

$$\int_0^{2\pi} e^{int}\, dt = \begin{cases} 0 & \text{if } n \neq 0, \\ 2\pi & \text{if } n = 0. \end{cases} \tag{2.8.5}$$

An interesting and advantageous turn of events is that certain real integrals can now be considered with complex methods using the relationships from Section 2.4.

2.8.7 Example. Using (2.4.13) and (2.8.5), we evaluate the trigonometric integral

$$\int_0^{2\pi} \cos^2 t \, dt = \frac{1}{4} \int_0^{2\pi} (e^{2it} + 2 + e^{-2it}) \, dt = \pi.$$

In calculus, this integral is typically found using the trigonometric identity (2.4.14). While this may not seem to be a great improvement, such techniques pay off at a higher rate for more complicated integrals. See the exercises for some examples.

We conclude with a frequently used inequality for the modulus of an integral.

2.8.8 Theorem. *If* $[a, b] \subseteq \mathbb{R}$ *and* φ *is a piecewise continuous complex-valued function on* $[a, b]$, *then*

$$\left| \int_a^b \varphi(t) \, dt \right| \leq \int_a^b |\varphi(t)| \, dt. \tag{2.8.6}$$

Proof. Consider the polar form

$$re^{i\theta} = \int_a^b \varphi(t) \, dt,$$

where $r \geq 0$. Then

$$\left| \int_a^b \varphi(t) \, dt \right| = r = \int_a^b e^{-i\theta} \varphi(t) \, dt.$$

Since the above quantities are real, we take the real part of both sides to find

$$\left| \int_a^b \varphi(t) \, dt \right| = \operatorname{Re} \int_a^b e^{-i\theta} \varphi(t) \, dt = \int_a^b \operatorname{Re}[e^{-i\theta} \varphi(t)] \, dt.$$

Now the inequality

$$\operatorname{Re}[e^{-i\theta} \varphi(t)] \leq |e^{-i\theta} \varphi(t)| = |\varphi(t)|$$

leads to

$$\int_a^b \operatorname{Re}[e^{-i\theta} \varphi(t)] \, dt \leq \int_a^b |\varphi(t)| \, dt,$$

proving (2.8.6). □

Summary and Notes for Section 2.8.

As we make our transition from real functions to complex functions, it is worthwhile to ask whether a difference we observe is because the domain is complex or because the range is complex. In the case of derivatives and integrals, we see that keeping a real domain and switching to a complex range has virtually no effect as we may simply consider the real and imaginary parts of the function. Furthermore, the chain rule works for compositions of functions real and complex.

   Despite the lack of novelty in this real-variable calculus, we note that the relationships between the exponential and trigonometric functions developed in Section 2.4 can be used here to evaluate real integrals by sidestepping into the complex realm. This idea will be revisited with more striking techniques later on, when we have developed a more robust complex integration theory.

Exercises for Section 2.8.

1. Use that $\tan^{-1} t = \mathrm{Arg}(1 + it)$ to find a formula for the derivative of $\tan^{-1} t$.

2. Let $a, b \in \mathbb{R} \setminus \{0\}$, and define $f(t) = e^{at} \cos bt$. Find the general antiderivative of f in two ways:

   (a) Integrate by parts twice and solve for the solution as is done in calculus.

   (b) Express f using the complex exponential function and use the results of this section.

3. Use the methods of this section to evaluate the following integrals. (Do not integrate by parts!)

   (a) $\displaystyle\int_0^{2\pi} e^{3t} \cos^2 2t\, dt$

   (b) $\displaystyle\int_0^{\pi} e^t \cos 3t \sin 4t\, dt$

4. Use the exponential forms of the trigonometric functions, the binomial formula (see Exercise 11 in Section 2.4), and (2.8.5) to evaluate the following integrals for any $n \in \mathbb{N}$.

$$\int_0^{2\pi} \cos^n t\, dt, \qquad \int_0^{2\pi} \sin^n t\, dt.$$

5. Let $[a, b] \subseteq \mathbb{R}$ and $\varphi \colon [a, b] \to \mathbb{C}$ be continuous such that $|\varphi(t)| \leq M$ for all $t \in [a, b]$, where $M > 0$. Show that if

$$\left| \int_a^b \varphi(t)\, dt \right| = M(b - a),$$

then $\varphi(t) = c$, where $c \in \mathbb{C}$ is a constant such that $|c| = M$.

6. Use the following steps to prove the fundamental theorem of calculus.

   (a) Prove the theorem in the case that f is continuous on $[a, b]$ by using real and imaginary parts and the version of the theorem seen in calculus.

   (b) Use part (a) on the partitioned integrals in (2.8.4) to complete the proof.

7. Prove that the algebraic derivative rules given in Exercise 8 of Section 2.7 also hold for functions of a real variable.

8. Give proofs of both versions of the chain rule presented in the section. (The hint for Exercise 11 in Section 2.7 may be helpful.)

9. Prove the integration-by-parts formula: If $f, g \colon [a, b] \to \mathbb{C}$ are continuous and piecewise continuously differentiable, then

$$\int_a^b f'(t)g(t)\, dt = f(b)g(b) - f(a)g(a) - \int_a^b g(t)f'(t)\, dt.$$

10. Show that if f is complex valued and piecewise continuously differentiable on $[a, b]$, then f is piecewise continuous on $[a, b]$. (*Hint*: Use the fundamental theorem of calculus.)

11. ▷ Let $\Omega \subseteq \mathbb{R}^2$ be open, $u, v \colon \Omega \to \mathbb{R}$, and $f = u + iv$. Explain how, as in Definition 2.8.1, the partial derivatives of f can be given by

$$\frac{\partial f}{\partial x} = \frac{\partial u}{\partial x} + i\frac{\partial v}{\partial x}, \qquad \frac{\partial f}{\partial y} = \frac{\partial u}{\partial y} + i\frac{\partial v}{\partial y}.$$

12. ▷ Let $R = [a, b] \times [c, d] \subseteq \mathbb{R}^2$ and $f \colon R \to \mathbb{C}$ be continuous. In the spirit of Definition 2.8.4, provide a definition for the double integral

$$\iint_R f(x, y)\, dA(x, y),$$

and show that *Fubini's theorem*

$$\iint_R f(x, y)\, dA(x, y) = \int_a^b \int_c^d f(x, y)\, dy\, dx = \int_c^d \int_a^b f(x, y)\, dx\, dy$$

holds. Use this to evaluate the following iterated integrals.

(a) $\displaystyle \int_0^2 \int_0^\pi y e^{2ixy}\, dy\, dx$

(b) $\displaystyle \int_0^1 \int_0^2 \frac{4xy}{(1 - ix^2y)^3}\, dy\, dx$

2.9 Contour Integrals

As with derivatives, there is also an integral calculus for functions of a complex variable. This extends significantly past a simple jump from real limits of integration to complex; complex integration has a distinct flavor and provides unique tools and insights to the study of function theory.

We saw in Section 2.8 how to define integrals of functions of a real variable with complex values. These are unsatisfactory, since most of the functions we are considering are of a complex variable. On occasion, using area integrals (double integrals) such as those seen in multivariable calculus is useful. However, the notion of a line integral, or contour integral, is much more so. What is especially nice is that in most practical circumstances, these integrals can be expressed using a real variable, allowing the work in Section 2.8 to apply.

2.9.1 Definition. Let $A \subseteq \mathbb{C}$. A *contour* in A is a function $\gamma \colon [a, b] \to A$, for some $a < b$, that is continuous and piecewise continuously differentiable. We often say that the contour is from $\gamma(a)$ to $\gamma(b)$. The contour is *closed* provided that $\gamma(a) = \gamma(b)$ and is *simple* if $\gamma(s) \neq \gamma(t)$ for distinct $s, t \in [a, b]$ such that $\{s, t\} \neq \{a, b\}$. The set of points $\gamma^* = \gamma([a, b])$ is called the *trace* of γ and may also be referred to as a contour, in which case γ is the *parameterization* of γ^*.

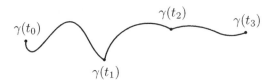

Figure 2.8   The trace of contour with partition $a = t_0 < t_1 < t_2 < t_3 = b$ in Definition 2.8.2

Without the requirement of piecewise continuous differentiability, γ would be called a *path* or *curve*. Observe that the real and imaginary parts of $z = \gamma(t)$ are parametric equations for a planar curve, as seen in multivariable calculus. Recall from Definition 2.8.2 that γ' may fail to exist at finitely many points. Figure 2.8 shows an illustration of this, although one cannot tell simply by looking at the trace whether γ' fails to exist or is equal to 0 at the "interesting" points.

Let us pause to remark on the imprecision of the definition of contour. On occasion, we will refer to a contour as a function defined on an interval, while later we may refer to a contour as a set of points. We invite this flexibility, as one point of view is usually more useful than the other depending on the circumstances. Consider the following examples.

2.9.2 Example. Define $\gamma_1 \colon [0, 2\pi] \to \mathbb{C}$ by $\gamma_1(t) = e^{it}$ and $\gamma_2 \colon [0, 4\pi] \to \mathbb{C}$ by $\gamma_2(t) = e^{it}$. The traces of these contours are the same, the unit circle. However, they are distinct contours; one traverses the circle once and the other twice. Notice that both contours are closed, but only the first is simple.

2.9.3 Example. Let E be a triangular region in the plane. Then $T = \partial E$ is a closed contour that we call a *triangular path*. While a function $\gamma \colon [a, b] \to \mathbb{C}$ can be defined such that $\gamma^* = T$, it is more natural to consider the set of points in this case. It is important to remark that in cases such as this, *we assume that the contour traverses the set once with a positive (counterclockwise) orientation unless otherwise noted!* Hence the contour is simple. That the starting/ending point remains ambiguous is not problematic, as will be seen in an exercise.

2.9.4 Example. Circles and line segments are commonly occurring contours, so we take note of their "standard" parameterizations. For $a \in \mathbb{C}$ and $r > 0$, we parameterize $\partial D(a; r)$ positively using $\gamma_1(t) = a + re^{it}$ for $0 \leq t \leq 2\pi$. For $\alpha, \beta \in \mathbb{C}$, we parameterize the line segment from α to β by $\gamma_2(t) = \alpha + t(\beta - \alpha) = (1 - t)\alpha + t\beta$ for $0 \leq t \leq 1$.

We now use contours for integration of complex functions.

2.9.5 Definition. Let $E \subseteq \mathbb{C}$ and $f \colon E \to \mathbb{C}$ be continuous. If $\gamma \colon [a, b] \to E$ is a contour, then we define the *contour integral* of f over γ by

$$\int_\gamma f(z)\, dz = \int_a^b f(\gamma(t))\gamma'(t)\, dt. \tag{2.9.1}$$

If C is a contour whose paramterization γ is understood, then we may write

$$\int_C f(z)\, dz \tag{2.9.2}$$

for the integral in (2.9.1).

These integrals form one of the primary tools for the study of functions of a complex variable. Note that the integrand on the right-hand side of (2.9.1) is piecewise continuous. The similarity to line integrals in vector calculus should be apparent and is addressed in an exercise.

Note that if $z = \gamma(t)$, then $dz = \gamma'(t)\, dt$ is the familiar relationship used for integral substitutions in calculus. When evaluating particular contour integrals, taking this point of view is often simpler than formally defining the parameterization of the contour.

2.9.6 Example. We use Example 2.9.4 to calculate

$$\int_{\partial \mathbb{D}} \frac{dz}{z} = \int_0^{2\pi} \frac{ie^{it}}{e^{it}}\, dt = i \int_0^{2\pi} dt = 2\pi i.$$

We can also think of using the integral substitution $z = e^{it}$, $dz = ie^{it}\, dt$, for $0 \le t \le 2\pi$.

The reader should recall that the definition of the integral in calculus is in terms of a limit of Riemann sums. Since Definition 2.8.4 depends upon real integrals from calculus and Definition 2.9.5 depends, in turn, upon Definition 2.8.4, our understanding of contour integrals then implicitly depends upon these Riemann sums. An alternative approach would be to define contour integrals directly as limits of Riemann sums and then prove equations (2.8.3) and (2.9.1). This less-typical approach would add nothing to our upcoming study of contour integration and hence is relegated to an exercise.

One point of consideration about integration over a contour C, described as a set of points, is that C could be parameterized in myriad ways.

2.9.7 Definition. Let $\gamma_1 \colon [a, b] \to \mathbb{C}$ and $\gamma_2 \colon [c, d] \to \mathbb{C}$ be contours. If there exists a continuous piecewise continuously differentiable increasing bijection $\sigma \colon [c, d] \to [a, b]$ such that $\gamma_2 = \gamma_1 \circ \sigma$, then γ_2 is a *reparameterization* of γ_1.

If γ_1 is a contour parameterizing C and γ_2 is a reparameterization of γ_1, then γ_2 also parameterizes C (the same number of times in the same direction with the same starting and ending points). Note that, although they have the same trace, the contours in Example 2.9.2 are not reparameterizations of one another. Depending on the nature of σ (for instance, if $\sigma'(t) = 0$ for some $t \in (c, d)$), γ_1 may not be a reparameterization of γ_2. Insisting that σ be continuously differentiable on (c, d) with $\sigma'(t) > 0$ for all t remedies this, but at the expense of generality.

The following makes us feel better about the notation (2.9.2).

2.9.8 Theorem. *Suppose that γ_1 is a contour in a set $E \subseteq \mathbb{C}$. If γ_2 is a reparameterization of γ_1, then*

$$\int_{\gamma_1} f(z)\, dz = \int_{\gamma_2} f(z)\, dz$$

for all continuous $f \colon E \to \mathbb{C}$.

Proof. Assume the notation in Definition 2.9.7. We use the method of substitution from calculus. If $s = \sigma(t)$, then $ds = \sigma'(t)\, dt$ wherever $\sigma'(t)$ exists. Since $\sigma(c) = a$ and $\sigma(d) = b$, we calculate

$$
\begin{aligned}
\int_{\gamma_2} f(z)\, dz &= \int_c^d f(\gamma_2(t))\gamma_2'(t)\, dt \\
&= \int_c^d f(\gamma_1(\sigma(t)))\gamma_1'(\sigma(t))\sigma'(t)\, dt \\
&= \int_a^b f(\gamma_1(s))\gamma_1'(s)\, ds \\
&= \int_{\gamma_1} f(z)\, dz,
\end{aligned}
$$

noting the integrands of the real-variable integrals are piecewise continuous because σ is a bijection and the substitution rule can be applied to real and imaginary parts on subintervals. $\qquad\square$

We now consider some "algebraic" operations on contours.

2.9.9 Definition. Suppose that C_1 and C_2 are the traces of contours $\gamma_1 \colon [a, b] \to \mathbb{C}$ and $\gamma_2 \colon [b, c] \to \mathbb{C}$ and that $\gamma_1(b) = \gamma_2(b)$. We define the contour $C = C_1 + C_2$ to be the trace of $\gamma \colon [a, c] \to \mathbb{C}$ given by

$$\gamma(t) = \begin{cases} \gamma_1(t) & \text{if } t \in [a, b], \\ \gamma_2(t) & \text{if } t \in [b, c]. \end{cases}$$

It is, of course, sufficient for the terminal point of C_1 to lie at the initial point of C_2, and then reparameterization can be used to get the intervals to match up.

If the above is "contour addition," then following provides for "subtraction."

2.9.10 Definition. Let C be the trace of the contour $\gamma \colon [a, b] \to \mathbb{C}$. Then define the contour $-C$ to be the trace of $\gamma_1 \colon [-b, -a] \to \mathbb{C}$ given by

$$\gamma_1(t) = \gamma(-t).$$

Then C and $-C$ contain the same points in $\mathbb{C}$, but they are parameterized in opposite directions.

These new contours affect integration in the predictable way. The proof of the following is left as an exercise.

2.9.11 Theorem. *Let C_1, C_2, and C be contours in the set $E \subseteq \mathbb{C}$ such that $C_1 + C_2$ exists. Then for any continuous $f\colon E \to \mathbb{C}$,*

$$\int_{C_1+C_2} f(z)\,dz = \int_{C_1} f(z)\,dz + \int_{C_2} f(z)\,dz$$

and

$$\int_{-C} f(z)\,dz = -\int_{C} f(z)\,dz.$$

We have carried on proving theorems about integrals without answering one important and obvious question: In calculus, the fundamental theorem of calculus allows us to easily evaluate integrals in certain circumstances. Can the same be done here? Yes.

2.9.12 Fundamental Theorem of Calculus for Contour Integrals. *Let $\Omega \subseteq \mathbb{C}$ be open, $f\colon \Omega \to \mathbb{C}$ be continuous, and $\gamma\colon [a, b] \to \Omega$ be a contour. If $F\colon \Omega \to \mathbb{C}$ is such that $F' = f$, then*

$$\int_{\gamma} f(z)\,dz = F(\gamma(b)) - F(\gamma(a)).$$

Proof. By the chain rule,

$$\frac{d}{dt} F(\gamma(t)) = f(\gamma(t))\gamma'(t)$$

for all t such that $\gamma'(t)$ exists. Hence

$$\int_{\gamma} f(z)\,dz = \int_{a}^{b} f(\gamma(t))\gamma'(t)\,dt = F(\gamma(t))]_{a}^{b}$$

by the real-variable fundamental theorem (Theorem 2.8.5). □

Note that Ω is assumed to be open in the above theorem because the definition of the derivative relies on the domain of the function being open. Here is an example when the fundamental theorem does not apply.

2.9.13 Example. Consider Example 2.9.6. Since $\gamma(0) = \gamma(2\pi)$, the fundamental theorem would seem to imply that the integral should be 0. The theorem does *not* apply, however, because $f(z) = 1/z$ does not have an antiderivative in an open set containing $\partial\mathbb{D}$. Any such antiderivative would be a branch of the logarithm, which cannot be defined as necessary.

The proof of the following useful corollary to the fundamental theorem is left as an exercise.

2.9.14 Corollary. *If C is a closed contour in the open set $\Omega \subseteq \mathbb{C}$, $f: \Omega \to \mathbb{C}$ is continuous, and $f = F'$ for some $F: \Omega \to \mathbb{C}$, then*

$$\int_C f(z)\, dz = 0.$$

Let us consider another example in the style of Example 2.9.6.

2.9.15 Example. Let $a \in \mathbb{C}$ and $n \in \mathbb{Z}$. If $n \geq 0$, then the function $z \mapsto (z-a)^n$ has an antiderivative on all of $\mathbb{C}$. If $n \leq -2$, then $z \mapsto (z-a)^n$ has an antiderivative on all of $\mathbb{C} \setminus \{a\}$. Specifically, if $r > 0$ then $z \mapsto (z-a)^n$ has an antiderivative on an open set containing $\partial D(a; r)$ as long as $n \neq -1$. Combining this observation with the technique of Example 2.9.6, we have that

$$\int_{\partial D(a;r)} (z-a)^n\, dz = \begin{cases} 0 & \text{if } n \neq -1, \\ 2\pi i & \text{if } n = -1. \end{cases}$$

2.9.16 Example. Here, we make use of Example 2.9.15 and power series to evaluate an integral. Using the series found in Example 2.4.13, we have

$$\int_{\partial D} \frac{\cos^2 z}{z^3}\, dz = \int_{\partial D} \frac{1}{z^3} \left(1 + \sum_{k=1}^{\infty} \frac{(-1)^k 2^{2k-1} z^{2k}}{(2k)!} \right) dz$$

$$= \int_{\partial D} z^{-3}\, dz - \int_{\partial D} z^{-1}\, dz + \int_{\partial D} \sum_{k=2}^{\infty} \frac{(-1)^k 2^{2k-1} z^{2k-3}}{(2k)!}\, dz.$$

Example 2.9.15 shows that the first integral is equal to 0 and the second is equal to $2\pi i$. The last integral is of a power series that converges on $\mathbb{C}$. Since the series has an antiderivative on $\mathbb{C}$ (Corollary 2.7.5), its integral is 0 by Corollary 2.9.14. Therefore

$$\int_{\partial D} \frac{\cos^2 z}{z^3}\, dz = -2\pi i.$$

As seen in calculus, the arc length of a curve is quite a useful quantity, especially when it comes to finding upper bounds on the values of integrals. Suppose that $\gamma: [a, b] \to \mathbb{C}$ is a contour, and write $\gamma = \alpha + i\beta$ for $\alpha, \beta: [a, b] \to \mathbb{R}$ in parametric form. From calculus, we know that the arc length of γ^* (in $\mathbb{R}^2$ terms) is given by

$$L(\gamma^*) = \int_a^b \sqrt{[\alpha'(t)]^2 + [\beta'(t)]^2}\, dt.$$

Note that

$$\sqrt{[\alpha'(t)]^2 + [\beta'(t)]^2} = |\alpha'(t) + i\beta'(t)| = |\gamma'(t)|,$$

and thus more simply,

$$L(\gamma^*) = \int_a^b |\gamma'(t)|\, dt. \tag{2.9.3}$$

This equation leads to the following definition, which may be of a familiar concept from the study of line integrals in multivariable calculus.

2.9.17 Definition. Let $E \subseteq \Omega$, $\gamma\colon [a,b] \to E$ be a contour, and $f\colon E \to \mathbb{C}$ be continuous. The *integral of f over γ with respect to arc length* is

$$\int_\gamma f(z)\,|dz| = \int_a^b f(\gamma(t))|\gamma'(t)|\,dt. \tag{2.9.4}$$

It is left as an exercise to see how the integral with respect to arc length can be alternatively defined using Riemann sums. Following Definition 2.9.5, we noted the integral substitution $z = \gamma(t)$, $dz = \gamma'(t)\,dt$. We can apply this to the above by using $|dz| = |\gamma'(t)|\,dt$.

We use integration with respect to arc length to obtain an important upper bound on the modulus of a contour integral.

2.9.18 Theorem. *If $E \subseteq \mathbb{C}$, $f\colon E \to \mathbb{C}$ is continuous, and $\gamma\colon [a,b] \to E$ is a contour, then*

$$\left| \int_\gamma f(z)\,dz \right| \leq \int_\gamma |f(z)|\,|dz|. \tag{2.9.5}$$

Proof. We have

$$\left| \int_\gamma f(z)\,dz \right| = \left| \int_a^b f(\gamma(t))\gamma'(t)\,dt \right| \leq \int_a^b |f(\gamma(t))||\gamma'(t)|\,dt = \int_\gamma |f(z)|\,|dz|,$$

by letting $\varphi(t) = f(\gamma(t))\gamma'(t)$ in Theorem 2.8.8. $\qquad\square$

Let us consider how uniform convergence interacts with contour integration.

2.9.19 Theorem. *Let $E \subseteq \mathbb{C}$, and suppose that $\gamma\colon [a,b] \to E$ is a contour. If $\{f_n\}_{n=1}^\infty$ is a sequence of continuous functions on E that converges uniformly to the function $f\colon E \to \mathbb{C}$, then*

$$\lim_{n\to\infty} \int_\gamma f_n(z)\,dz = \int_\gamma f(z)\,dz.$$

Proof. We know that f is continuous by Theorem 2.2.5. Let $\varepsilon > 0$. Set $L = \int_\gamma |dz|$, the arc length of γ^*. There exists $N \in \mathbb{N}$ such that $|f_n(z) - f(z)| < \varepsilon/L$ for all $n \geq N$ and $z \in E$. And thus for $n \geq N$, we have

$$\left| \int_\gamma f_n(z)\,dz - \int_\gamma f(z)\,dz \right| \leq \int_\gamma |f_n(z) - f(z)|\,|dz| < \int_\gamma \frac{\varepsilon}{L}\,|dz| = \varepsilon.$$

This establishes the desired convergence. $\qquad\square$

Using inequalities with integrals with respect to arc length as above is a standard technique. The nervous reader should consult the exercises.

The next corollary follows by applying Theorem 2.9.19 to the sequence of partial sums. Its proof is left as an exercise.

2.9.20 Corollary. *Let $E \subseteq \mathbb{C}$, and suppose that $\gamma \colon [a, b] \to E$ is a contour. If $\{f_n\}_{n=1}^{\infty}$ is a sequence of continuous functions on E such that $\sum_{n=1}^{\infty} f_n$ converges uniformly on E, then*

$$\int_{\gamma} \sum_{n=1}^{\infty} f_n(z)\, dz = \sum_{n=1}^{\infty} \int_{\gamma} f_n(z)\, dz,$$

implying convergence of the right-hand side.

The contour integral in the following example reappears in a proof in Section 3.2.

2.9.21 Example. For fixed $z \in \mathbb{D}$, we calculate the integral

$$\int_{\partial \mathbb{D}} \frac{d\zeta}{\zeta - z}.$$

We use the geometric series to expand

$$\frac{1}{\zeta - z} = \frac{1}{\zeta} \frac{1}{1 - z/\zeta} = \frac{1}{\zeta} \sum_{n=0}^{\infty} \frac{z^n}{\zeta^n} = \sum_{n=0}^{\infty} \frac{z^n}{\zeta^{n+1}}$$

for $\zeta \in \partial \mathbb{D}$. The series converges uniformly for such ζ by an application of the Weierstrass M-test (Theorem 2.2.7). Corollary 2.9.20 then gives

$$\int_{\partial \mathbb{D}} \frac{d\zeta}{\zeta - z} = \sum_{n=0}^{\infty} z^n \int_{\partial \mathbb{D}} \zeta^{-n-1}\, d\zeta = 2\pi i.$$

(Example 2.9.15 gives the values of the individual integrals in the series, most of which are 0.) Note the integral has the same value for all $z \in \mathbb{D}$.

Summary and Notes for Section 2.9.

Moving to a complex domain is similar to moving from $\mathbb{R}$ to $\mathbb{R}^2$ in multivariable calculus, and we turn to contour integrals when considering functions defined in planar regions. (We note that our restriction to contours instead of something more general is a convenient choice that does not diminish the results.) These integrals are actually simpler in $\mathbb{C}$ than line integrals are in $\mathbb{R}^2$ because there is only one planar variable, so their definition is analogous to making an integral substitution. Furthermore, we see that integrating over a real segment is a special case of this broader notion.

Two things are worth special attention. Observe how the fundamental theorem of calculus "fails" in Example 2.9.6 because the logarithm cannot be appropriately defined. This will be a common theme in upcoming sections. Also, note how uniform convergence is once again central to interchanging limit processes in Theorem 2.9.19

as was discussed at the end of Section 2.2, where here we recall that the definition of the integral is a limit.

The concept of complex contour integration was explored by Gauss and Cauchy (independently) in the early part of the 19th century. It is with complex integration that Cauchy really made his mark, as we will soon be convinced by the number of times his name appears in the following chapters.

Exercises for Section 2.9.

1. Evaluate the following contour integrals, where the contour C is as specified. Assume a counterclockwise orientation when appropriate.

   (a) $\displaystyle\int_C e^z \, dz,$ C is the segment from $1 + i$ to $2 + i$

   (b) $\displaystyle\int_C \frac{dz}{z-2},$ C is the top half of $\partial D(2; 1)$

   (c) $\displaystyle\int_C \bar{z} \, dz,$ C is the triangular path with vertices 0, 2, and $1 + i$

   (d) $\displaystyle\int_C \operatorname{Re} z \, dz,$ $C = \partial D(1; 1)$

   (e) $\displaystyle\int_C z^2 \, |dz|,$ $C = \partial D(i; 2)$

2. Let C be the segment from 1 to i. Use Theorem 2.9.18 to determine an upper bound on

$$\left| \int_C \cos^2 z \, dz \right|.$$

3. Let $\gamma\colon [0, 1] \to \mathbb{C}$ be given by $\gamma(t) = t + it^2$. Sketch γ^*. If $z = x + iy$, evaluate

$$\int_\gamma \left(2xy - ix^2\right) dz.$$

4. Use power series to calculate the following contour integrals.

   (a) $\displaystyle\int_\gamma \frac{\sin^2 z}{z^6} \, dz,$ $\gamma(t) = e^{it},\ 0 \leq t \leq 2\pi$

   (b) $\displaystyle\int_\gamma \frac{e^z + e^{2z}}{z^4} \, dz,$ $\gamma(t) = 2e^{it},\ 0 \leq t \leq 4\pi$

5. Let $\gamma\colon [0, 2\pi] \to \mathbb{C}$ be given by $\gamma(t) = e^{it}$. Calculate

$$\int_\gamma \frac{z^2 - 1}{z(z^2 + 4)} \, dz.$$

   (*Hint*: Use partial fractions.)

6. Let C be a contour in $\mathbb{C}$ with initial point α and terminal point β. Evaluate

$$\int_C z^2 \, dz.$$

7. Prove Theorem 2.9.11.

8. Prove Corollary 2.9.14.

9. Prove Corollary 2.9.20.

10. Let $E \subseteq \mathbb{C}$, $C \subseteq E$ be a closed contour, and $\alpha, \beta \in C$ be distinct. Suppose that γ_1 parameterizes C beginning and ending at α and γ_2 parameterizes C beginning and ending at β, both once in the same direction. Show that

$$\int_{\gamma_1} f(z)\, dz = \int_{\gamma_2} f(z)\, dz$$

for any continuous $f: E \to \mathbb{C}$. This shows that the choice of starting/ending point of the parameterization of a closed contour does not affect integration over the contour.

11. ▷ Let $E \subseteq \mathbb{C}$ and $\gamma: [a, b] \to E$ be a contour. Show that if $f, g: E \to \mathbb{R}$ are continuous and satisfy $f(z) \le g(z)$ on γ^*, then

$$\int_{\gamma} f(z)\, |dz| \le \int_{\gamma} g(z)\, |dz|.$$

This was used in the proof of Theorem 2.9.19 and will be freely used in upcoming sections.

12. Let $a \in \mathbb{C}$ and $r > 0$. Suppose that $f: D(a;r) \to \mathbb{C}$ is continuously differentiable and $|f'(z)| \le M$ for some $M > 0$ and all $z \in D(a;r)$. Prove that

$$|f(w) - f(z)| \le M|w - z|$$

for all $z, w \in D(a;r)$.

13. Let $f: \partial\mathbb{D} \to \mathbb{C}$ be continuous such that $|f(z)| \le M$ for all $z \in \partial\mathbb{D}$, where $M > 0$. Prove that if

$$\left| \int_{\partial\mathbb{D}} f(z)\, dz \right| = 2\pi M,$$

then $f(z) = c\bar{z}$, where $c \in \mathbb{C}$ is a constant such that $|c| = M$. (*Hint*: Consider Exercise 5 in Section 2.8.)

14. Let $f: \mathbb{C} \to \mathbb{C}$ be given by

$$f(z) = \int_0^1 e^{-zt} t^2\, dt.$$

(a) Calculate a power series for f based at 0. What is its radius of convergence?

(b) Find $f'(0)$.

15. Fill in the details, including partitioning intervals and breaking into real and imaginary parts, of how the substitution rule is used in the proof of Theorem 2.9.8.

16. Prove *Fubini's theorem* for contour integrals: Let $A, B \subseteq \mathbb{C}$, and let $C_1 \subseteq A$ and $C_2 \subseteq B$ be contours. If $\varphi: A \times B \to \mathbb{C}$ is continuous (see Exercise 16 in Section 2.1), then

$$\int_{C_1} \int_{C_2} \varphi(z, w)\, dw\, dz = \int_{C_2} \int_{C_1} \varphi(z, w)\, dz\, dw.$$

(*Hint*: Parameterize the contours using γ_1, γ_2 to make a double integral over a rectangle $R = [a, b] \times [c, d]$. Use the piecewise continuous differentiability of γ_1, γ_2 to subdivide

$R = \bigcup_{j=1}^{m} \bigcup_{k=1}^{n} [s_{j-1}, s_j] \times [t_{k-1}, t_k]$ and apply the real-variable Fubini's theorem [Exercise 12 of Section 2.8] to each sub-rectangle.)

17. Relate the definition of contour integrals in $\mathbb{C}$ to line integrals of real-valued functions in $\mathbb{R}^2$ in the following way. Suppose that C is a contour in $\mathbb{R}^2 \approx \mathbb{C}$. If $f \colon C \to \mathbb{C}$ is continuous, write $u, v \colon C \to \mathbb{R}$ as the real and imaginary parts of f (where $z = x + iy$). Show that

$$\int_C f(z)\, dz = \int_C u(x,y)\, dx - v(x,y)\, dy + i \int_C v(x,y)\, dx + u(x,y)\, dy.$$

Explain how this suggests the relationship $dz = dx + i\, dy$.

18. Let $\Omega \subseteq \mathbb{C}$ be open. Prove that Ω is connected if and only if for every $\alpha, \beta \in \Omega$, there is a contour $\gamma \colon [a,b] \to \Omega$ such that $\gamma(a) = \alpha$ and $\gamma(b) = \beta$. (*Hint*: Consider Exercise 16 in Section 1.4.)

19. In this exercise, we consider directly defining contour integrals in terms of Riemann sums: Let $E \subseteq \mathbb{C}$, $\gamma \colon [a,b] \to E$ be a contour, $f \colon E \to \mathbb{C}$, and $I \in \mathbb{C}$. If for all $\varepsilon > 0$, there exists $\delta > 0$ such that for any partition

$$a = t_0 < t_1 < \cdots < t_n = b$$

with $\Delta t_k = t_k - t_{k-1} < \delta$ for all k and for all choices of sample points $t_k^* \in [t_{k-1}, t_k]$ for all k,

$$\left| I - \sum_{k=1}^{n} f(z_k^*) \Delta z_k \right| < \varepsilon, \tag{2.9.6}$$

where $z_k^* = \gamma(t_k^*)$ and $\Delta z_k = \gamma(t_k) - \gamma(t_{k-1})$, then I is said to be the *contour integral of f over γ*, denoted

$$I = \int_\gamma f(z)\, dz.$$

Answer the following.

(a) In the special case that $\gamma(t) = t$, we write

$$I = \int_a^b f(t)\, dt.$$

Appeal to the definition of the integral from calculus to show that if $u, v \colon [a,b] \to \mathbb{R}$ are Riemann integrable and $f = u + iv$, then f is integrable and (2.8.3) holds.

(b) If f is continuous and γ is continuously differentiable over (a,b), prove that the contour integral of f exists and (2.9.1) holds. (*Hint*: Given $\varepsilon > 0$, use the mean value theorem from calculus and uniform continuity [see Appendix B] to show that there exists $\delta_1 > 0$ such that

$$\left| \sum_{k=1}^{n} f(z_k^*)[\Delta z_k - \gamma'(t_k^*)\Delta t_k] \right| < \frac{\varepsilon}{2}$$

holds for all partitions satisfying $\Delta t_k < \delta_1$ and sample points $t_k^* \in [t_{k-1}, t_k]$.)

(c) Let $c \in (a, b)$, and define $\gamma_1 \colon [a, c] \to \mathbb{C}$ and $\gamma_2 \colon [c, b] \to \mathbb{C}$ by $\gamma_1(t) = \gamma(t)$ and $\gamma_2(t) = \gamma(t)$. Show that if the contour integrals of f over γ_1 and γ_2 exist, then

$$\int_\gamma f(z)\, dz = \int_{\gamma_1} f(z)\, dz + \int_{\gamma_2} f(z)\, dz,$$

implying the existence of the left-hand side. Conclude, using induction, that if f is continuous then the contour integral of f exists for any contour γ and is given by (2.9.1).

(d) If we replace (2.9.6) with

$$\left| I - \sum_{k=1}^n f(z_k^*)\, |\Delta z_k| \right| < \varepsilon,$$

then I is the arc length integral

$$I = \int_\gamma f(z)\, |dz|.$$

By modifying the argument in (b), show that (2.9.4) holds.

CHAPTER 3

ANALYTIC FUNCTIONS

The study of analytic functions is the fundamental purpose of complex analysis. Sections 3.3–3.7 contain a wealth of strong properties of these functions which will no doubt seem surprising, and possibly unbelievable, to those familiar only with the calculus of real functions.

In Sections 2.7 and 2.8, we contrasted the definitions of the real and complex derivative. We are on the verge of seeing just how different they are!

3.1 The Principle of Analyticity

Now that we have considered the basics of the complex plane and the calculus of functions of a complex variable, we are ready to address the primary object of study in function theory.

3.1.1 Definition. Let $\Omega \subseteq \mathbb{C}$ be open. A function $f\colon \Omega \to \mathbb{C}$ is *analytic* provided that for every $a \in \Omega$, there exists an $r > 0$ and a sequence $\{c_n\}_{n=0}^{\infty}$ of complex numbers such that

$$f(z) = \sum_{n=0}^{\infty} c_n(z - a)^n \qquad (3.1.1)$$

87

Complex Analysis: A Modern First Course in Function Theory, First Edition. Jerry R. Muir, Jr.
© 2015 John Wiley & Sons, Inc. Published 2015 by John Wiley & Sons, Inc.

for all $z \in D(a; r) \subseteq \Omega$. The set of all analytic functions on Ω is denoted by $H(\Omega)$.

Since Ω is open, there is an open disk in Ω about any given point $a \in \Omega$, and therefore the convergence of a power series to f on some $D(a; r)$ is the significant portion of the above definition.

It may seem strange that $H(\Omega)$ is used to denote the set of analytic functions on Ω instead of $A(\Omega)$. It is common to reserve $A(\Omega)$ for the set of all continuous functions $f : \overline{\Omega} \to \mathbb{C}$ that are analytic on Ω. Often the word *holomorphic* is used in place of analytic, explaining the notation. In older publications, the word *regular* is also used for analytic.

3.1.2 Example. Let $\Omega = \mathbb{C} \setminus \{1\}$, and define $f : \Omega \to \mathbb{C}$ by

$$f(z) = \frac{1}{1 - z}.$$

We know that f is equal to a power series based at $a = 0$ converging on $\mathbb{D}$. (It is the geometric series.) To see that $f \in H(\Omega)$, let $a \in \Omega$ and use the geometric series to calculate

$$\begin{aligned}
f(z) &= \frac{1}{1 - a} \frac{1}{1 - (z - a)/(1 - a)} \\
&= \frac{1}{1 - a} \sum_{n=0}^{\infty} \left(\frac{z - a}{1 - a} \right)^n \\
&= \sum_{n=0}^{\infty} \frac{(z - a)^n}{(1 - a)^{n+1}},
\end{aligned}$$

valid for $z \in D(a; |1 - a|) \subseteq \Omega$.

Corollaries 2.3.5 and 2.7.5 provide the following elementary properties of analytic functions.

3.1.3 Theorem. *Let $\Omega \subseteq \mathbb{C}$ be open and $f \in H(\Omega)$. Then f is continuous and infinitely differentiable on Ω, and all derivatives of f are analytic on Ω. Furthermore, f has infinitely many antiderivatives on a neighborhood of (disk centered at) any point in Ω.*

Summary and Notes for Section 3.1.

The idea of an analytic function is central to complex analysis and will occupy us for the next few chapters. Our definition of analyticity is local representation by a convergent power series. We will see that there are other conditions on a function equivalent to this that could be used as alternate definitions and will point these out as they are encountered.

Clearly, analytic functions of a real variable could be (and are) similarly defined. However, since power series must converge in disks, these functions are nothing more than restrictions of complex analytic functions to real domains. Furthermore, almost all of the profound properties of analytic functions we are about to see are tied to planar domains and fail to hold for functions of a real variable.

Exercises for Section 3.1.

1. Suppose that $p \colon \mathbb{C} \to \mathbb{C}$ is a polynomial. Show that $p \in H(\mathbb{C})$. (*Hint*: To expand p in a series about a given $a \in \mathbb{C}$, recognize that for each $n \in \mathbb{N}$, $z^n = (z - a + a)^n$ and use the binomial formula, which is stated in Exercise 11 of Section 2.4.)

2. Manipulate the power series of the exponential function (2.4.1) to show that the following functions are analytic on $\mathbb{C}$.

    (a) $f(z) = e^z$

    (b) $f(z) = \sin z$

    (c) $f(z) = \cos z$

3. Let $\alpha \in \mathbb{R}$ and $f(z) = \log z$ be the branch of the logarithm corresponding to $\alpha < \arg z < \alpha + 2\pi$. Manipulate the geometric series to show that f is analytic on its domain.

4. Let $\Omega \subseteq \mathbb{C}$ be open, $f, g \in H(\Omega)$, and $c \in \mathbb{C}$. Show that following functions are analytic on Ω.

    (a) cf

    (b) $f + g$

    (c) fg (*Hint*: Consider Exercise 6 in Section 2.3.)

3.2   Differentiable Functions are Analytic

We now address the first significant milestone in our development of analytic functions, revealing a remarkable difference between the function theories of real and complex variables. Specifically, in the theory of functions of a real variable, the notion of a differentiable function is common, and certainly analytic functions are naturally defined, as well. The important point is that there are many differentiable real-variable functions that are not analytic, which is why the following theorem is so spectacular. Results like this give complex analysis its identity as an area of study. Notice that it is a converse to Theorem 3.1.3.

3.2.1 Goursat's Theorem. *Let $\Omega \subseteq \mathbb{C}$ be open. If $f \colon \Omega \to \mathbb{C}$ is differentiable, then f is analytic on Ω.*

This amazing theorem will take some care to prove. In fact, we will develop some helpful results first, each of which is important in its own right. This first lemma and subsequent corollary will be generalized later in Section 4.2.

3.2.2 Lemma. *Let $\Omega \subseteq \mathbb{C}$ be open such that $\overline{\mathbb{D}} \subseteq \Omega$. If $f \colon \Omega \to \mathbb{C}$ is differentiable and f' is continuous, then*

$$f(z) = \frac{1}{2\pi i} \int_{\partial \mathbb{D}} \frac{f(\zeta)}{\zeta - z} \, d\zeta$$

for all $z \in \mathbb{D}$.

A first glimpse at a big theme in complex integration, this lemma indicates that the values of $f(z)$ for $z \in \mathbb{D}$ are completely determined by the values of f on $\partial \mathbb{D}$. The proof makes use of Leibniz's rule from advanced calculus, stated and proved in Appendix B, which allows for differentiation to move inside of a partial integral.

Proof. For fixed $z \in \mathbb{D}$, define $\varphi \colon [0,1] \times [0, 2\pi] \to \mathbb{C}$ by

$$\varphi(s,t) = \frac{e^{it} f((1-s)z + se^{it})}{e^{it} - z}.$$

For each $t \in [0, 2\pi]$, $\{(1-s)z + se^{it} : s \in [0,1]\}$ is the line segment between z and e^{it} and therefore lies in $\overline{\mathbb{D}}$; hence φ is well defined. In addition, φ is continuous, as is

$$\frac{\partial \varphi}{\partial s}(s,t) = e^{it} f'((1-s)z + se^{it}).$$

By Leibniz's rule, the function $g \colon [0,1] \to \mathbb{C}$ given by

$$g(s) = \int_0^{2\pi} \varphi(s,t) \, dt$$

is continuously differentiable with derivative

$$g'(s) = \int_0^{2\pi} e^{it} f'((1-s)z + se^{it}) \, dt = \left. \frac{f((1-s)z + se^{it})}{is} \right]_{t=0}^{2\pi} = 0$$

for $s \in (0,1)$. Thus g is constant. We now apply the integral in Example 2.9.21 to calculate

$$g(0) = \int_0^{2\pi} \varphi(0,t) \, dt = \int_0^{2\pi} \frac{e^{it} f(z)}{e^{it} - z} \, dt = \frac{f(z)}{i} \int_{\partial \mathbb{D}} \frac{d\zeta}{\zeta - z} = 2\pi f(z),$$

where we used the substitution $\zeta = e^{it}$. It now follows that $g(s) = 2\pi f(z)$ for all $s \in [0,1]$. Therefore

$$f(z) = \frac{g(1)}{2\pi} = \frac{1}{2\pi} \int_0^{2\pi} \frac{e^{it} f(e^{it})}{e^{it} - z} \, dt = \frac{1}{2\pi i} \int_{\partial \mathbb{D}} \frac{f(\zeta)}{\zeta - z} \, d\zeta,$$

using the same substitution. $\qquad \square$

The following generalization comes from substitution. Its proof is left as an exercise.

3.2.3 Corollary. *Let $\Omega \subseteq \mathbb{C}$ be open, $a \in \Omega$, and $r > 0$ be such that $\overline{D}(a; r) \subseteq \Omega$. If $f \colon \Omega \to \mathbb{C}$ is differentiable and f' is continuous, then*

$$f(z) = \frac{1}{2\pi i} \int_{\partial D(a;r)} \frac{f(\zeta)}{\zeta - z} \, d\zeta \tag{3.2.1}$$

for all $z \in D(a; r)$.

We close in on Goursat's theorem with the following result.

3.2.4 Theorem. *Let $\Omega \subseteq \mathbb{C}$ be open. If $f \colon \Omega \to \mathbb{C}$ is differentiable and f' is continuous, then f is analytic on Ω.*

Proof. Fix $a \in \Omega$, and let $r > 0$ be such that $\overline{D}(a; r) \subseteq \Omega$. Now f can be written in the form (3.2.1) for all $z \in D(a; r)$.

Fix $z \in D(a; r)$. For $\zeta \in \partial D(a; r)$, we use the geometric series to calculate

$$
\begin{aligned}
\frac{1}{\zeta - z} &= \frac{1}{\zeta - a} \frac{1}{1 - [(z - a)/(\zeta - a)]} \\
&= \frac{1}{\zeta - a} \sum_{n=0}^{\infty} \left(\frac{z - a}{\zeta - a} \right)^n \\
&= \sum_{n=0}^{\infty} \frac{(z - a)^n}{(\zeta - a)^{n+1}}.
\end{aligned}
\tag{3.2.2}
$$

Therefore

$$
\frac{f(\zeta)}{\zeta - z} = \sum_{n=0}^{\infty} \frac{f(\zeta)(z - a)^n}{(\zeta - a)^{n+1}}.
\tag{3.2.3}
$$

The left-hand side of (3.2.3) is the integrand in (3.2.1). We will see that the right-hand side of (3.2.3) converges uniformly for $\zeta \in \partial D(a; r)$.

Let

$$
M = \sup\{|f(\zeta)| : \zeta \in \partial D(a; r)\}.
$$

(The supremum exists because f is continuous and $\partial D(a; r)$ is compact.) Note that

$$
\frac{|f(\zeta)||z - a|^n}{|\zeta - a|^{n+1}} \leq \frac{M|z - a|^n}{r^{n+1}} = \frac{M}{r} \left(\frac{|z - a|}{r} \right)^n
$$

for $\zeta \in \partial D(a; r)$. Since $|z - a|/r < 1$, the right-hand side of the above inequality is the nth term of a convergent geometric series. The Weierstrass M-test (Theorem 2.2.7) now implies that the right-hand side of (3.2.3) converges uniformly for $\zeta \in \partial D(a; r)$. We may therefore use Corollary 2.9.20 to see that

$$
\begin{aligned}
f(z) &= \frac{1}{2\pi i} \int_{\partial D(a; r)} \sum_{n=0}^{\infty} \frac{f(\zeta)(z - a)^n}{(\zeta - a)^{n+1}} \, d\zeta \\
&= \sum_{n=0}^{\infty} \left[\frac{1}{2\pi i} \int_{\partial D(a; r)} \frac{f(\zeta)}{(\zeta - a)^{n+1}} \, d\zeta \right] (z - a)^n.
\end{aligned}
$$

Letting

$$
c_n = \frac{1}{2\pi i} \int_{\partial D(a; r)} \frac{f(\zeta)}{(\zeta - a)^{n+1}} \, d\zeta,
\tag{3.2.4}
$$

shows that

$$
f(z) = \sum_{n=0}^{\infty} c_n (z - a)^n.
$$

Since $z \in D(a;r)$ was arbitrarily chosen and c_n is independent of z, we have that f is analytic on Ω. $\square$

Most results in complex integration theory are stated for functions assumed to be analytic on some open set. In contrast, the following theorem provides a way in which integrals can be used to *conclude* a function is analytic.

3.2.5 Morera's Theorem. *Let $\Omega \subseteq \mathbb{C}$ be open and $f: \Omega \to \mathbb{C}$ be continuous. If*

$$\int_T f(z)\,dz = 0$$

for every triangular path $T \subseteq \Omega$, then f is analytic on Ω.

In the following proof, we use the notation $[a, b]$ for distinct points $a, b \in \mathbb{C}$ to indicate the directed line segment from a to b.

Proof. It suffices to show that f is analytic on every open disk in Ω, so suppose that $D(a;r) \subseteq \Omega$ for some $a \in \Omega$ and $r > 0$. We will see that the function $F: D(a;r) \to \mathbb{C}$ given by

$$F(z) = \int_{[a,z]} f(\zeta)\,d\zeta$$

satisfies $F' = f$ on $D(a;r)$. With that in place, the continuity of f implies F is analytic by Theorem 3.2.4. But then f is analytic on $D(a;r)$ by Theorem 3.1.3, completing the proof.

Fix $z_0 \in D(a;r)$. For any given $z \in D(a;r) \setminus \{z_0\}$, the segments $[a, z]$, $[z, z_0]$, and $[z_0, a]$ sum (in the sense of Theorem 2.9.11) to a triangular path T (unless a, z_0, and z lie on the same line, which presents no complication). By hypothesis,

$$0 = \int_T f(\zeta)\,d\zeta$$

$$= \int_{[a,z]} f(\zeta)\,d\zeta + \int_{[z,z_0]} f(\zeta)\,d\zeta + \int_{[z_0,a]} f(\zeta)\,d\zeta$$

$$= F(z) - F(z_0) + \int_{[z,z_0]} f(\zeta)\,d\zeta.$$

Now

$$F(z) - F(z_0) = \int_{[z_0,z]} f(\zeta)\,d\zeta.$$

Subtracting $(z - z_0)f(z_0) = \int_{[z_0,z]} f(z_0)\,d\zeta$ and dividing by $z - z_0$ gives

$$\frac{F(z) - F(z_0)}{z - z_0} - f(z_0) = \frac{1}{z - z_0} \int_{[z_0,z]} (f(\zeta) - f(z_0))\,d\zeta \tag{3.2.5}$$

for all $z \in D(a;r) \setminus \{z_0\}$.

Let $\varepsilon > 0$. Since f is continuous, there is some $\delta > 0$ such that $|f(z) - f(z_0)| < \varepsilon$ whenever $z \in \Omega$ satisfies $|z - z_0| < \delta$. Therefore let $z \in D(a;r)$ be such that

$0 < |z - z_0| < \delta$ and apply the modulus to both sides of (3.2.5). An application of Theorem 2.9.18 yields

$$\left| \frac{F(z) - F(z_0)}{z - z_0} - f(z_0) \right| \leq \frac{1}{|z - z_0|} \int_{[z_0, z]} \varepsilon \, |d\zeta| = \varepsilon.$$

This proves that

$$F'(z_0) = \lim_{z \to z_0} \frac{F(z) - F(z_0)}{z - z_0} = f(z_0),$$

as needed. □

We can now prove Goursat's theorem. In the proof, we use the concept of the *diameter* of a bounded set $E \subseteq \mathbb{C}$, which is the value

$$\operatorname{diam} E = \sup\{|z - w| : z, w \in E\}. \tag{3.2.6}$$

Proof of Goursat's theorem. It is sufficient to show that f is analytic on an arbitrary open disk $D \subseteq \Omega$. We intend to use Morera's theorem, and therefore let $T \subseteq D$ be a triangular path, and set $\varepsilon > 0$. Let E be the set formed by T and its inside, and let $\alpha = \operatorname{diam} E$.

Use the midpoints of the sides of T to divide E into four subtriangles, and let $S_1, \ldots, S_4$ denote their boundaries. Assuming counterclockwise orientations, use Theorem 2.9.11 on the sides of $S_1, \ldots, S_4$ to calculate

$$\int_T f(z) \, dz = \sum_{k=1}^{4} \int_{S_k} f(z) \, dz.$$

Define T_1 to be one of $S_1, \ldots, S_4$ that achieves the maximum

$$\max\left\{ \left| \int_{S_k} f(z) \, dz \right| : k = 1, \ldots, 4 \right\}.$$

Then

$$\left| \int_T f(z) \, dz \right| \leq \sum_{k=1}^{4} \left| \int_{S_k} f(z) \, dz \right| \leq 4 \left| \int_{T_1} f(z) \, dz \right|.$$

We follow the above method inductively to form triangular paths T_n, for all $n \in \mathbb{N}$, such that if E_n is the set formed by taking T_n together with its inside, T_{n+1} is the boundary of the subtriangle of E_n with maximum integral modulus. (See Figure 3.1.) Now since

$$\left| \int_{T_n} f(z) \, dz \right| \leq 4 \left| \int_{T_{n+1}} f(z) \, dz \right|$$

for all n, we have

$$\left| \int_T f(z) \, dz \right| \leq 4^n \left| \int_{T_n} f(z) \, dz \right|$$

for all n. It is also evident that diam $E_{n+1} = (\text{diam } E_n)/2$ for all n, and thus

$$\text{diam } E_n = \frac{\alpha}{2^n}$$

for all n.

Since each E_n is compact and $E_{n+1} \subseteq E_n$ for all n, Theorem 1.6.15 implies that there exists

$$a \in \bigcap_{n=1}^{\infty} E_n.$$

Since f is differentiable at a, there exists some $\delta > 0$ such that for all $z \in \Omega$ satisfying $0 < |z - a| < \delta$,

$$\left| \frac{f(z) - f(a)}{z - a} - f'(a) \right| < \frac{\varepsilon}{3\alpha^2}.$$

Assume δ is chosen small enough so that $D(a; \delta) \subseteq D$. Then

$$|f(z) - f(a) - f'(a)(z - a)| \leq \frac{\varepsilon}{3\alpha^2} |z - a|$$

for all $z \in D(a; \delta)$.

Choose $n \in \mathbb{N}$ such that $\alpha/2^n < \delta$. Then diam $E_n < \delta$. Since $a \in E_n$, $E_n \subseteq D(a; \delta)$. Constant functions and linear functions have antiderivatives in all of $\mathbb{C}$, and therefore Corollary 2.9.14 gives

$$\int_{T_n} f(a)\, dz = 0, \qquad \int_{T_n} f'(a)(z - a)\, dz = 0.$$

Hence

$$
\begin{aligned}
\left| \int_{T_n} f(z)\, dz \right| &= \left| \int_{T_n} [f(z) - f(a) - f'(a)(z - a)]\, dz \right| \\
&\leq \int_{T_n} |f(z) - f(a) - f'(a)(z - a)|\, |dz| \\
&\leq \frac{\varepsilon}{3\alpha^2} \int_{T_n} |z - a|\, |dz| \\
&< \frac{\varepsilon}{3\alpha^2} \text{ diam } E_n \int_{T_n} |dz|.
\end{aligned}
$$

Now diam $E_n = \alpha/2^n$, and the arc length of T_n is less than or equal to 3 diam E_n, and thus

$$\left| \int_{T_n} f(z)\, dz \right| < \frac{\varepsilon}{3\alpha^2} \frac{\alpha}{2^n} \frac{3\alpha}{2^n} = \frac{\varepsilon}{4^n}.$$

It follows that

$$\left| \int_T f(z)\, dz \right| \leq 4^n \left| \int_{T_n} f(z)\, dz \right| < \varepsilon.$$

Since $\varepsilon > 0$ was arbitrary, it must be that

$$\int_T f(z)\, dz = 0.$$

Therefore by Morera's theorem, f is analytic on D, and hence on Ω. $\qquad\square$

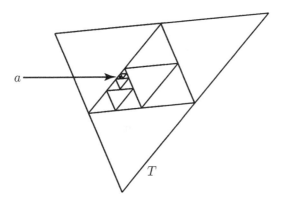

Figure 3.1   Possible triangular subdivisions used in the proof of Goursat's theorem

In the following corollary, we summarize some observations made over the course of this section. Notice that, amongst other things, we find that *certain integrals can be calculated using derivatives!* This will be generalized in Section 4.2.

3.2.6 Corollary. *Let $\Omega \subseteq \mathbb{C}$ be open, $f \in H(\Omega)$, and $a \in \Omega$. Let R be the radius of the largest disk centered at a within Ω. ($R = \infty$ if $\Omega = \mathbb{C}$.) Then f has the power series expansion*

$$f(z) = \sum_{n=0}^{\infty} c_n(z-a)^n$$

convergent on $D(a;R)$ (on $\mathbb{C}$ if $R = \infty$), where for all $n = 0,1,\ldots,$

$$c_n = \frac{f^{(n)}(a)}{n!} = \frac{1}{2\pi i} \int_{\partial D(a;r)} \frac{f(z)}{(z-a)^{n+1}}\, dz \qquad (3.2.7)$$

for any $r \in (0,R)$. If $\{f(z_n)\}$ diverges for a sequence $\{z_n\}_{n=1}^{\infty} \subseteq D(a;R)$ converging to a point on $\partial D(a;R)$, then R is the radius of convergence of the series. Furthermore, this series is unique.

By uniqueness, we mean that f has no power series expansion centered at a with coefficients other than $\{c_n\}$.

Proof. The coefficients $\{c_n\}$ are calculated in (3.2.4) in Theorem 3.2.4 by integrating over $\partial D(a;r)$, where r is any positive number such that $\overline{D}(a;r) \subseteq \Omega$. Further, the power series converges on $D(a;r)$. Therefore the series will converge on the largest disk centered at a in Ω, namely $D(a; R)$. If $\{f(z_n)\}$ diverges for a sequence $\{z_n\}_{n=1}^{\infty} \subseteq D(a; R)$ converging to a point on $\partial D(a; R)$, then the series cannot converge in a bigger disk, and hence R is the radius of convergence.

If $\sum_{n=0}^{\infty} d_n(z - a)^n$ converges to f on some disk $D(a; \rho) \subseteq \Omega$, then successive differentiation reveals that the coefficients must satisfy

$$f^{(n)}(a) = n! d_n$$

for any n. This implies uniqueness and (3.2.7).   □

3.2.7 Example. We will evaluate the contour integral

$$I = \int_{\partial D(-1;2)} \frac{\sin(\pi z^2)}{(z + 1)^3} \, dz.$$

We use $f(z) = \sin(\pi z^2)$, $a = -1$, and $n = 2$ in (3.2.7). A direct calculation reveals

$$f''(z) = 2\pi \cos(\pi z^2) - 4\pi^2 z^2 \sin(\pi z^2),$$

and hence $I = 2\pi i f''(-1)/2! = -2\pi^2 i$.

3.2.8 Example. Consider the function

$$f(z) = \frac{\text{Log } z}{z^2 + 4}.$$

Since f is differentiable, it is analytic on its natural domain by Goursat's theorem. That is, on the set of all $z \in \mathbb{C}$ except for $z = \pm 2i$ and $z \leq 0$. It follows that $f(z)$ is equal to a power series based at $1 + i$ for z in a disk centered at $1 + i$, and Corollary 3.2.6 gives that this disk can have the largest radius possible to lie in the domain of f, $R = \sqrt{2}$. Furthermore, since $f(z) \to \infty$ as $z \to 2i$, R is the radius of convergence of the series.

Now consider a similar analysis for a series based at $-2 + i$. The largest disk centered at $-2 + i$ in the domain of f has radius 1, and hence $f(z)$ is equal to a power series converging on $D(-2 + i; 1)$. As f does not have an infinite limit at a point on $\partial D(-2+i; 1)$, we cannot conclude that the series has radius of convergence 1. And it does not! Indeed, consider the function $g(z) = (\log z)/(z^2 + 4)$, where the branch of the logarithm corresponds to $-\pi/2 < \arg z < 3\pi/2$. Then $f = g$ on $D(-2 + i; 1)$, and hence they have the same power series expansion on that disk. However the series converges to g on $D(-2 + i; \sqrt{5})$. Since $g(z) \to \infty$ as $z \to 2i$, the radius of convergence of the series converging to f on $D(-2 + i; 1)$ is $R = \sqrt{5}$.

Summary and Notes for Section 3.2.

Goursat's theorem gives the striking result that if a function has a complex derivative, then it is analytic, meaning that it can be expressed as a power series in a disk about any given point in the domain of the function. The existence of one derivative implies the existence of all derivatives! To see how significantly this differs from the calculus of functions on $\mathbb{R}$, consider that the function $f_0(x) = x^{1/3}$ is not differentiable at $x = 0$. Therefore $f_1(x) = x^{4/3}$ has one, but not two, derivatives at $x = 0$. Inductively, $f_n(x) = x^{n+1/3}$ has n, but not $n+1$, derivatives at $x = 0$ for each $n \in \mathbb{N}$. From this, we see that for an open interval $I \subseteq \mathbb{R}$, the family of functions with n derivatives on I is a *proper subset* of the family of functions with m derivatives on I if $n, m \in \mathbb{N}$ with $n > m$. Adding more complexity is the observation that a real function can be differentiable without the derivative being continuous, and hence we can place the family of functions with n continuous derivatives properly between the family with n derivatives and the family with $n + 1$ derivatives. Even more, the family of functions with infinitely many derivatives is bigger than the family of real analytic functions (functions that are equal to power series in a neighborhood of each point). But due to Goursat's theorem, all of these families are the same in an open subset of $\mathbb{C}$.

To prove Goursat's theorem, we first proved that it was true with the additional hypothesis that f' is continuous. On the surface, this seems like a small addition, but it turns out to be very important. For Morera's theorem is proven by showing that the continuous function in its statement has an antiderivative, making it analytic. Goursat's theorem follows from this.

As a result of this work, we learned that a power series representing an analytic function converges to the function in the largest possible disk; that is, the largest disk centered at the expansion point lying in the domain of the function. This is certainly not true for real power series, although this complex interpretation helps us understand why some real series converge where they do.

Augustin-Louis Cauchy is largely responsible for most of the main results involving complex integration, and he proved what is essentially Theorem 3.2.4 through this work in the 19th century. Edouard Jean-Baptiste Goursat's contribution was to show that the hypothesis that f' be continuous is not necessary, which he did at the turn of the 20th century.

Giacinto Morera proved his theorem in the late 19th century. It is commonly stated with the hypothesis that

$$\int_C f(z)\, dz = 0$$

for all closed contours (or some more general type of closed curve) $C \subseteq \Omega$. Considering only triangular paths makes for a stronger result and also nicely fits our proof of Goursat's theorem.

In many books, a function is defined to be analytic if it has a complex derivative, and in others, if it has a continuous complex derivative. We now see that these definitions are equivalent to ours.

Exercises for Section 3.2.

1. The real function $f : \mathbb{R} \to \mathbb{R}$ given by

$$f(x) = \frac{1}{1 + x^2}$$

is infinitely differentiable for all $x \in \mathbb{R}$. However, it is seen in calculus that its power series representation based at 0 has radius of convergence 1. What is an explanation of this?

2. Write out the terms of degree $n \leq 2$ of the power series converging to $f(z) = e^z \sec z$ based at 0. What is the radius of convergence of the series?

3. For each of the following functions f and points $a \in \mathbb{C}$, determine the radius of convergence of the series converging to f in a disk centered at a.

   (a) $f(z) = \dfrac{\sin z}{1 + z^4}$, $a = 2$

   (b) $f(z) = \dfrac{z^2}{e^z - 1}$, $a = i$

   (c) $f(z) = \dfrac{\operatorname{Log} z}{z^2 - iz + 2}$, $a = -2 - 2i$

4. Evaluate the following integrals using (3.2.7).

   (a) $\displaystyle\int_{\partial D(i;1)} \dfrac{z^2 + 2z}{z^2 - 2iz - 1}\, dz$

   (b) $\displaystyle\int_{\partial D(0;2\pi)} \dfrac{\cos z}{z^5}\, dz$

   (c) $\displaystyle\int_{\partial D(0;\pi/4)} \dfrac{\tan z}{z^2}\, dz$

   (d) $\displaystyle\int_{\partial D(1;1)} \dfrac{e^{2iz}}{(z-1)^4}\, dz$

5. For each $n \in \mathbb{N}$, evaluate the integral

$$\int_{\partial D(1;1)} \left(\frac{z}{z-1}\right)^n dz.$$

6. By considering the integral $\int_{\partial D} e^z/z\, dz$, prove that

$$\int_0^\pi e^{\cos t} \cos(\sin t)\, dt = \pi.$$

7. Evaluate the integral

$$\int_0^{2\pi} \frac{\sin(2e^{it})}{e^{3it}}\, dt$$

   by recognizing it as a contour integral.

8. Let $a \in \mathbb{C}$. We define the generalized binomial coefficients by $\binom{a}{0} = 1$ and

$$\binom{a}{n} = \frac{a(a-1)\cdots(a-n+1)}{n!}, \qquad n \in \mathbb{N}.$$

   Prove the binomial series representation

$$(1+z)^a = \sum_{n=0}^\infty \binom{a}{n} z^n, \qquad z \in \mathbb{D},$$

where the principal branch of the power is used on the left-hand side. How does this compare to the binomial formula?

9. Verify the following *mean value theorems*. For each, suppose that $\Omega \subseteq \mathbb{C}$ is open, $f \in H(\Omega)$, and $\overline{D}(a; r) \subseteq \Omega$ for some $a \in \Omega$ and $r > 0$.

   (a) The value of f at a is equal to the average of its values on the circle $\partial D(a; r)$. That is,

   $$f(a) = \frac{1}{2\pi} \int_0^{2\pi} f(a + re^{it})\, dt.$$

   (b) If we consider $D(a; r)$ as a subset of $\mathbb{R}^2$, then the value of f at a is equal to the average of its values over $D(a; r)$. That is,

   $$f(a) = \frac{1}{\pi r^2} \iint\limits_{D(a;r)} f(x + iy)\, dA(x, y).$$

   (Consider Exercise 16 in Section 2.9.)

10. Use a substitution in the integral formula given in Lemma 3.2.2 to verify the formula in Corollary 3.2.3.

11. Let $f \in H(D(0; r))$ for some $r > 0$. Define $g \colon D(0; r) \to \mathbb{C}$ by

    $$g(z) = \int_0^1 f(tz)\, dt.$$

    (a) Show that $g \in H(D(0; r))$.

    (b) Show that for all $z \in D(0; r)$,

    $$\frac{d}{dz}[zg(z)] = f(z).$$

12. The Fibonacci numbers are defined by $c_0 = c_1 = 1$ and $c_n = c_{n-1} + c_{n-2}$ for $n \geq 2$. Suppose that

    $$f(z) = \sum_{n=0}^{\infty} c_n z^n$$

    for z in the disk of convergence of the series.

    (a) Show that

    $$f(z) = \frac{1}{1 - z - z^2}$$

    for z in the disk of convergence.

    (b) Find the radius of convergence of the series.

13. Let $[a, b] \subseteq \mathbb{R}$ and $f \colon [a, b] \to \mathbb{C}$ be continuous. The *Fourier transform* of f is the function $F \colon \mathbb{C} \to \mathbb{C}$ given by

    $$F(z) = \int_a^b e^{-itz} f(t)\, dt.$$

    (a) Prove that $F \in H(\mathbb{C})$.

(b) Show that there are constants $M, \alpha \geq 0$ such that

$$|F(z)| \leq Me^{\alpha|\operatorname{Im} z|}$$

for all $z \in \mathbb{C}$.

14. A set $E \subseteq \mathbb{C}$ is *starlike* with respect to $a \in E$ (or just starlike) if $[a, z] \subseteq E$ for all $z \in E \setminus \{a\}$. Show that if $\Omega \subseteq \mathbb{C}$ is a starlike domain and $f \in H(\Omega)$, then there is $F \in H(\Omega)$ such that $F' = f$.

15. Suppose that $f, g \in H(\mathbb{D})$ are written as $f(z) = \sum_{n=0}^{\infty} c_n z^n$ and $g(z) = \sum_{n=0}^{\infty} d_n z^n$. Define the *Hadamard product* of f and g to be the function

$$(f * g)(z) = \sum_{n=0}^{\infty} c_n d_n z^n.$$

(a) Show that $f * g \in H(\mathbb{D})$.

(b) Find a function $I \in H(\mathbb{D})$ that serves as an identity for the Hadamard product. (That is, $I * f = f$ for all $f \in H(\mathbb{D})$.)

(c) Which functions $f \in H(\mathbb{D})$ are not invertible under the Hadamard product? (That is, there is no $g \in H(\mathbb{D})$ such that $g * f = I$.)

(d) Let $r \in (0, 1)$. Verify that $f * g$ can be calculated with the convolution integral

$$(f * g)(z) = \frac{1}{2\pi i} \int_{\partial D(0;r)} \frac{f(\zeta)}{\zeta} g\left(\frac{z}{\zeta}\right) d\zeta$$

for all $z \in D(0; r)$. (*Hint:* Try using a series and Corollary 3.2.6.)

3.3   Consequences of Goursat's Theorem

Goursat's theorem gives the amazing result that for an open set $\Omega \subseteq \mathbb{C}$, a function $f : \Omega \to \mathbb{C}$ is analytic if and only if it is differentiable. The following is now immediate because of differentiation rules. (Compare with Exercise 4 in Section 3.1.)

3.3.1 Theorem. *Let $\Omega \subseteq \mathbb{C}$ be open. If $f, g \in H(\Omega)$ and $c \in \mathbb{C}$, then*

(a) $cf \in H(\Omega)$,

(b) $f + g \in H(\Omega)$,

(c) $fg \in H(\Omega)$, *and*

(d) $f/g \in H(\Omega)$ *if $g(z) \neq 0$ for all $z \in \Omega$.*

Furthermore, if $f(\Omega)$ is open and $h \in H(f(\Omega))$, then $h \circ f \in H(\Omega)$.

The reader with an interest in algebra may gladly note that, together with the accompanying algebraic properties, items (a) and (b) show $H(\Omega)$ is a *vector space* and items (b) and (c) show $H(\Omega)$ is a *ring*, which means $H(\Omega)$ is an *algebra*.

The following is an immediate consequence of Corollary 3.2.6.

3.3.2 Cauchy's Estimates. *Suppose that $\Omega \subseteq \mathbb{C}$ is open, $f \in H(\Omega)$, $a \in \Omega$, and $R > 0$ is such that $D(a; R) \subseteq \Omega$. If $|f(z)| \leq M$ for all $z \in D(a; R)$ and some $M \geq 0$, then*

$$|f^{(n)}(a)| \leq \frac{n!M}{R^n}$$

for all $n = 0, 1, \ldots$.

If it is known that the closure of the disk $D(a; R)$ lies in Ω, then the existence of such an $M \geq 0$ is guaranteed because f is continuous and $\overline{D}(a; R)$ is compact.

Proof. Using Theorem 2.9.18 and Corollary 3.2.6, we have that

$$|f^{(n)}(a)| \leq \frac{n!}{2\pi} \int_{\partial D(a;r)} \frac{|f(z)|}{|z-a|^{n+1}} \, |dz| \leq \frac{n!}{2\pi} \int_{\partial D(a;r)} \frac{M}{r^{n+1}} \, |dz| = \frac{n!M}{r^n}$$

for all $r \in (0, R)$. The limit

$$|f^{(n)}(a)| = \lim_{r \to R^-} |f^{(n)}(a)| \leq \lim_{r \to R^-} \frac{n!M}{r^n} = \frac{n!M}{R^n}$$

gives the result. $\qquad\square$

The following corollary is a slight alteration of the hypotheses of Cauchy's estimates, and its proof follows through a similar calculation.

3.3.3 Corollary. *Suppose that $\Omega \subseteq \mathbb{C}$ is open, $f \in H(\Omega)$, $a \in \Omega$, and $R > 0$ is such that $\overline{D}(a; R) \subseteq \Omega$. If $|f(z)| \leq M$ for all $z \in \partial D(a; R)$ and some $M \geq 0$, then*

$$|f^{(n)}(a)| \leq \frac{n!M}{R^n}$$

for all $n = 0, 1, \ldots$.

The first, and arguably most important, application of Cauchy's estimates deals with the following special type of analytic function.

3.3.4 Definition. The members of $H(\mathbb{C})$ are called *entire functions*.

It follows from Corollary 3.2.6 that an entire function f has a series expansion of the form

$$f(z) = \sum_{n=0}^{\infty} c_n z^n \tag{3.3.1}$$

for all $z \in \mathbb{C}$. All analytic functions are, in a sense, polynomials with an infinite number of terms. This point of view is most evident in the case of entire functions because of the equality of domains.

Let us consider a property of polynomials on $\mathbb{C}$. If $p \colon \mathbb{C} \to \mathbb{C}$ is a polynomial of degree $n \in \mathbb{N}$, then we may write

$$p(z) = \sum_{k=0}^{n} a_k z^k$$

for all $z \in \mathbb{C}$, where $a_n \neq 0$. If $z \neq 0$, then

$$p(z) = z^n \left(a_n + \sum_{k=0}^{n-1} a_k z^{k-n} \right). \tag{3.3.2}$$

Since $z^{k-n} \to 0$ as $z \to \infty$ for $k < n$, the quantity in parentheses in (3.3.2) tends to $a_n \neq 0$ as $z \to \infty$. Thus

$$\lim_{z \to \infty} p(z) = \infty.$$

The above is not generally true for entire functions, but we can say the following.

3.3.5 Liouville's Theorem.   *A bounded entire function must be constant.*

Proof. Suppose that f is entire and bounded by the value $M \geq 0$. Write f as in (3.3.1). Cauchy's estimates and Corollary 3.2.6 imply that for all $n \in \mathbb{N}$ and $R > 0$,

$$|c_n| = \left| \frac{f^{(n)}(0)}{n!} \right| \leq \frac{M}{R^n}.$$

It follows that $c_n = 0$ for all $n \in \mathbb{N}$, which implies the result. $\square$

It comes as a pleasant surprise that we can prove the following so simply (given that we are studying analysis, after all).

3.3.6 Fundamental Theorem of Algebra.   *Every nonconstant polynomial over $\mathbb{C}$ has a root (or zero).*

Proof. We prove the contrapositive. If $p \colon \mathbb{C} \to \mathbb{C}$ is a polynomial with no root, then the function

$$f(z) = \frac{1}{p(z)}$$

is entire. Since $p(z) \to \infty$ as $z \to \infty$, $f(z) \to 0$ as $z \to \infty$. Specifically, there exists $R > 0$ such that $|z| > R$ implies $|f(z)| < 1$. Since $\overline{D}(0; R)$ is compact, f is bounded by some $M \geq 0$ on $\overline{D}(0; R)$. Therefore f is bounded on $\mathbb{C}$, and hence is constant by Liouville's theorem. Thus p is constant. $\square$

Summary and Notes for Section 3.3.

Goursat's theorem shows that functions of a complex variable behave quite differently than functions of a real variable. Some consequences of that theorem are simultaneously striking and simple. Cauchy's estimates give crude bounds on the derivatives of an analytic function, and Liouville's theorem shows that a nonconstant analytic function on $\mathbb{C}$ must be unbounded, a property far from true for functions on $\mathbb{R}$.

It is nice that complex analysis provides such an elementary proof of the fundamental theorem of algebra. This theorem was proposed as early as the 1600s and many proofs were attempted by a great number of mathematicians. Jean-Robert Argand had the first rigorous proof. Carl Friedrich Gauss actually gave four proofs, not all perfect, the last of which fully embraced complex numbers.

Exercises for Section 3.3.

1. Give an example of a nonconstant entire function f such that $f(z)$ fails to tend to ∞ as $z \to \infty$.

2. Prove Corollary 3.3.3.

3. Show that if $f \colon \mathbb{D} \to \mathbb{D}$ is analytic, then every coefficient in the power series representation of f based at 0 is bounded by 1.

4. Suppose that f is an entire function and that for some $M \geq 0$ and $n \in \mathbb{N}$, $|f(z)| \leq M|z|^n$ for all $z \in \mathbb{C}$. Verify that $f(z) = cz^n$ for some $c \in \mathbb{C}$ such that $|c| \leq M$.

5. Let f be an entire function. Prove that if $f(\mathbb{C})$ has an exterior point, then f is constant.

6. Let $p \colon \mathbb{C} \to \mathbb{C}$ be a nonconstant polynomial. Prove that $p(\mathbb{C}) = \mathbb{C}$.

7. Let $f \colon \mathbb{D} \to \mathbb{C}$ be given by

$$f(z) = \frac{1}{(1-z)^2}.$$

(a) Show that for $r \in (0,1)$,

$$\sup_{|z|=r} |f(z)| = \frac{1}{(1-r)^2}.$$

(b) Use Cauchy's estimates to prove that for all $r \in (0,1)$,

$$n + 1 \leq \frac{1}{r^n(1-r)^2}.$$

8. Suppose that $f \in H(\mathbb{D})$ satisfies

$$|f(z)| \leq \frac{1}{1-|z|}$$

for all $z \in \mathbb{D}$. Using techniques similar to Exercise 7, find the "best" upper bound on $|f^{(n)}(0)|$ that Cauchy's estimates can provide.

3.4   The Zeros of Analytic Functions

As we continue our analysis of analytic functions, we turn to the zeros (roots) of the functions. We shall see that there are strict rules governing their behavior, leading to several remarkable consequences.

3.4.1 Definition. Let $\Omega \subseteq \mathbb{C}$ be open and $f \in H(\Omega)$. The *zero set* of f is

$$Z(f) = \{a \in \Omega : f(a) = 0\}. \tag{3.4.1}$$

Suppose $a \in Z(f)$ for some $f \in H(\Omega)$, where $\Omega \subseteq \mathbb{C}$ is open. Since $a \in \Omega$, there exists $r > 0$ such that $D(a; r) \subseteq \Omega$, and f has the expansion

$$f(z) = \sum_{n=0}^{\infty} c_n(z - a)^n$$

for all $z \in D(a; r)$. Since $f(a) = 0$, $c_0 = 0$. Therefore the first nonzero term in the series (if there is one) has positive index. This prompts the next definition.

3.4.2 Definition. Suppose that $\Omega \subseteq \mathbb{C}$ is open, $f \in H(\Omega)$, and $a \in Z(f)$. Suppose that for a given $m \in \mathbb{N}$, $c_m \neq 0$ and

$$f(z) = \sum_{n=m}^{\infty} c_n(z - a)^n$$

is the expansion of f in a disk $D(a; r) \subseteq \Omega$. Then f is said to have a zero at a of *order m*. If $m = 1$, the zero is *simple*.

The uniqueness of the power series expansion of f at a removes any ambiguity from the above definition.

3.4.3 Example. Consider the entire function $f(z) = e^{z^2} - 1$, for which $0 \in Z(f)$. The order of this zero can be found by calculating the series

$$f(z) = \sum_{n=0}^{\infty} \frac{(z^2)^n}{n!} - 1 = \sum_{n=1}^{\infty} \frac{z^{2n}}{n!}$$

and observing that the z^2 term is the first nonzero term of the series. Therefore the zero is of order 2.

Alternatively, observe that $f'(z) = 2ze^{z^2}$ so that $f'(0) = 0$, and $f''(z) = 2(2z^2 + 1)e^{z^2}$ so that $f''(0) = 2 \neq 0$. Thus, without calculating the entire series, we see that its first nonzero term has degree 2.

Our first result is a factorization theorem that will help establish an important characteristic of the zero set of an analytic function.

3.4.4 Theorem. *Let $\Omega \subseteq \mathbb{C}$ be open. A function $f \in H(\Omega)$ has a zero of order $m \in \mathbb{N}$ at the point $a \in \Omega$ if and only if there exists $g \in H(\Omega)$ such that $g(a) \neq 0$ and*

$$f(z) = (z - a)^m g(z)$$

for all $z \in \Omega$.

Proof. Assume that f has a zero of order m at a. In a disk $D(a; r) \subseteq \Omega$, f has the power series expansion

$$f(z) = \sum_{n=m}^{\infty} c_n (z - a)^n = (z - a)^m \sum_{n=0}^{\infty} c_{m+n}(z - a)^n, \qquad (3.4.2)$$

where $c_m \neq 0$. This prompts defining $g \colon \Omega \to \mathbb{C}$ by

$$g(z) = \begin{cases} \dfrac{f(z)}{(z - a)^m} & \text{if } z \in \Omega \setminus \{a\}, \\[2mm] \dfrac{f^{(m)}(a)}{m!} & \text{if } z = a. \end{cases}$$

Since $g(a) = c_m$, g agrees with the series on the right-hand side of (3.4.2) on $D(a; r)$, and hence is analytic there. But g is analytic on $\Omega \setminus \{a\}$ by definition, and so g is as desired.

The converse follows immediately. $\qquad \qquad \square$

The following results involve *domains* (connected open sets) in $\mathbb{C}$.

3.4.5 Theorem. *Let $\Omega \subseteq \mathbb{C}$ be a domain and $f \in H(\Omega)$. Either $f(z) = 0$ for all $z \in \Omega$ or $Z(f)$ has no limit points in Ω.*

Proof. Since both instances cannot occur simultaneously, assume that $a \in \Omega$ is a limit point of $Z(f)$. We shall show that $f(z) = 0$ for all $z \in \Omega$.

For each $k \in \mathbb{N}$, there exists $z_k \in D(a; 1/k) \cap Z(f) \setminus \{a\}$. Since $z_k \to a$, continuity gives $a \in Z(f)$. Suppose that a is a zero of order $m \in \mathbb{N}$. Then by Theorem 3.4.4, there is $g \in H(\Omega)$ such that $g(a) \neq 0$ and

$$f(z) = (z - a)^m g(z), \qquad z \in \Omega.$$

But for all k, $(z_k - a)^m g(z_k) = 0$ implies $g(z_k) = 0$, and continuity gives $g(a) = 0$, a contradiction.

It must therefore be the case that $f^{(n)}(a) = 0$ for all $n \in \mathbb{N} \cup \{0\}$. Define the set

$$E = \{z \in \Omega : f^{(n)}(z) = 0 \text{ for all } n \in \mathbb{N} \cup \{0\}\} \neq \varnothing.$$

Let $z_0 \in E$, and let $r > 0$ such that $D(z_0; r) \subseteq \Omega$. Now f has a series expansion on $D(z_0; r)$, and since $f^{(n)}(z_0) = 0$ for all n, all of the series coefficients are 0. Thus f is identically 0 on $D(z_0; r)$. But then for any $z \in D(z_0; r)$, $f^{(n)}(z) = 0$ for all n, proving $z \in E$. Therefore $D(z_0; r) \subseteq E$, showing E is open.

Now choose $z_0 \in \Omega \setminus E$. There exists n such that $f^{(n)}(z_0) \neq 0$. Since $f^{(n)}$ is continuous, $(f^{(n)})^{-1}(\mathbb{C} \setminus \{0\})$ is open by Theorem 2.1.9 and thus contains a disk $D(z_0; r)$ for some $r > 0$. Since $D(z_0; r) \subseteq \Omega \setminus E$, we have $\Omega \setminus E$ open.

By Theorem 1.4.16, $\Omega \setminus E = \varnothing$. Specifically, $f(z) = 0$ for all $z \in \Omega$. □

The statement of the previous theorem can be rephrased to say that unless an analytic function f is equivalently 0 on a domain Ω, the zeros of f are *isolated*.

Here are some immediate consequences. The first is evident from the proof of Theorem 3.4.5.

3.4.6 Corollary. *Let $\Omega \subseteq \mathbb{C}$ be a domain. If $f \in H(\Omega)$ is not equivalently 0 and $a \in Z(f)$, then a is a zero of order m for some $m \in \mathbb{N}$.*

3.4.7 Corollary. *Suppose that $\Omega \subseteq \mathbb{C}$ is a domain and that $f, g \in H(\Omega)$. If the set*

$$E = \{z \in \Omega : f(z) = g(z)\}$$

has a limit point in Ω, then $f = g$.

Proof. Since $f - g \in H(\Omega)$ and $E = Z(f - g)$, Theorem 3.4.5 gives the result. □

3.4.8 Remark. Corollary 3.4.7 implies that our definitions of the exponential, sine, and cosine functions in Section 2.4 are the only analytic functions on $\mathbb{C}$ that agree with their real counterparts on $\mathbb{R}$.

If we combine Theorem 3.4.4 with the fundamental theorem of algebra, we see that polynomials can be factored as we would hope.

3.4.9 Corollary. *If $p \colon \mathbb{C} \to \mathbb{C}$ is a polynomial of degree $n \in \mathbb{N}$, then there exist distinct $a_1, \ldots, a_k \in \mathbb{C}$ and $m_1, \ldots, m_k \in \mathbb{N}$ such that $\sum_{j=1}^{k} m_j = n$, $Z(p) = \{a_1, \ldots, a_k\}$, the order of the zero of p at each a_j is m_j, and p has the factorization*

$$p(z) = c \prod_{j=1}^{k} (z - a_j)^{m_j}$$

for all $z \in \mathbb{C}$, where $c \in \mathbb{C}$ is a constant.

Proof. The fundamental theorem of algebra states that p has a zero, say a_1. Let m_1 be the order of that zero. Then there is some $q \in H(\mathbb{C})$ such that

$$p(z) = (z - a_1)^{m_1} q(z).$$

Now q must be a polynomial of degree $n - m_1$ and will have, other than a_1, the same zeros as p. We may then continue inductively to get the result. □

Summary and Notes for Section 3.4.

Because analytic functions are locally expressible as power series, which are, in a sense, infinite polynomials, we wonder what properties of polynomials extend to general analytic functions. Here, we see that we can factor out $(z - a)^m$ from an analytic function when a is a zero of f of order m, where the order of the zero is the degree of the nonzero term of lowest power in the power series expansion of f about a. This will be quite advantageous down the road.

Because zeros of nontrivial analytic functions defined on a domain are isolated, we see that any two such functions equal on a set with a limit point in the domain must be the same. To see just how restrictive this characteristic of analytic functions is, note that the values of an entire function everywhere in the plane are, in essence, determined by its values on a given tiny line segment.

Exercises for Section 3.4.

1. For each entire function f, determine the order of each zero of f.

   (a) $f(z) = z^4 + 2z^2 + 1$

   (b) $f(z) = z^3 \cos^2 z$

   (c) $f(z) = z^2 \sin z$

   (d) $f(z) = (1 - e^{iz}) \sin z$

2. Show how the results of this section imply that $f(z) = \operatorname{Re} z$ and $g(z) = \operatorname{Im} z$ are not analytic on any nonempty open subset of $\mathbb{C}$.

3. Show that the function $f \colon \mathbb{C} \to \mathbb{C}$ given by $f(z) = e^z - e^{i \operatorname{Im} z}$ is not analytic.

4. Give an example showing why Ω must be a domain in the statement of Theorem 3.4.5.

5. Let $\Omega \subseteq \mathbb{C}$ be open, $a \in \Omega$, and $f, g \in H(\Omega)$. Suppose that f and g have zeros at a of order $n \in \mathbb{N}$ and $m \in \mathbb{N}$, respectively.

   (a) What can be said about the zero of $f + g$ at a?

   (b) What can be said about the zero of fg at a?

6. Let $R > 0$, and set $\Omega = \mathbb{C} \setminus \overline{D}(0; R)$. We say that a function $f \in H(\Omega)$ has a *zero at* ∞ of order $m \in \mathbb{N}$ if the function $g(z) = f(1/z)$ can be extended to have a zero of order m at 0. Prove that for such f, there exists $\{c_n\}_{n=m}^{\infty} \subseteq \mathbb{C}$ such that $c_m \neq 0$ and

$$f(z) = \sum_{n=m}^{\infty} \frac{c_n}{z^n}$$

   for all $z \in \Omega$.

7. Let $\Omega \subseteq \mathbb{C}$ be a domain and $f \in H(\Omega)$. Show that if $f'(z) = 0$ for all $z \in \Omega$, then f is a constant function.

8. Suppose $\Omega \subseteq \mathbb{C}$ is a domain and $f \in H(\Omega)$ is not equivalently 0. Use the following steps to show that $Z(f)$ is countable. (See Exercise 17 in Appendix A.)

   (a) Let $\{K_n\}_{n=1}^{\infty}$ be an exhaustion of Ω. (See Exercise 17 in Section 1.4.) Use the Bolzano–Weierstrass theorem to argue that $Z(f) \cap K_n$ is finite for all n.

(b) Finish the argument using the aforementioned exercise in Appendix A.

9. Explain how Corollary 3.4.6 follows from the proof of Theorem 3.4.5.

10. Let $\Omega \subseteq \mathbb{C}$ be a domain and $f \in H(\Omega)$. Suppose that the power series expansion of f about any point $a \in \Omega$ is such that one of its coefficients is equal to 0. Prove that f must be a polynomial.

11. Suppose that $f, g \in H(\mathbb{D})$ have no zeros in $\mathbb{D}$. Set $x_n = 1/n$ for $n = 2, 3, \ldots$. Show that if

$$\frac{f'(x_n)}{f(x_n)} = \frac{g'(x_n)}{g(x_n)}$$

for all n, then $f = cg$ for some nonzero constant $c \in \mathbb{C}$.

3.5 The Open Mapping Theorem and Maximum Principle

The following important theorem relies subtly on the isolation of zeros.

3.5.1 Open Mapping Theorem. *Let $\Omega \subseteq \mathbb{C}$ be a domain. If $f \in H(\Omega)$ is nonconstant, then $f(U)$ is open for every open set $U \subseteq \Omega$.*

Functions f satisfying that $f(U)$ is open whenever U is open are called *open functions*. Compare this with the topological notion of continuity (Theorem 2.1.9).

Proof. Let $U \subseteq \Omega$ be open and $\alpha \in f(U)$. We must show that there is an open disk centered at α contained in $f(U)$. Set $a \in U$ such that $f(a) = \alpha$. Define $g: \Omega \to \mathbb{C}$ by $g(z) = f(z) - \alpha$. Now g is analytic and not equivalently 0, and so a is an isolated zero of g. There is then $r > 0$ such that $\overline{D}(a; r) \subseteq U$ and $g(z) \neq 0$ for $z \in \overline{D}(a; r) \setminus \{a\}$. For these z, $f(z) \neq \alpha$.

Let $\Gamma = f(\partial D(a; r))$. Then Γ is compact and $\alpha \notin \Gamma$. Therefore α is an exterior point of Γ, and thus there exists $\varepsilon > 0$ such that $D(\alpha; \varepsilon) \subseteq \mathbb{C} \backslash \Gamma$. If $D(\alpha; \varepsilon) \subseteq f(U)$, then we are done. Suppose instead that $w \in D(\alpha; \varepsilon) \setminus f(U)$.

Define $h: U \to \mathbb{C}$ by

$$h(z) = \frac{1}{f(z) - w}.$$

Evidently, h is well defined and analytic. If $z \in \partial D(a; r)$, then $f(z) \in \Gamma$, and hence $|f(z) - w| \geq \varepsilon - |w - \alpha| > 0$. It follows that

$$|h(z)| \leq \frac{1}{\varepsilon - |w - \alpha|}$$

for all $z \in \partial D(a; r)$. The corollary to Cauchy's estimates (Corollary 3.3.3) gives that

$$\frac{1}{|w - \alpha|} = |h(a)| \leq \frac{1}{\varepsilon - |w - \alpha|}.$$

This solves to give

$$|w - \alpha| \geq \frac{\varepsilon}{2}.$$

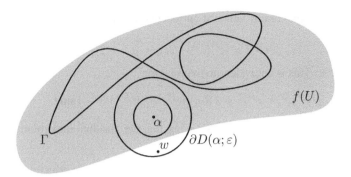

Figure 3.2 Constructions in the proof of Theorem 3.5.1

This shows that if $w \in D(\alpha; \varepsilon)$ fails to lie in $f(U)$, then $w \notin D(\alpha; \varepsilon/2)$. (See Figure 3.2.) Thus $D(\alpha; \varepsilon/2) \subseteq f(U)$. $\qquad \square$

We can now remove some hypotheses from the statement of Theorem 2.7.7.

3.5.2 Corollary. *If $\Omega \subseteq \mathbb{C}$ is open and $f \in H(\Omega)$ is one-to-one, then $f(\Omega)$ is open and $f^{-1} \in H(f(\Omega))$.*

Proof. We first prove that f is an open function. Let $U \subseteq \Omega$ be open and $\alpha \in f(U)$. Then $\alpha = f(a)$ for some $a \in U$. Let $r > 0$ such that $D(a; r) \subseteq U$. Since $D(a; r)$ is a domain and f is nonconstant on $D(a; r)$, $f(D(a; r))$ is open. Hence there is $\varepsilon > 0$ such that $D(\alpha; \varepsilon) \subseteq f(D(a; r)) \subseteq f(U)$, as desired.

We now have that $f(\Omega)$ is open and $f^{-1} : f(\Omega) \to \Omega$ is well defined. If $U \subseteq \mathbb{C}$ is open, then $U \cap \Omega$ is open, and $(f^{-1})^{-1}(U) = f(U \cap \Omega)$ is open. Theorem 2.1.9 gives that f^{-1} is continuous, and analyticity follows from Theorem 2.7.7 and Goursat's theorem. $\qquad \square$

The following theorem is a direct result of the open mapping theorem and continues the theme of fantastic consequences of Goursat's theorem.

3.5.3 Maximum Principle. *Let $\Omega \subseteq \mathbb{C}$ be a domain and $f \in H(\Omega)$. If there exists $a \in \Omega$ such that $|f(a)| \geq |f(z)|$ for all $z \in \Omega$, then f is constant.*

In other words, $|f|$ cannot attain a maximum in Ω unless f is constant. Of course, a similar argument shows that $|f|$ cannot even have a relative maximum unless f is constant. This is also referred to as the *maximum modulus theorem*.

Proof. Suppose that f is nonconstant and $a \in \Omega$. Since $f(\Omega)$ is open (open mapping theorem), there is a disk $D(f(a); \varepsilon) \subseteq f(\Omega)$ for some $\varepsilon > 0$. This disk contains a point of modulus greater than $|f(a)|$. $\qquad \square$

3.5.4 Example. Consider the polynomial $p(z) = z^2 - 2iz$ on the set $\overline{\mathbb{D}}$. Since $\overline{\mathbb{D}}$ is compact, $|p|$ must attain its absolute maximum value on $\overline{\mathbb{D}}$. By the maximum principle, this maximum cannot occur at a point in the domain $\mathbb{D}$ as p is nonconstant.

It follows that the maximum value of $|p|$ occurs at a point of $\partial\mathbb{D}$. We therefore maximize

$$|p(e^{it})| = |e^{it}(e^{it} - 2i)| = |e^{it} - 2i|$$

over $0 \le t \le 2\pi$. This is equivalent to maximizing

$$|p(e^{it})|^2 = |e^{it}|^2 + 2\operatorname{Re}(2ie^{it}) + |2i|^2 = 5 - 4\sin t.$$

The maximum occurs at $t = 3\pi/2$, showing that the absolute maximum value attained by $|p|$ on $\overline{\mathbb{D}}$ is $|p(-i)| = 3$.

It is surprising how restrictive and enlightening the maximum principle can be, as the next result suggests. Referring to this as a lemma is almost criminal, given its importance in function theory, but it is tradition.

3.5.5 Schwarz's Lemma. *Let $f : \mathbb{D} \to \mathbb{D}$ be an analytic function satisfying $f(0) = 0$. Then*

(a) $|f(z)| \le |z|$ *for all $z \in \mathbb{D}$ and*

(b) $|f'(0)| \le 1$.

If equality occurs in (a) for any $z \in \mathbb{D} \setminus \{0\}$ or in (b), then there is $c \in \partial\mathbb{D}$ such that $f(z) = cz$ for all $z \in \mathbb{D}$.

Proof. Since f has a zero at 0, there exists $g \in H(\mathbb{D})$ such that

$$f(z) = zg(z)$$

for all $z \in \mathbb{D}$. It is immediate that $g(0) = f'(0)$. If $r \in (0, 1)$ and $|z| = r$, then

$$|g(z)| = \frac{|f(z)|}{r} < \frac{1}{r}.$$

The maximum principle implies that $|g(z)| < 1/r$ for all $z \in D(0; r)$. Since this is true for all $r \in (0, 1)$, we have that $|g(z)| \le 1$ for all $z \in \mathbb{D}$. This gives (a) for $z \ne 0$ and (b) for $z = 0$. (Note that (a) holds for $z = 0$ by hypothesis.)

If $|g(z)| = 1$ for some $z \in \mathbb{D}$, then g is constant, specifically $g(z) = c \in \partial\mathbb{D}$ for all $z \in \mathbb{D}$, by the maximum principle. But $|g(z)| = 1$ for some $z \in \mathbb{D}$ if and only if equality occurs in (a) for some $z \in \mathbb{D} \setminus \{0\}$ or in (b), giving the result. $\square$

We will now explore just how powerful this theory is by providing an unlikely classification theorem. Specifically, we will describe *all* analytic bijections of the unit disk onto itself, the so-called analytic automorphisms.

3.5.6 Definition. Let $\Omega \subseteq \mathbb{C}$ be open. The one-to-one analytic functions taking Ω onto Ω are called the *analytic automorphisms* of Ω. The set of all such functions is denoted $\operatorname{Aut}\Omega$.

Corollary 3.5.2 gives that $f^{-1} \in \operatorname{Aut}\Omega$ when $f \in \operatorname{Aut}\Omega$. It is worth noting that $\operatorname{Aut}\Omega$ is a *group* under the operation of function composition.

The methods of Section 2.6, specifically Example 2.6.21, show that the functions

$$\varphi_a(z) = \frac{z - a}{1 - \overline{a}z}, \tag{3.5.1}$$

for $a \in \mathbb{D}$, are in Aut $\mathbb{D}$. Of course, multiplying φ_a by a constant of modulus 1 yields another automorphism. Are there any more?

3.5.7 Classification of Disk Automorphisms. *A function f is in* Aut $\mathbb{D}$ *if and only if there are $a \in \mathbb{D}$ and $c \in \partial\mathbb{D}$ such that $f = c\varphi_a$.*

Proof. We know that $c\varphi_a \in$ Aut $\mathbb{D}$ for any $a \in \mathbb{D}$ and $c \in \partial\mathbb{D}$, thus we need only address the converse.

For $f \in$ Aut $\mathbb{D}$, let $g = f^{-1}$, and set $a = g(0)$. Define $\psi = f \circ \varphi_a^{-1} \in$ Aut $\mathbb{D}$. Then $\psi(0) = 0$. Since $\varphi_a^{-1} = \varphi_{-a}$ (check this), direct calculation yields

$$\psi'(0) = f'(a)\varphi'_{-a}(0) = (1 - |a|^2)f'(a).$$

Schwarz's lemma implies that $|\psi'(0)| \leq 1$, and therefore

$$|f'(a)| \leq \frac{1}{1 - |a|^2}.$$

Now set $\omega = \varphi_a \circ g \in$ Aut $\mathbb{D}$. Then $\omega(0) = 0$. Direct calculation gives

$$\omega'(0) = \varphi'_a(a)g'(0) = \frac{g'(0)}{1 - |a|^2}.$$

Again, Schwarz's lemma gives that

$$|g'(0)| \leq 1 - |a|^2.$$

Now $(f \circ g)(z) = z$ for all $z \in \mathbb{D}$. Differentiating both sides at $z = 0$ gives $f'(a)g'(0) = 1$. But then

$$1 = |f'(a)||g'(0)| \leq \frac{1 - |a|^2}{1 - |a|^2} = 1.$$

Therefore the inequality in the above line must be an equality, and accordingly, $|f'(a)| = 1/(1 - |a|^2)$. Hence $|\psi'(0)| = 1$, and Schwarz's lemma implies that there is $c \in \partial\mathbb{D}$ such that $\psi(z) = cz$ for all $z \in \mathbb{D}$. But then $f = \psi \circ \varphi_a = c\varphi_a$. $\square$

In a sense, we now have an idea of how many analytic bijections there are from $\mathbb{D}$ onto $\mathbb{D}$ – exactly one for each $c \in \partial\mathbb{D}$ and $a \in \mathbb{D}$.

Summary and Notes for Section 3.5.

The open mapping theorem and maximum principle are two strong results that certainly do not hold for the calculus of real functions. Indeed, a major topic in single-variable calculus is the identification of relative extrema of functions on an interval, and we now see that nonconstant analytic functions have no relative maxima (in modulus) inside their domains.

Schwarz's lemma should not be taken lightly, for it has many important applications. Indeed, we see from it that all one-to-one analytic mappings of $\mathbb{D}$ onto $\mathbb{D}$ are linear fractional transformations of a specific type. One gets the sense that there are a lot fewer analytic functions than there are real differentiable functions because of the many strong properties that they satisfy. Those in doubt should try to find an algebraic formula for every one-to-one differentiable function from $[0, 1]$ onto $[0, 1]$!

Exercises for Section 3.5.

1. Show that the open mapping theorem fails for real-valued functions of a real variable. (That is, find an example.)

2. For each of the following, a continuous function f is given together with a compact set E in its domain. Find the maximum value of $|f|$ on E and the point (or points) where the maximum occurs.

   (a) $f(z) = e^z$, $E = \overline{D}(1; 1)$
   (b) $f(z) = z^2 + 1$, $E = \overline{D}(i; 1)$
   (c) $f(z) = z^2 + iz - 1$, $E = \overline{\mathbb{D}}$

3. Prove the *minimum principle*: If $\Omega \subseteq \mathbb{C}$ is a domain, $f \in H(\Omega)$, and $a \in \Omega$ is such that $0 < |f(a)| \leq |f(z)|$ for all $z \in \Omega$, then f is constant. Why must $|f(a)| > 0$ be assumed?

4. Let $\Omega \subseteq \mathbb{C}$ be a domain, and let $f \in H(\Omega)$. Show that if one of $\text{Re } f$ or $\text{Im } f$ has a relative maximum or minimum at a point $a \in \Omega$, then f is constant.

5. Let $\Omega \subseteq \mathbb{C}$ be a bounded domain, and let $f : \overline{\Omega} \to \mathbb{C}$ be continuous and be analytic on Ω. Further, suppose that $|f(z)| = \alpha$ for all $z \in \partial\Omega$, where $\alpha \geq 0$ is fixed. Prove that either f is constant or f has a zero in Ω.

6. Find all analytic automorphisms of the right half-plane $H = \{z \in \mathbb{C} : \text{Re } z > 0\}$, i.e. the elements of $\text{Aut } H$.

7. Let $H = \{z \in \mathbb{C} : \text{Re } z > 0\}$ be the right half-plane. Suppose that $f : \mathbb{D} \to H$ is analytic and $f(0) = 1$. Verify that for all $z \in \mathbb{D}$,

$$\frac{1 - |z|}{1 + |z|} \leq |f(z)| \leq \frac{1 + |z|}{1 - |z|}$$

   using linear fractional transformations and Schwarz's lemma.

8. Prove the *Schwarz–Pick theorem*: If $f : \mathbb{D} \to \mathbb{D}$ is analytic, then for all $z \in \mathbb{D}$,

$$\frac{|f'(z)|}{1 - |f(z)|^2} \leq \frac{1}{1 - |z|^2},$$

   where equality for some $z \in \mathbb{D}$ implies that f is an automorphism. (*Hint*: Fix z and compose f with two automorphisms.)

9. Let $C \subseteq \mathbb{D}$ be a contour. Define the *hyperbolic length* of C to be

$$L_h(C) = \int_C \frac{|dz|}{1 - |z|^2}.$$

(a) Prove that composition with an analytic function $f: \mathbb{D} \to \mathbb{D}$ does not increase the hyperbolic length of C. That is, show that $L_h(f(C)) \le L_h(C)$.

(b) If C has positive arc length (i.e. C is not a single point), then show that $L_h(f(C)) = L_h(C)$ if and only if $f \in \operatorname{Aut} \mathbb{D}$.

3.6   The Cauchy–Riemann Equations

Suppose that $\Omega \subseteq \mathbb{C}$ is open. It is convenient to recall that $z = x + iy \in \mathbb{C}$ can be written as an ordered pair $z = (x, y)$, and hence we can think of Ω as also belonging to $\mathbb{R}^2$. If $f \in H(\Omega)$, there are then $u, v: \Omega \to \mathbb{R}$ such that

$$f(z) = u(x, y) + iv(x, y), \qquad z = x + iy \in \Omega.$$

Consider the derivative of f at some $z \in \Omega$:

$$f'(z) = \lim_{h \to 0} \frac{f(z + h) - f(z)}{h}.$$

In the limit, h is complex. Since the limit exists as $h \to 0$, it exists as $h \to 0$ along any path. Assume $h = s$ for $s \in \mathbb{R}$. Then

$$
\begin{aligned}
f'(z) &= \lim_{s \to 0} \frac{f(z + s) - f(z)}{s} \\
&= \lim_{s \to 0} \frac{u(x + s, y) - u(x, y)}{s} + i \lim_{s \to 0} \frac{v(x + s, y) - v(x, y)}{s} \\
&= \frac{\partial u}{\partial x}(x, y) + i \frac{\partial v}{\partial x}(x, y).
\end{aligned}
\tag{3.6.1}
$$

Similarly, suppose that $h = it$ for $t \in \mathbb{R}$. Then

$$
\begin{aligned}
f'(z) &= \lim_{t \to 0} \frac{f(z + it) - f(z)}{it} \\
&= -i \lim_{t \to 0} \frac{u(x, y + t) - u(x, y)}{t} + \lim_{t \to 0} \frac{v(x, y + t) - v(x, y)}{t} \\
&= -i \frac{\partial u}{\partial y}(x, y) + \frac{\partial v}{\partial y}(x, y).
\end{aligned}
\tag{3.6.2}
$$

Setting the real and imaginary parts of (3.6.1) and (3.6.2) equal gives the system of partial differential equations on Ω,

$$\frac{\partial u}{\partial x} = \frac{\partial v}{\partial y}, \qquad \frac{\partial u}{\partial y} = -\frac{\partial v}{\partial x}. \tag{3.6.3}$$

3.6.1 Definition. The partial differential equations in (3.6.3) are called the *Cauchy–Riemann equations*.

We see that the Cauchy–Riemann equations hold for any analytic function. The converse, that satisfaction of the Cauchy–Riemann equations is sufficient for analyticity, is the more substantial result.

3.6.2 Theorem. *Let $\Omega \subseteq \mathbb{C}$ be open and $f: \Omega \to \mathbb{C}$ be written $f = u + iv$ for $u, v: \Omega \to \mathbb{R}$. Then $f \in H(\Omega)$ if and only if u and v are differentiable functions of two real variables that satisfy the Cauchy–Riemann equations on Ω.*

We know from multivariable calculus that u and v are differentiable at a point (x, y) if

$$u(x + s, y + t) = u(x, y) + \frac{\partial u}{\partial x}(x, y)s + \frac{\partial u}{\partial y}(x, y)t + A(s, t)$$

$$v(x + s, y + t) = v(x, y) + \frac{\partial v}{\partial x}(x, y)s + \frac{\partial v}{\partial y}(x, y)t + B(s, t),$$

$\qquad(3.6.4)$

where A and B are defined in a neighborhood of $(0, 0)$ and satisfy

$$\lim_{(s,t)\to(0,0)} \frac{A(s, t)}{\sqrt{s^2 + t^2}} = \lim_{(s,t)\to(0,0)} \frac{B(s, t)}{\sqrt{s^2 + t^2}} = 0. \qquad(3.6.5)$$

Recall that existence of continuous partial derivatives implies differentiability and differentiability implies existence of partial derivatives, but these implications are not reversible.

Proof. If $f \in H(\Omega)$, we have established that the Cauchy–Riemann equations hold for u and v. Since f' is continuous, it follows from (3.6.1) and (3.6.2) that u and v have continuous partial derivatives and hence are differentiable.

For the converse, let $z = x + iy \in \Omega$. Using (3.6.4), the Cauchy–Riemann equations, and writing $h = s + it$, we have

$$\begin{aligned}
f(z + h) - f(z) &= u(x + s, y + t) + iv(x + s, y + t) - u(x, y) - iv(x, y) \\
&= \frac{\partial u}{\partial x}(x, y)s + \frac{\partial u}{\partial y}(x, y)t + i\frac{\partial v}{\partial x}(x, y)s + i\frac{\partial v}{\partial y}(x, y)t \\
&\quad + A(s, t) + iB(s, t) \\
&= \frac{\partial u}{\partial x}(x, y)s - \frac{\partial v}{\partial x}(x, y)t + i\frac{\partial v}{\partial x}(x, y)s + i\frac{\partial u}{\partial x}(x, y)t \\
&\quad + A(s, t) + iB(s, t) \\
&= \left(\frac{\partial u}{\partial x}(x, y) + i\frac{\partial v}{\partial x}(x, y) \right)(s + it) + A(s, t) + iB(s, t).
\end{aligned}$$

Then for $h \neq 0$,

$$\frac{f(z + h) - f(z)}{h} = \frac{\partial u}{\partial x}(x, y) + i\frac{\partial v}{\partial x}(x, y) + \frac{A(s, t)}{s + it} + \frac{iB(s, t)}{s + it}.$$

Applying (3.6.5) (using moduli), we have

$$\lim_{h\to 0} \frac{f(z + h) - f(z)}{h} = \frac{\partial u}{\partial x}(x, y) + i\frac{\partial v}{\partial x}(x, y).$$

Therefore $f'(z)$ exists for all $z \in \Omega$, and f is analytic by Goursat's theorem. $\qquad\square$

3.6.3 Example. Consider the function $f \colon \mathbb{C} \to \mathbb{C}$ given by $f(z) = e^{ix} + y^2$, where $z = x + iy$. Note that $u(x, y) = \cos x + y^2$ and $v(x, y) = \sin x$ have continuous partial derivatives and hence are differentiable. From the above work, it is clear that f has a complex derivative for each $z \in \mathbb{C}$ at which the Cauchy–Riemann equations hold. Now (3.6.3) becomes

$$- \sin x = 0, \qquad 2y = - \cos x.$$

We solve this to see that $f'(z)$ exists precisely for $z = n\pi + (-1)^{n+1} i/2$ for $n \in \mathbb{Z}$, and its value is $(-1)^n i$ at these points.

It is important to note that f is not analytic anywhere, for analyticity is equivalent to complex differentiability on an open set.

Summary and Notes for Section 3.6.

Functions from a set in $\mathbb{C}$ into $\mathbb{C}$ can naturally be thought of as functions from the analogous set in $\mathbb{R}^2$ into $\mathbb{R}^2$. If u and v are the coordinate functions, then u and v are functions of two real variables, as are studied in multivariable calculus. As we know by now, there are many such functions u and v that are differentiable, but it takes a special relationship between u and v for the function $f = u + iv$ to be analytic. The Cauchy–Riemann equations give the exact conditions on u and v for this to occur.

Because satisfaction of the Cauchy–Riemann equations on an open set is equivalent to analyticity, they are often used to define analyticity. Our choice to cast analyticity in the language of series allows us to postpone considering these equations until they are used, as we will in the next section, to understand some geometric properties of analytic functions. One should not underestimate the importance of the Cauchy–Riemann equations to the development of function theory just because they fall at this position in this text.

The equations first appeared in the work of Jean le Rond d'Alembert in 1752, a century earlier than the work of Cauchy and Bernhard Riemann, on hydrodynamics. Leonhard Euler also considered these equations in the following decades in his work on evaluating real integrals with complex methods. Cauchy and Riemann, of course, both (independently) saw the usefulness of these equations, but it was Riemann who fully saw their value.

Exercises for Section 3.6.

1. Use the Cauchy–Riemann equations to show that $f(z) = e^z$ is analytic on $\mathbb{C}$.

2. In each of the following, $f \colon \mathbb{C} \to \mathbb{C}$ and $z = x + iy$. Determine the values of $z \in \mathbb{C}$ for which $f'(z)$ exists.

   (a) $f(z) = x^2 + y^2 - 2xyi$

   (b) $f(z) = x^3 + 3xy^2 + i(y^3 + 3x^2 y)$

   (c) $f(z) = e^y + (1 + iy)e^x$

   (d) $f(z) = (y + 1)^2 + i(x + 1)^2$

3. Define $u \colon \mathbb{R}^2 \to \mathbb{R}$ by $u(x, y) = e^x \sin y + 2xy$. Find a function $v \colon \mathbb{R}^2 \to \mathbb{R}$ such that u and v satisfy the Cauchy–Riemann equations.

4. Let $\Omega \subseteq \mathbb{C}$ be open, $u, v \colon \Omega \to \mathbb{R}$ be differentiable, and $f = u + iv$. The *Jacobian* of f (thought of as a transformation from $\Omega \subseteq \mathbb{R}^2$ into $\mathbb{R}^2$, as in multivariable calculus) is the determinant

$$J_f(x, y) = \det \begin{bmatrix} \dfrac{\partial u}{\partial x}(x, y) & \dfrac{\partial u}{\partial y}(x, y) \\[2mm] \dfrac{\partial v}{\partial x}(x, y) & \dfrac{\partial v}{\partial y}(x, y) \end{bmatrix}.$$

(a) Show that if $f \in H(\Omega)$, then $J_f(x, y) = |f'(z)|^2$ for $z = x + iy \in \Omega$.

(b) Suppose $f \in H(\Omega)$ is one-to-one, $E \subseteq \Omega$ is a set over which double integrals of continuous functions are well defined, and $\varphi \colon f(E) \to \mathbb{C}$ is continuous. Prove that

$$\iint_{f(E)} \varphi(u, v)\, dA(u, v) = \iint_E \varphi(f(x, y))|f'(x, y)|^2\, dA(x, y),$$

where $f'(x, y) = f'(z)$ for $z = x + iy \in \Omega$.

(c) For f and E as in part (b), verify the area formula

$$A(f(E)) = \iint_E |f'(x, y)|^2\, dA(x, y).$$

5. Let $\Omega \subseteq \mathbb{C}$ be open. If $\varphi \colon \Omega \to \mathbb{R}$ has partial derivatives, then the *gradient* of φ is the vector-valued function $\nabla \varphi \colon \Omega \to \mathbb{R}^2$ given by

$$\nabla \varphi(x, y) = \left(\frac{\partial \varphi}{\partial x}(x, y), \frac{\partial \varphi}{\partial y}(x, y) \right).$$

Suppose that $f = u + iv \in H(\Omega)$.

(a) Show that

$$\|\nabla u(x, y)\| = \|\nabla v(x, y)\| = |f'(z)|$$

for all $z = x + iy \in \Omega$.

(b) Show that for any $z = x + iy \in \Omega$, the vector $\nabla v(x, y)$ is a rotation of the vector $\nabla u(x, y)$ by $\pi/2$ radians in the counterclockwise direction.

6. Let $\Omega \subseteq \mathbb{C}$ be open. Let $C^1(\Omega)$ be the set of all functions $f \colon \Omega \to \mathbb{C}$ with continuous partial derivatives. On $C^1(\Omega)$, define the operators

$$\frac{\partial}{\partial z} = \frac{1}{2}\left(\frac{\partial}{\partial x} - i\frac{\partial}{\partial y} \right), \qquad \frac{\partial}{\partial \bar{z}} = \frac{1}{2}\left(\frac{\partial}{\partial x} + i\frac{\partial}{\partial y} \right).$$

(a) Show that $f \in C^1(\Omega)$ is analytic if and only if

$$\frac{\partial f}{\partial \bar{z}} = 0.$$

(For this reason, $\partial/\partial\bar{z}$ is called the *Cauchy–Riemann operator*. This shows that $H(\Omega)$ is the kernel (nullspace) of the linear transformation $\partial/\partial\bar{z}$ defined on the vector space $C^1(\Omega)$.)

(b) Let $f \in C^1(\Omega)$. Show that

$$\frac{\partial f}{\partial z} = 0$$

if and only if $\bar{f} \in H(\Omega)$. (Such functions are sometimes called *anti-analytic*.)

(c) Show that if $f \in H(\Omega)$, then

$$\frac{\partial f}{\partial z} = f'.$$

(d) Show how $\partial/\partial z$ and $\partial/\partial\bar{z}$ can be derived by *formally* applying the chain rule with x and y as "functions" of z and $\bar{z}$ as follows:

$$x = \frac{z + \bar{z}}{2}, \qquad y = \frac{z - \bar{z}}{2i}.$$

7. Let $\Omega \subseteq \mathbb{C} \setminus \{0\}$ be open. Suppose $u, v \colon \Omega \to \mathbb{R}$ are given in the polar coordinates (r, θ) instead of the rectangular coordinates (x, y). Show that the polar form of the Cauchy–Riemann equations is

$$\frac{\partial u}{\partial r} = \frac{1}{r}\frac{\partial v}{\partial \theta}, \qquad \frac{\partial v}{\partial r} = -\frac{1}{r}\frac{\partial u}{\partial \theta}.$$

8. Keep with the notation of Exercise 7. Show that if $f'(z)$ exists for some $z = re^{i\theta}$, then

$$f'(z) = \left(\frac{\partial u}{\partial r}(r, \theta) + i\frac{\partial v}{\partial r}(r, \theta)\right)(\cos\theta - i\sin\theta)$$

$$= -\frac{i}{r}\left(\frac{\partial u}{\partial \theta}(r, \theta) + i\frac{\partial v}{\partial \theta}(r, \theta)\right)(\cos\theta - i\sin\theta).$$

3.7 Conformal Mapping and Local Univalence

As the reader will observe, the title of this section could be, "Fun Things that Occur When an Analytic Function Has a Nonzero Derivative." The demand that a complex function be analytic has already been seen as a restrictive enough condition to imply some very strong results. This is especially so when the derivative is nonzero.

3.7.1 Definition. Let $\Omega \subseteq \mathbb{C}$ be open and $f \in H(\Omega)$. If f is one-to-one on Ω, then f is said to be *univalent*. If for any $z \in \Omega$, there exists an open set $U \subseteq \Omega$ such that $z \in U$ and f is univalent on U, then f is *locally univalent* on Ω.

The reader may be curious why one would define a new word for "one-to-one." This is based in part on tradition, and in part on language. First note that univalent implies analytic. If a function is not univalent, it can be referred to as *multivalent*. If the function is m-to-one, then it is sometimes called m-*valent*. (A function $f \colon E \to \mathbb{C}$ is m-to-one on E if for all $w \in f(E)$, the set $f^{-1}(\{w\})$ has exactly m elements.)

3.7.2 Theorem. *Let $\Omega \subseteq \mathbb{C}$ be open and $f \in H(\Omega)$. For each $a \in \Omega$ such that $f'(a) \neq 0$, there is an open set $U \subseteq \Omega$ such that $a \in U$ and f is univalent on U.*

Of course, it then follows that if $f'(z) \neq 0$ for all $z \in \Omega$, then f is locally univalent on Ω.

Proof. Since f' is continuous and $|f'(a)| > 0$, there exists $r > 0$ such that for all $z \in D(a;r) \subseteq \Omega$, $|f'(z) - f'(a)| < |f'(a)|$. Let $z_1, z_2 \in D(a;r)$ be distinct, and let $L \subseteq D(a;r)$ denote the directed line segment from z_1 to z_2. Then

$$\int_L (f'(z) - f'(a)) \, dz = f(z_2) - f(z_1) - f'(a)(z_2 - z_1)$$

by the fundamental theorem of calculus for contour integrals (Theorem 2.9.12). But

$$\left| \int_L (f'(z) - f'(a)) \, dz \right| \leq \int_L |f'(z) - f'(a)| \, |dz| < |f'(a)||z_2 - z_1|,$$

which gives

$$|f'(a)| - \left| \frac{f(z_2) - f(z_1)}{z_2 - z_1} \right| \leq \left| \frac{f(z_2) - f(z_1)}{z_2 - z_1} - f'(a) \right| < |f'(a)|.$$

This shows that $(f(z_2) - f(z_1))/(z_2 - z_1) \neq 0$. Therefore $f(z_1) \neq f(z_2)$, proving that f is univalent on $D(a;r)$. □

From the following theorem, we will obtain the converse to Theorem 3.7.2.

3.7.3 Theorem. *Suppose that $\Omega \subseteq \mathbb{C}$ is a domain, $f \in H(\Omega)$ is nonconstant, and $a \in \Omega$. Set $\alpha = f(a)$. If $m \in \mathbb{N}$ is the order of the zero of $F(z) = f(z) - \alpha$ at a, then there is an open set $U \subseteq \Omega$ such that $a \in U$ and f is m-to-one on $U \setminus \{a\}$.*

The careful reader might note that Theorem 3.7.2 is essentially a consequence of Theorem 3.7.3. This is not an oversight; Theorem 3.7.2 is needed in the proof of Theorem 3.7.3.

Proof. Since F has a zero at a and is nonconstant, its zero at a is of order m for some $m \in \mathbb{N}$. Due to Theorem 3.4.4, there is $g \in H(\Omega)$ such that $g(a) \neq 0$ and

$$F(z) = (z - a)^m g(z).$$

By continuity, there exists $r > 0$ such that $g(z) \neq 0$ for all $z \in D(a;r) \subseteq \Omega$. (Note that $g^{-1}(\mathbb{C} \setminus \{0\})$ is open and contains a.) The function g'/g is analytic on $D(a;r)$ and analytic functions can be antidifferentiated in disks. By adding an appropriate constant, there is $\varphi \in H(D(a;r))$ such that $\varphi' = g'/g$ on $D(a;r)$ and $\varphi(a) = \log g(a)$ for some allowable branch of the logarithm. Note that

$$\frac{d}{dz} g(z) e^{-\varphi(z)} = g'(z) e^{-\varphi(z)} - \varphi'(z) g(z) e^{-\varphi(z)} = 0, \qquad z \in D(a;r).$$

There is thus a constant $c \in \mathbb{C}$ such that $g(z)e^{-\varphi(z)} = c$ for all $z \in D(a;r)$. Evaluating at $z = a$ shows $c = 1$ and $g(z) = e^{\varphi(z)}$.

Now define $h: D(a;r) \to \mathbb{C}$ by

$$h(z) = (z - a)e^{\varphi(z)/m}.$$

Then h is analytic and $h'(a) = e^{\varphi(a)/m} \neq 0$. Therefore there is an open set $V \subseteq D(a;r)$ such that $a \in V$ and h is univalent on V by Theorem 3.7.2. Since $h(a) = 0$, the open mapping theorem (Theorem 3.5.1) provides some $\varepsilon > 0$ such that $D(0;\varepsilon) \subseteq h(V)$. Let $U = h^{-1}(D(0;\varepsilon)) \cap V$. Then U is open and $a \in U$. Since $w \mapsto w^m$ is m-to-one on $D(0;\varepsilon) \setminus \{0\}$, $z \mapsto [h(z)]^m$ is m-to-one on $U \setminus \{a\}$. But for such z,

$$f(z) = \alpha + F(z) = \alpha + [h(z)]^m \tag{3.7.1}$$

is then m-to-one on $U \setminus \{a\}$. □

Here is the promised converse to Theorem 3.7.2.

3.7.4 Corollary. *If* $\Omega \subseteq \mathbb{C}$ *is open,* $a \in \Omega$, *and* $f \in H(\Omega)$ *is such that* $f'(a) = 0$, *then* f *fails to be univalent on any open set* $U \subseteq \Omega$ *such that* $a \in U$.

Proof. Let $U \subseteq \Omega$ be an open set containing a. There is $r > 0$ such that $D(a;r) \subseteq U$. The result is trivial if f is constant on the domain $D(a;r)$, so assume otherwise. The function $F: D(a;r) \to \mathbb{C}$ given by $F(z) = f(z) - f(a)$ then has a zero at a of order $m \geq 2$. Therefore there is an open set $V \subseteq D(a;r)$ containing a such that f is m-to-one on $V \setminus \{a\} \subseteq U$. □

3.7.5 Example. As an illustration of the above concepts, consider $f \in H(\mathbb{C}\setminus\{-1\})$ given by

$$f(z) = \frac{z^2}{1+z}.$$

Note that

$$f'(z) = \frac{2z + z^2}{(1+z)^2}, \qquad f''(z) = \frac{2}{(1+z)^3}$$

so that $f'(z) = 0$ for $z = -2, 0$. Since f'' is nonzero on the domain of f, there must be open sets $U, V \subseteq \mathbb{C} \setminus \{-1\}$ such that $-2 \in U$, $0 \in V$, and f is two-to-one on $U \setminus \{-2\}$ and on $V \setminus \{0\}$ by Theorem 3.7.3.

To study this further, consider setting $f(z) = f(w)$ for $z \neq w$. This leads to

$$0 = z^2 - w^2 + z^2 w - zw^2 = (z - w)(z + w + zw).$$

Since $z - w \neq 0$, we obtain

$$w = \frac{-z}{1+z}.$$

Let $T(z)$ denote the right-hand side of the above. Then T is a linear fractional transformation, -2 and 0 are (the only) fixed points of T, and $T^{-1} = T$. Since T

exchanges -1 and ∞, T maps the line $\{z \in \mathbb{C} : \mathrm{Re}\, z = -1\} \cup \{\infty\}$ onto itself, as seen by using the symmetric points -2 and 0. Let

$$U = \{z \in \mathbb{C} : \mathrm{Re}\, z < -1\}, \qquad V = \{z \in \mathbb{C} : \mathrm{Re}\, z > -1\}.$$

Then $T(U) = U$, and $U \setminus \{-2\}$ is decomposed into pairs of distinct points $\{a, b\}$ such that $T(a) = b$ and $T(b) = a$. Each such pair of points has the same image under f. A similar result holds for V, showing they are as concluded from Theorem 3.7.3.

We now move to a second concept closely connected to nonzero derivatives. Consider an open set $\Omega \subseteq \mathbb{C}$, and fix a point $a \in \Omega$. If $\gamma \colon [-r, r] \to \Omega$, $r > 0$, is a contour such that $\gamma(0) = a$ and γ is differentiable at 0 with $\gamma'(0) \neq 0$, then $\gamma'(0)$ is a tangent vector to γ^* at a when thought of as a planar vector based at a. (Recall tangent vectors to curves in multivariable calculus.) The angle of the tangent vector at a with respect to the positive real axis is $\arg \gamma'(0)$. *To avoid choosing branches, there is an understood "modulo 2π" here and in what follows for the remainder of the section.* The angle can be expressed

$$\arg \gamma'(0) = \lim_{t \to 0} \arg\left(\frac{\gamma(t) - \gamma(0)}{t} \right) = \lim_{t \to 0^+} \arg\left(\frac{\gamma(t) - a}{t} \right).$$

Because the argument is unchanged by multiplication by a positive number, the above angle can be written

$$\Theta_a(\gamma) = \lim_{t \to 0^+} \arg(\gamma(t) - a). \tag{3.7.2}$$

This limit may exist even when γ fails to be differentiable at 0 or is such that $\gamma'(0) = 0$, and hence serves as a more flexible definition of the angle of γ^* at a. (Note that existence of the limit implies there is $\delta \in (0, r]$ such that $\gamma(t) \neq a$ for all $t \in (0, \delta)$.) Moreover, this permits us to consider γ defined only on $[0, r]$.

If $\gamma_1, \gamma_2 \colon [0, r] \to \Omega$ are such that $\gamma_1(0) = \gamma_2(0) = a$ and $\Theta_a(\gamma_1)$ and $\Theta_a(\gamma_2)$ exist, then we consider the angle between γ_1^* and γ_2^* at a to be $\theta = \Theta_a(\gamma_2) - \Theta_a(\gamma_1)$. This is the angle between the tangent vectors to γ_1^* and γ_2^* at a, if they exist. (See Figure 3.3.) Note that orientation of the contours manifests itself in the order of subtraction.

3.7.6 Definition. Let $\Omega \subseteq \mathbb{C}$ be open and $a \in \Omega$. A continuous function $f \colon \Omega \to \mathbb{C}$ is *conformal* at a if it preserves angles and orientation. That is, if $\gamma_1, \gamma_2 \colon [0, r] \to \Omega$ are contours such that $\gamma_1(0) = \gamma_2(0) = a$ and $\Theta_a(\gamma_1)$ and $\Theta_a(\gamma_2)$ exist, then $\Theta_{f(a)}(f \circ \gamma_1)$ and $\Theta_{f(a)}(f \circ \gamma_2)$ exist and

$$\Theta_{f(a)}(f \circ \gamma_2) - \Theta_{f(a)}(f \circ \gamma_1) = \Theta_a(\gamma_2) - \Theta_a(\gamma_1) \tag{3.7.3}$$

Furthermore, f is *conformal* on Ω if it is conformal at each point of Ω.

Note that $f \circ \gamma_1$ and $f \circ \gamma_2$ are contours in $f(\Omega)$ emanating from $f(a)$. The following theorem shows how the nature of the derivative of an analytic function

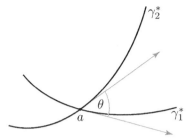

Figure 3.3 θ is the angle between γ_1^* and γ_2^* at a when both contours have tangent vectors

at a point affects the angle between contours, and its proof conveniently uses the construction in the proof of Theorem 3.7.3.

3.7.7 Theorem. *Assume the hypotheses of Theorem 3.7.3. Then if $\gamma_1, \gamma_2 \colon [0, r] \to \Omega$ are contours such that $\gamma_1(0) = \gamma_2(0) = a$ and $\Theta_a(\gamma_1)$ and $\Theta_a(\gamma_2)$ exist, then $\Theta_{f(a)}(f \circ \gamma_1)$ and $\Theta_{f(a)}(f \circ \gamma_2)$ exist and*

$$\Theta_{f(a)}(f \circ \gamma_2) - \Theta_{f(a)}(f \circ \gamma_1) = m[\Theta_a(\gamma_2) - \Theta_a(\gamma_1)].$$

In particular, f is conformal at a if and only if $f'(a) \neq 0$ (so that $m = 1$).

In other words, the angle between the contours is magnified by m, the order of the zero of $f(z) - f(a)$ at a. There is no loss of generality assuming Ω is a domain, since the theorem can be applied to a disk about a given point.

Proof. Let α and h be as in the proof of Theorem 3.7.3. For γ equal to either γ_1 or γ_2, we have

$$\begin{aligned}
\Theta_\alpha(f \circ \gamma) &= \lim_{t \to 0^+} \arg(f(\gamma(t)) - \alpha) \\
&= \lim_{t \to 0^+} \arg([h(\gamma(t))]^m) \\
&= m \lim_{t \to 0^+} \arg h(\gamma(t)) \\
&= m \lim_{t \to 0^+} \arg\left(\frac{h(\gamma(t)) - h(a)}{\gamma(t) - a}(\gamma(t) - a) \right) \\
&= m \arg h'(a) + m\Theta_a(\gamma),
\end{aligned}$$

where we used that the argument of a product of complex numbers is the sum of their arguments. Applying this identity to contours γ_1 and γ_2 and subtracting gives the result. $\qquad\square$

3.7.8 Remark. The curious reader may wonder what advantage lies in defining the angle between contours in the way we have instead of more simply using the angle between tangent vectors of contours with nonzero derivatives. Suppose $\Omega \subseteq \mathbb{C}$ is open, $f \in H(\Omega)$, and $\gamma \colon [-r, r] \to \Omega$ is a contour with $\gamma(0) = a$ and $\gamma'(0) \neq 0$.

Then the image contour $f \circ \gamma$ has derivative $f'(a)\gamma'(0)$ at 0. If $f'(a) = 0$, then this contour has a zero tangent vector. Thus the magnification of angles observed in Theorem 3.7.7 would not be available to us with the tangent vector definition.

The following important theorem shows how the assumption of conformality actually *implies* analyticity!

3.7.9 Conformal Mapping Theorem. *Let $\Omega \subseteq \mathbb{C}$ be open and $f: \Omega \to \mathbb{C}$ be written $f = u + iv$ for $u, v: \Omega \to \mathbb{R}$. Then u and v are differentiable on Ω and f is conformal on Ω if and only if $f \in H(\Omega)$ and $f'(z) \neq 0$ for all $z \in \Omega$.*

Proof. Suppose that u and v are differentiable and f is conformal. As in Exercise 11 of Section 2.8, we write

$$\frac{\partial f}{\partial x} = \frac{\partial u}{\partial x} + i\frac{\partial v}{\partial x}, \qquad \frac{\partial f}{\partial y} = \frac{\partial u}{\partial y} + i\frac{\partial v}{\partial y}.$$

The Cauchy–Riemann equations are the real and imaginary parts of

$$\frac{\partial f}{\partial x} = -i\frac{\partial f}{\partial y}.$$

Suppose $a \in \Omega$ is such that the Cauchy–Riemann equations do not hold for f at a.

Let $r > 0$ so that $\overline{D}(a; r) \subseteq \Omega$. For all $\theta \in \mathbb{R}$, let $\gamma_\theta: [0, r] \to \Omega$ be given by $\gamma_\theta(t) = a + e^{i\theta}t$. Since f is conformal at a, it must be that

$$\Theta_{f(a)}(f \circ \gamma_{\theta_2}) - \Theta_{f(a)}(f \circ \gamma_{\theta_1}) = \Theta_a(\gamma_{\theta_2}) - \Theta_a(\gamma_{\theta_2}) = \theta_2 - \theta_1$$

for all $\theta_1, \theta_2 \in \mathbb{R}$. But then

$$\Theta_{f(a)}(f \circ \gamma_{\theta_2}) - \theta_2 = \Theta_{f(a)}(f \circ \gamma_{\theta_1}) - \theta_1,$$

showing that $\Theta_{f(a)}(f \circ \gamma_\theta) - \theta$ is constant over all values of θ.

By using (3.6.4), we see that

$$f(\gamma_\theta(t)) = f(a) + \frac{\partial f}{\partial x}(a)(t\cos\theta) + \frac{\partial f}{\partial y}(a)(t\sin\theta) + A(te^{i\theta})$$

for $t \in (0, r]$, where

$$\lim_{t \to 0^+} \frac{A(te^{i\theta})}{t} = \lim_{t \to 0^+} \frac{A(t\cos\theta, t\sin\theta)}{\sqrt{t^2\cos^2\theta + t^2\sin^2\theta}} = 0.$$

Because $\Theta_{f(a)}(f \circ \gamma_\theta)$ exists, there is $\delta \in (0, r]$ such that $f(\gamma_\theta(t)) \neq f(a)$ for $t \in (0, \delta)$. For such t, we multiply the inside of the argument by the positive number $2/t$ and use properties of the argument in the calculation

$\arg(f(\gamma_\theta(t)) - f(a)) - \theta$

$$= \arg\left(\frac{\partial f}{\partial x}(a)(e^{i\theta} + e^{-i\theta}) - i\frac{\partial f}{\partial y}(a)(e^{i\theta} - e^{-i\theta}) + \frac{2A(te^{i\theta})}{t}\right) - \theta$$

$$= \arg\left(\frac{\partial f}{\partial x}(a) - i\frac{\partial f}{\partial y}(a) + e^{-2i\theta}\left[\frac{\partial f}{\partial x}(a) + i\frac{\partial f}{\partial y}(a)\right] + \frac{2e^{-i\theta}A(te^{i\theta})}{t}\right).$$

By our assumption, the set

$$C = \left\{ \frac{\partial f}{\partial x}(a) - i\frac{\partial f}{\partial y}(a) + e^{-2i\theta}\left[\frac{\partial f}{\partial x}(a) + i\frac{\partial f}{\partial y}(a)\right] : 0 \le \theta < \pi \right\} \qquad (3.7.4)$$

is a circle with radius

$$\left|\frac{\partial f}{\partial x}(a) + i\frac{\partial f}{\partial y}(a)\right| > 0.$$

If $0 \in C$, only one value of θ corresponds to it in the set (3.7.4). For all other θ, we may take the limit needed to obtain

$$\Theta_{f(a)}(f \circ \gamma_\theta) - \theta = \arg\left(\frac{\partial f}{\partial x}(a) - i\frac{\partial f}{\partial y}(a) + e^{-2i\theta}\left[\frac{\partial f}{\partial x}(a) + i\frac{\partial f}{\partial y}(a)\right]\right).$$

A contradiction ensues, as the argument of the nonzero points on C cannot be constant. We conclude that $f \in H(\Omega)$ by Theorem 3.6.2.

The remainder of the proof follows from Theorem 3.7.7.   □

3.7.10 Example. The functions in Aut $\mathbb{D}$ are now seen to be the only one-to-one conformal mappings of $\mathbb{D}$ onto $\mathbb{D}$ (whose real and imaginary parts are differentiable). In Figure 3.4, we visualize the mapping $\varphi_a \in$ Aut $\mathbb{D}$, given by (3.5.1) with $a = e^{i\pi/4}/2$, as was done in Section 2.5. Our work in Section 2.6 gives that line segments in $\mathbb{D}$ map to segments or circular arcs and circles in $\mathbb{D}$ map to circles, and symmetric points can be used to determine their characteristics. Observe that the images of segments through 0 and circles centered at 0 intersect at right angles.

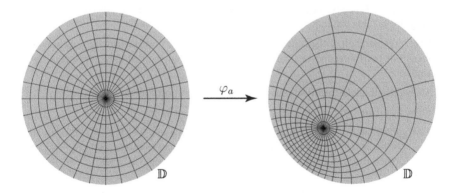

Figure 3.4   Mapping of segments and concentric circles in $\mathbb{D}$ into $\mathbb{D}$ by φ_a, $a = e^{i\pi/4}/2$

Summary and Notes for Section 3.7.

We see that an analytic function is conformal (angle and orientation preserving) at any point at which its derivative is nonzero and, conversely, every (differentiable) conformal mapping on the plane is actually a complex analytic function. This is a significant geometric result that follows from the Cauchy–Riemann equations. We also see that an analytic function is one-to-one on a neighborhood of (open set containing) any point at which the derivative is nonzero and not in any neighborhood of a point at which the derivative is zero.

Laplace's equation, a partial differential equation related to such things as electromagnetic potential and fluid flow that will be considered in Section 6.1, is invariant under a conformal transformation of domain. Therefore, when defined in a complicated planar region, it can be solved in a nicer region that is the conformal image of the original. In particular, the mappings we considered in Sections 2.5 and 2.6 could be used to this end. This is a key application of complex analysis to the sciences.

The connection between conformal mapping and the Cauchy–Riemann equations is what led to the discovery of the equations prior to the study of functions of a complex variable. For instance, Joseph Louis Lagrange in the 18th century and Gauss in the 19th derived the Cauchy–Riemann equations when considering conformal mapping in cartography. Riemann's work drew largely from Gauss's.

Exercises for Section 3.7.

1. Let $f: \mathbb{C} \to \mathbb{C}$ be given by $f(z) = z^2 - 2z + 2$. Find a point $a \in \mathbb{C}$ such that $f'(a) = 0$.

   (a) Demonstrate that f is not one-to-one in any open set containing a.

   (b) Demonstrate that f is not conformal at a.

2. Perform an analysis as in Example 3.7.5 for the function

$$f(z) = \frac{1+z}{1+z^2}.$$

3. Let $f: \mathbb{C} \to \mathbb{C}$ be given by $f(z) = |z|z$. Show that $f'(0) = 0$ and f is conformal at 0. Why doesn't this contradict Theorem 3.7.7?

4. Give an illustration of the conformal nature of $f(z) = e^z$ everywhere by sketching the images of horizontal and vertical lines under f and showing that they intersect at right angles.

5. Show that all linear fractional transformations are conformal on their domains.

6. Show that there is no one-to-one conformal mapping of $\mathbb{D}$ onto $\mathbb{C}$.

7. Verify the analytic version of the *inverse function theorem*: Let $\Omega \subseteq \mathbb{C}$ be open and $f \in H(\Omega)$. If $a \in \Omega$ is such that $f'(a) \neq 0$, then there exist open sets $U \subseteq \Omega$ and $V \subseteq \mathbb{C}$ such that $a \in U$, f is a one-to-one mapping of U onto V, and $f^{-1}: V \to U$ is analytic.

8. Let $\Omega \subseteq \mathbb{C}$ be open, $f \in H(\Omega)$, and $a \in \Omega$ such that f is conformal at a. Show that there is a constant angle α, called the *angle of rotation* at a, such that for every contour $\gamma: [0, r] \to \Omega$ such that $\gamma(0) = a$ and $\Theta_a(\gamma)$ exists,

$$\alpha = \Theta_{f(a)}(f \circ \gamma) - \Theta_a(\gamma).$$

   Calculate a simple formula for α.

9. Let $\Omega \subseteq \mathbb{C}$ be open and $a \in \Omega$. A continuous function $f: \Omega \to \mathbb{C}$ is *anti-conformal* at a if it preserves angles and reverses orientation. With the notation of Definition 3.7.6, this

means

$$\Theta_{f(a)}(f \circ \gamma_2) - \Theta_{f(a)}(f \circ \gamma_1) = \Theta_a(\gamma_1) - \Theta_a(\gamma_2).$$

Adapt the proof of the conformal mapping theorem to show that if $f = u + iv$, then u and v are differentiable on Ω and f is anti-conformal on Ω if and only if $f = \bar{g}$ with $g \in H(\Omega)$ and $g'(z) \neq 0$ for all $z \in \Omega$. (See Exercise 6 in Section 3.6.)

10. Let $\Omega \subseteq \mathbb{C}$ be open, $a \in \Omega$, and $f : \Omega \to \mathbb{C}$. If the limit

$$M_a(f) = \lim_{z \to a} \frac{|f(z) - f(a)|}{|z - a|}$$

exists, then its value $M_a(f)$ is called the *magnification factor* of f at a.

(a) Explain why the term "magnification factor" is appropriate.

(b) Show that if f is conformal, then f has a positive magnification factor at each point in Ω.

(c) Suppose Ω is a domain. Show that if f satisfies the hypotheses of the conformal mapping theorem, u and v are continuously differentiable, and the magnification factor of f exists and is positive at each point of Ω, then $f \in H(\Omega)$ or $\bar{f} \in H(\Omega)$. (*Hint*: Modify the proof of the conformal mapping theorem, including using the contours γ_θ. Consider Exercise 6 in Section 3.6.)

11. Let $f(z) = \sin z$ for $z \in \mathbb{C}$.

(a) Show that the points $z = x + iy \in \mathbb{C}$ at which f has magnification factor 1 satisfy $\cos 2x + \cosh 2y = 2$.

(b) Find all $z \in \mathbb{C}$ such that the angle of rotation of f at z is 0. (In other words, direction is preserved as a vector at such points.) Sketch the solution set.

12. Let $g : \mathbb{D} \to \mathbb{C}$ be univalent. Show that there is a linear function $h(z) = az + b$ with $a \neq 0$ so that the composition $f = h \circ g$ is a univalent function on the disk with series expansion of the form

$$f(z) = z + \sum_{n=2}^{\infty} c_n z^n.$$

Such functions f are called *schlicht* mappings of the disk.

13. Find all of the schlicht mappings that take $\mathbb{D}$ onto a half-plane. What do the half-planes have in common?

CHAPTER 4

CAUCHY'S INTEGRAL THEORY

A critical step in the proof of Goursat's theorem is Corollary 3.2.3, in which it is shown that the values of an analytic function in a disk can be expressed as an integral involving the function on the disk's boundary. In Corollary 3.2.6, this type of representation is also seen to hold for derivatives of the function at the center of the disk. In this chapter, we significantly extend the representation to integrals over arbitrary closed contours in the famous Cauchy integral formula. This formula and its consequences will greatly expand our ability to evaluate complex contour integrals.

4.1 The Index of a Closed Contour

Before addressing the main topic of this section, we require a lemma that we will utilize on a number of occasions.

4.1.1 Lemma. *Let C be a contour in $\mathbb{C}$ and $\varphi \colon C \to \mathbb{C}$ be continuous. For each $m \in \mathbb{N}$, define $f_m \colon \mathbb{C} \setminus C \to \mathbb{C}$ by*

$$f_m(z) = \int_C \frac{\varphi(\zeta)}{(\zeta - z)^m} \, d\zeta.$$

127

Complex Analysis: A Modern First Course in Function Theory, First Edition. Jerry R. Muir, Jr.
© 2015 John Wiley & Sons, Inc. Published 2015 by John Wiley & Sons, Inc.

Then for all m, $f_m \in H(\mathbb{C} \setminus C)$ and $f'_m = m f_{m+1}$.

Proof. Fix $m \in \mathbb{N}$. Let $a \in \mathbb{C} \setminus C$ and $r > 0$ be such that $D(a; r) \subseteq \mathbb{C} \setminus C$. Let $\zeta \in C$. The calculation giving (3.2.2) holds here, and so

$$\frac{1}{\zeta - z} = \sum_{n=0}^{\infty} \frac{(z - a)^n}{(\zeta - a)^{n+1}}$$

for all $z \in D(a; r)$. Differentiating $m - 1$ times with respect to z gives

$$\frac{(m-1)!}{(\zeta - z)^m} = \sum_{n=m-1}^{\infty} \frac{n!}{(n - m + 1)!} \frac{(z - a)^{n-m+1}}{(\zeta - a)^{n+1}}$$

for such z. Accordingly,

$$\frac{\varphi(\zeta)}{(\zeta - z)^m} = \sum_{n=m-1}^{\infty} \frac{n! \varphi(\zeta)}{(m-1)!(n - m + 1)!} \frac{(z - a)^{n-m+1}}{(\zeta - a)^{n+1}} \tag{4.1.1}$$

holds for all $\zeta \in C$ and $z \in D(a; r)$.

Now fix $z \in D(a; r)$. We know that there is some $M > 0$ such that $|\varphi(\zeta)| \le M$ for all $\zeta \in C$ because φ is continuous and C is compact. We then have the bounds

$$\left| \frac{n! \varphi(\zeta)}{(m-1)!(n - m + 1)!} \frac{(z - a)^{n-m+1}}{(\zeta - a)^{n+1}} \right|$$
$$\le \frac{M}{(m-1)! |z - a|^m} \left[\frac{n!}{(n - m + 1)!} \left(\frac{|z - a|}{r} \right)^{n+1} \right] \tag{4.1.2}$$

for $n = m - 1, m, \ldots$. For such n, write M_n for the right-hand side of (4.1.2). Then

$$\lim_{n \to \infty} \frac{M_{n+1}}{M_n} = \lim_{n \to \infty} \frac{n + 1}{n - m + 2} \frac{|z - a|}{r} = \frac{|z - a|}{r} < 1.$$

Therefore $\sum_{n=m-1}^{\infty} M_n$ converges by the ratio test. The Weierstrass M-test (Theorem 2.2.7) then implies that the series (4.1.1) converges uniformly for $\zeta \in C$.

Due to uniform convergence, we may integrate inside the series in the following calculation:

$$f_m(z) = \int_C \frac{\varphi(\zeta)}{(\zeta - z)^m} \, d\zeta$$
$$= \sum_{n=m-1}^{\infty} \left[\frac{n!}{(m-1)!(n - m + 1)!} \int_C \frac{\varphi(\zeta)}{(\zeta - a)^{n+1}} \, d\zeta \right] (z - a)^{n-m+1}.$$

Since z was arbitrarily chosen in $D(a; r)$ and the expression in brackets is independent of z, f_m is equal to a power series on each disk in $\mathbb{C} \setminus C$ and is hence analytic. Differentiation of the series and simplification of factorials shows that $f'_m = m f_{m+1}$. $\qquad \square$

We now define a useful function closely tied to the geometry of a closed contour in $\mathbb{C}$.

4.1.2 Definition. Let γ be a closed contour in $\mathbb{C}$. The *index* of γ is the function $\mathrm{Ind}_\gamma \colon \mathbb{C} \setminus \gamma^* \to \mathbb{C}$ given by

$$\mathrm{Ind}_\gamma z = \frac{1}{2\pi i} \int_\gamma \frac{d\zeta}{\zeta - z}. \tag{4.1.3}$$

We first address some analytic aspects of the index and then consider its geometric interpretation.

4.1.3 Theorem. *Let $\gamma \colon [a, b] \to \mathbb{C}$ be a closed contour. Then for all $z \in \mathbb{C} \setminus \gamma^*$, $\mathrm{Ind}_\gamma z \in \mathbb{Z}$. Furthermore, Ind_γ is analytic and hence is constant on any connected subset of $\mathbb{C} \setminus \gamma^*$. Specifically, $\mathrm{Ind}_\gamma z = 0$ if z is in the unbounded component of $\mathbb{C} \setminus \gamma^*$.*

Proof. Fix $z \in \mathbb{C} \setminus \gamma^*$. Let $\varphi \colon [a, b] \to \mathbb{C}$ be defined by

$$\varphi(t) = \int_a^t \frac{\gamma'(s)}{\gamma(s) - z} \, ds.$$

By the fundamental theorem of calculus (Theorem 2.8.5), φ is continuous, and for all t such that $\gamma'(t)$ exists (for all but a possible finite number of points in $[a, b]$),

$$\varphi'(t) = \frac{\gamma'(t)}{\gamma(t) - z},$$

and accordingly,

$$\frac{d}{dt} e^{-\varphi(t)} (\gamma(t) - z) = e^{-\varphi(t)} \gamma'(t) - \varphi'(t) e^{-\varphi(t)} (\gamma(t) - z) = 0.$$

Therefore $e^{-\varphi(t)} (\gamma(t) - z)$ is constant for all values of t. Since $\varphi(a) = 0$,

$$e^{-\varphi(b)} (\gamma(b) - z) = e^{-\varphi(a)} (\gamma(a) - z) = \gamma(a) - z.$$

Now $\gamma(b) = \gamma(a)$ implies $e^{-\varphi(b)} = 1$. This shows that $\varphi(b) = 2n\pi i$ for some $n \in \mathbb{Z}$. It follows that

$$\mathrm{Ind}_\gamma z = \frac{1}{2\pi i} \int_a^b \frac{\gamma'(s)}{\gamma(s) - z} \, ds = \frac{\varphi(b)}{2\pi i} = n.$$

Lemma 4.1.1 gives that Ind_γ is analytic, analytic functions are continuous, and the continuous image of a connected set is connected by Theorem 2.1.13. Singleton sets are the only connected subsets of $\mathbb{Z}$, proving Ind_γ is constant on any connected set.

Since γ^* is compact, $\gamma^* \subseteq D(0; R)$ for some $R > 0$. Let $z \in \mathbb{C} \setminus \overline{D}(0; R)$, and choose a branch of the logarithm such that $\log(\zeta - z)$ is defined for all $\zeta \in D(0; R)$. Then Corollary 2.9.14 implies

$$\operatorname{Ind}_\gamma z = \frac{1}{2\pi i} \int_\gamma \frac{d\zeta}{\zeta - z} = 0,$$

as $\log(\zeta - z)$ is an antiderivative of the integrand. This proves the final claim. $\square$

Let us consider how the index of a closed contour can be interpreted geometrically. With the above hypotheses, suppose $z \in \mathbb{C} \setminus \gamma^*$ is fixed. Since $\mathbb{C} \setminus \gamma^*$ is open, there is some $r > 0$ such that $D(z; r) \subseteq \mathbb{C} \setminus \gamma^*$. For any $t \in [a, b]$, $z \notin D(\gamma(t); r)$, and therefore a branch of $\arg(\zeta - z)$ can be defined for $\zeta \in D(\gamma(t); r)$.

Partition $[a, b]$ by

$$a = t_0 < t_1 < \cdots < t_n = b$$

so that for any $k = 1, \ldots, n$, $|\gamma(t_k) - \gamma(t)| < r$ for all $t \in [t_{k-1}, t_k]$. (Since the arc length of γ^* is finite, this can be done by partitioning γ^* into a finite number of pieces of arc length less than r.) For each k, $\gamma([t_{k-1}, t_k]) \subseteq D(\gamma(t_k); r)$, and using the branch of $\arg(\zeta - z)$ defined in that disk, we write

$$\Delta_k \arg(\gamma(t) - z) = \arg(\gamma(t_k) - z) - \arg(\gamma(t_{k-1}) - z) \qquad (4.1.4)$$

for the total net change in the argument of $\gamma(t) - z$ as t ranges from t_{k-1} to t_k. Using the corresponding branch of $\log(\zeta - z)$, we have

$$\int_{t_{k-1}}^{t_k} \frac{\gamma'(t)}{\gamma(t) - z}\, dt = \log(\gamma(t_k) - z) - \log(\gamma(t_{k-1}) - z)$$

$$= \ln|\gamma(t_k) - z| - \ln|\gamma(t_{k-1}) - z| + i\Delta_k \arg(\gamma(t) - z).$$

Now consider

$$2\pi i \operatorname{Ind}_\gamma z = \sum_{k=1}^n \int_{t_{k-1}}^{t_k} \frac{\gamma'(t)}{\gamma(t) - z}\, dt$$

$$= \sum_{k=1}^n [\ln|\gamma(t_k) - z| - \ln|\gamma(t_{k-1}) - z|] + i\sum_{k=1}^n \Delta_k \arg(\gamma(t) - z).$$

The sum of the differences of logs telescopes to $\ln|\gamma(b) - z| - \ln|\gamma(a) - z| = 0$ because $\gamma(a) = \gamma(b)$. The sum of the changes in argument represents *the total net change in the argument of $\gamma(t) - z$ as t ranges from a to b*. (It will not telescope to 0 because *distinct branches of the argument are used for distinct k in (4.1.4)*.)

From this discussion, we see that $\operatorname{Ind}_\gamma z$ is the net number of full turns in the counterclockwise direction made by γ around z. For this reason, $\operatorname{Ind}_\gamma z$ is often called the *winding number* of γ about z. See Figure 4.1.

4.1.4 Remark. If γ is a simple closed contour in $\mathbb{C}$, it may seem self-evident that $\mathbb{C} \setminus \gamma^*$ consists of two connected components, one bounded (the *inside* of γ^*) and

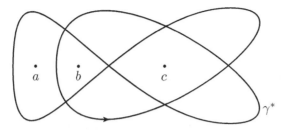

Figure 4.1 $\text{Ind}_\gamma\, a = -1$, $\text{Ind}_\gamma\, b = 0$, and $\text{Ind}_\gamma\, c = 2$

one unbounded (the *outside* of γ^*). In fact, this is a consequence of the *Jordan curve theorem* from topology, a great example of a theorem that is easy to state yet difficult to prove. We could carefully avoid assuming this when stating some upcoming results, but only at the expense of clarity and intuition. Hence we take it for granted and observe that if γ is positively oriented, the above argument gives that $\text{Ind}_\gamma\, z = 1$ if z lies inside of γ^* and $\text{Ind}_\gamma\, z = 0$ if z lies outside of γ^*.

4.1.5 Example. Fix $z \in \mathbb{D}$ and $n \in \mathbb{Z}$, and consider the contour $\gamma \colon [0, 2\pi] \to \mathbb{C}$ given by $\gamma(t) = e^{int}$. Evidently, γ parameterizes the unit circle n times in the positive direction if $n \in \mathbb{N}$, $-n$ times in the negative direction if $-n \in \mathbb{N}$, and is the constant value 1 if $n = 0$. It follows from the above reasoning that $\text{Ind}_\gamma\, z = n$. For a direct calculation, we have

$$
\begin{aligned}
\text{Ind}_\gamma\, z &= \frac{1}{2\pi i} \int_0^{2\pi} \frac{ine^{int}}{e^{int} - z}\, dt \\
&= \frac{n}{2\pi} \int_0^{2\pi} \frac{dt}{1 - e^{-int}z} \\
&= \frac{n}{2\pi} \int_0^{2\pi} \sum_{k=0}^{\infty} e^{-inkt} z^k\, dt.
\end{aligned}
$$

The series inside the last integral converges uniformly for $t \in [0, 2\pi]$ by an application of the Weierstrass M-test, so integration can be done term by term. The result then follows by applying (2.8.5) to the separate integrals in

$$
\text{Ind}_\gamma\, z = \frac{n}{2\pi} \sum_{k=0}^{\infty} z^k \int_0^{2\pi} e^{-inkt}\, dt.
$$

Summary and Notes for Section 4.1.

Geometrically, the index of a closed contour about a point is the net number of times the curve turns around the point in the counterclockwise direction. Its analytic definition, as the integral of a simple function whose antiderivative involves the logarithm, will be used in some upcoming proofs. It is the multivalued nature of the logarithm, particularly the argument, that makes this work. As one traverses the contour, the real part of the logarithm returns to its original value, but the imaginary part changes based upon how the curve winds about the point.

Although technical, Lemma 4.1.1 is a handy result that will pop up on a number of upcoming occasions. It essentially says that we can differentiate f_m by differentiating inside the integral. This is yet another example of interchanging limit processes and is a (albeit not general) complex result similar to Leibniz's rule (in Appendix B).

That (what is now) the Jordan curve theorem was nontrivial and thus needed proof was argued by Bernard Bolzano in the early 19th century. There is some debate as to whether Camille Jordan's proof of 1887 is complete. Some see his proof as missing details that are not too hard to fill in, while others credit Oswald Veblen with providing the first complete proof 18 years later. Although the complicated proof is somewhat simplified by assuming the curve to be a contour, it is still too great a digression for us.

Exercises for Section 4.1.

1. Let $n \in \mathbb{Z}$, $z \in \mathbb{C} \setminus \overline{\mathbb{D}}$, and γ be the contour in Example 4.1.5. Use a similar manipulation of the geometric series to that in the example so show that $\mathrm{Ind}_\gamma\, z = 0$.

2. Let C_1 and C_2 be closed contours in $\mathbb{C}$ such that $C = C_1 + C_2$ is defined. If γ, γ_1, and γ_2 parameterize C, C_1, and C_2, respectively, then prove that

$$\mathrm{Ind}_{\gamma_1}\, z + \mathrm{Ind}_{\gamma_2}\, z = \mathrm{Ind}_\gamma\, z$$

   for all $z \in \mathbb{C} \setminus C$.

3. Let C be a closed contour in $\mathbb{C}$ parameterized by γ. If $\hat{\gamma}$ parameterizes $-C$, show that

$$\mathrm{Ind}_{\hat{\gamma}}\, z = -\,\mathrm{Ind}_\gamma\, z$$

   for all $z \in \mathbb{C} \setminus C$.

4. Sketch the traces of closed contours γ that satisfy the given conditions.

   (a) $\mathrm{Ind}_\gamma(-1) = 1$ and $\mathrm{Ind}_\gamma\, 1 = 3$

   (b) $\mathrm{Ind}_\gamma\, 0 = 2$, $\mathrm{Ind}_\gamma\, 1 = 2$, and $\mathrm{Ind}_\gamma\, i = -1$

5. Let γ be a closed contour whose trace is the sum of the segments $[2 - i, 2 + i]$, $[2 + i, -2 - i]$, $[-2 - i, -2 + i]$, and $[-2 + i, 2 - i]$. Calculate $\mathrm{Ind}_\gamma(-1)$ and $\mathrm{Ind}_\gamma\, 1$ by directly evaluating the integrals in the definition. Compare this with the geometric interpretation.

6. Let $p\colon \mathbb{C} \to \mathbb{C}$ be a polynomial, and choose $R > 0$ large enough so that all zeros of p lie in $D(0; R)$. Show that

$$\int_{\partial D(0;R)} \frac{p'(z)}{p(z)}\, dz = 2n\pi i,$$

   where n is the degree of p.

7. Verify the final step in the proof of Lemma 4.1.1. That is, differentiate the series representation of f_m to arrive at $f_m' = m f_{m+1}$ for all $m \in \mathbb{N}$.

4.2 The Cauchy Integral Formula

Both theoretically and computationally useful, the Cauchy integral formula is a significant improvement over Corollary 3.2.3. Our proof is a modern version (see [8]) and requires the following lemma that deals with continuity in two complex variables, a concept analogous to continuity of a function of two real variables and defined in Exercise 16 of Section 2.1.

4.2.1 Lemma. *Let $\Omega \subseteq \mathbb{C}$ be open and $f \in H(\Omega)$. The function $\varphi \colon \Omega \times \Omega \to \mathbb{C}$ defined by*

$$\varphi(z,w) = \begin{cases} \dfrac{f(z) - f(w)}{z - w} & \text{if } z \neq w, \\[2ex] f'(z) & \text{if } z = w \end{cases}$$

is continuous. Furthermore, if $w \in \Omega$ and $\varphi_w \colon \Omega \to \mathbb{C}$ is given by $\varphi_w(z) = \varphi(z,w)$, then $\varphi_w \in H(\Omega)$.

Proof. It is left as an exercise to show that φ is continuous at points (z, w) for $z, w \in \Omega$, $z \neq w$. Here, we address the more complicated argument that φ is continuous at (a, a) for $a \in \Omega$. Let $\varepsilon > 0$. There exists $\delta > 0$ such that $D(a; \delta) \subseteq \Omega$ and $|f'(z) - f'(a)| < \varepsilon$ for all $z \in D(a; \delta)$. Fix $z, w \in \Omega$ such that $\sqrt{|z - a|^2 + |w - a|^2} < \delta$. Then $z, w \in D(a; \delta)$. If $z = w$, then

$$|\varphi(z,w) - \varphi(a,a)| = |f'(z) - f'(a)| < \varepsilon.$$

Otherwise, let $L \subseteq D(a; \delta)$ be the directed line segment from w to z. We use the fundamental theorem of calculus for contour integrals to calculate

$$\begin{aligned} |\varphi(z,w) - \varphi(a,a)| &= \frac{1}{|z - w|} |f(z) - f(w) - f'(a)(z - w)| \\ &= \frac{1}{|z - w|} \left| \int_L (f'(\zeta) - f'(a)) \, d\zeta \right| \\ &\leq \frac{1}{|z - w|} \int_L |f'(\zeta) - f'(a)| \, |d\zeta| \\ &< \frac{1}{|z - w|} \int_L \varepsilon \, |d\zeta| = \varepsilon. \end{aligned}$$

Therefore φ is continuous at (a, a).

Now if $w \in \Omega$ is fixed, there is a disk $D(w; r) \subseteq \Omega$ for some $r > 0$ and expansion

$$f(z) = \sum_{n=0}^{\infty} c_n (z - w)^n, \qquad z \in D(w; r).$$

Direct calculation verifies that for all $z \in D(w; r)$,

$$\varphi_w(z) = \sum_{n=0}^{\infty} c_{n+1}(z - w)^n.$$

Since $\varphi'_w(z)$ exists if $z \neq w$ by the very definition of φ, $\varphi_w \in H(\Omega)$. $\square$

4.2.2 Cauchy's Integral Formula. *Let $\Omega \subseteq \mathbb{C}$ be open and $f \in H(\Omega)$. If γ is a closed contour in Ω such that $\mathrm{Ind}_\gamma z = 0$ for all $z \in \mathbb{C} \setminus \Omega$, then*

$$f(z)\,\mathrm{Ind}_\gamma z = \frac{1}{2\pi i} \int_\gamma \frac{f(\zeta)}{\zeta - z}\,d\zeta \tag{4.2.1}$$

for all $z \in \Omega \setminus \gamma^*$.

The hypothesis that $\mathrm{Ind}_\gamma z = 0$ for all $z \in \mathbb{C} \setminus \Omega$, seen here for the first time, will be a regular occurrence in upcoming theorems. It is equivalent to saying that Ω contains all points $z \in \mathbb{C} \setminus \gamma^*$ for which $\mathrm{Ind}_\gamma z \neq 0$. See Figure 4.2.

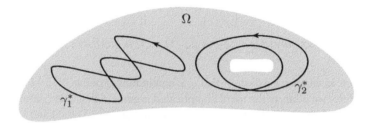

Figure 4.2 γ_1 satisfies the hypotheses of Cauchy's integral formula, but γ_2 does not

Proof. Let φ be as defined in Lemma 4.2.1. Define $g_1 \colon \Omega \to \mathbb{C}$ by

$$g_1(z) = \int_\gamma \varphi(z, \zeta)\,d\zeta.$$

The continuity of φ implies the continuity of g_1. (See the exercises.) We will show $g_1 \in H(\Omega)$ by using Morera's theorem (Theorem 3.2.5) to see that g_1 is analytic on an arbitrary open disk $D \subseteq \Omega$. Given $\zeta \in \gamma^*$, we know $\varphi_\zeta \in H(\Omega)$, and therefore, since φ_ζ can be antidifferentiated on D (Corollary 2.7.5),

$$\int_T \varphi_\zeta(z)\,dz = 0$$

for every triangular path $T \subseteq D$. Now for such a path, we calculate

$$\int_T g_1(z)\,dz = \int_T \int_\gamma \varphi(z, \zeta)\,d\zeta\,dz = \int_\gamma \int_T \varphi(z, \zeta)\,dz\,d\zeta = 0,$$

as desired. Note that the reversal of integrals is valid using Fubini's theorem (see Exercise 16 of Section 2.9) because of the continuity of φ.

Now set $U = \{z \in \mathbb{C} \setminus \gamma^* : \mathrm{Ind}_\gamma z = 0\}$. Then U is open due to the properties of the index. Define $g_2 \colon U \to \mathbb{C}$ by

$$g_2(z) = \int_\gamma \frac{f(\zeta)}{\zeta - z}\,d\zeta.$$

Now g_2 is analytic by Lemma 4.1.1. If $z \in \Omega \cap U$, then

$$
\begin{aligned}
g_1(z) &= \int_\gamma \frac{f(\zeta) - f(z)}{\zeta - z} \, d\zeta \\
&= \int_\gamma \frac{f(\zeta)}{\zeta - z} \, d\zeta - f(z) \int_\gamma \frac{d\zeta}{\zeta - z} \\
&= g_2(z) - 2\pi i f(z) \operatorname{Ind}_\gamma z \\
&= g_2(z).
\end{aligned}
$$

By hypothesis, $\Omega \cup U = \mathbb{C}$, and we use the above reasoning to define $g \colon \mathbb{C} \to \mathbb{C}$ by

$$
g(z) = \begin{cases} g_1(z) & \text{if } z \in \Omega, \\ g_2(z) & \text{if } z \in U. \end{cases}
$$

We see that g is analytic, and hence is entire. Let $\varepsilon > 0$. Since γ^* is compact, $\gamma^* \subseteq D(0; R)$ for some $R > 0$. Furthermore, $|f|$ is bounded by some $M \geq 0$ on γ^*. If L is the arc length of γ^*, then choose $z \in \mathbb{C}$ such that $|z| > R + ML/\varepsilon$. Then for all $\zeta \in \gamma^*$, $|z - \zeta| \geq |z| - |\zeta| > ML/\varepsilon$. This implies that

$$
|g(z)| = |g_2(z)| = \left| \int_\gamma \frac{f(\zeta)}{\zeta - z} \, d\zeta \right| \leq M \int_\gamma \frac{|d\zeta|}{|\zeta - z|} < ML \frac{\varepsilon}{ML} = \varepsilon.
$$

Since g is continuous, it is bounded on the compact set $\overline{D}(0; R + ML/\varepsilon)$, and therefore is bounded on $\mathbb{C}$. Liouville's theorem (Theorem 3.3.5) implies that g is constant. Further, since ε was arbitrary, $g(z) = 0$ for all $z \in \mathbb{C}$. Specifically, for $z \in \Omega \setminus \gamma^*$ this gives that

$$
0 = \int_\gamma \frac{f(\zeta) - f(z)}{\zeta - z} \, d\zeta.
$$

That is,

$$
f(z) \int_\gamma \frac{d\zeta}{\zeta - z} = \int_\gamma \frac{f(\zeta)}{\zeta - z} \, d\zeta.
$$

Dividing by $2\pi i$ proves the theorem. $\qquad\square$

The following corollary, a substantial upgrade of Corollary 3.2.6, is particularly nice, even as it undermines our calculus intuition. *It shows how derivatives can be calculated using integrals and vice versa!*

4.2.3 Cauchy's Integral Formula for Derivatives. *Let $\Omega \subseteq \mathbb{C}$ be open and $f \in H(\Omega)$. If γ is a closed contour in Ω such that $\operatorname{Ind}_\gamma z = 0$ whenever $z \in \mathbb{C} \setminus \Omega$, then*

$$
f^{(m)}(z) \operatorname{Ind}_\gamma z = \frac{m!}{2\pi i} \int_\gamma \frac{f(\zeta)}{(\zeta - z)^{m+1}} \, d\zeta \tag{4.2.2}
$$

for all $m = 0, 1, \ldots$ and $z \in \Omega \setminus \gamma^*$.

Proof. We will use induction on m since the Cauchy integral formula provides the base case. For $m \in \mathbb{N}$, set $f_m \colon \mathbb{C} \setminus \gamma^* \to \mathbb{C}$ by

$$f_m(z) = \int_\gamma \frac{f(\zeta)}{(\zeta - z)^m}\, d\zeta,$$

and assume that

$$f^{(m-1)}(z) \operatorname{Ind}_\gamma z = \frac{(m-1)!}{2\pi i} \int_\gamma \frac{f(\zeta)}{(\zeta - z)^m}\, d\zeta = \frac{(m-1)!}{2\pi i} f_m(z) \qquad (4.2.3)$$

for some $m \in \mathbb{N}$ and all $z \in \Omega \setminus \gamma^*$. Lemma 4.1.1 gives that f_m is analytic and $f_m' = m f_{m+1}$. Now $\operatorname{Ind}_\gamma$ has derivative 0 in $\Omega \setminus \gamma^*$, and therefore differentiating both sides of (4.2.3) gives

$$f^{(m)}(z) \operatorname{Ind}_\gamma z = \frac{m!}{2\pi i} f_{m+1}(z) = \frac{m!}{2\pi i} \int_\gamma \frac{f(\zeta)}{(\zeta - z)^{m+1}}\, d\zeta.$$

This completes the inductive step. □

Many common applications of the Cauchy integral formula and its corollary for derivatives involve simple closed contours. The following corollary gives the subsequent simpler form in this situation. (See Remark 4.1.4.)

4.2.4 Corollary. *Let $\Omega \subseteq \mathbb{C}$ be open, $f \in H(\Omega)$, and γ be a simple closed contour in Ω such that the inside of γ^* lies in Ω. Then for all z inside of γ^*,*

$$f^{(m)}(z) = \frac{m!}{2\pi i} \int_\gamma \frac{f(\zeta)}{(\zeta - z)^{m+1}}\, d\zeta$$

for all $m = 0, 1, \dots$.

We complete this section with a couple of examples.

4.2.5 Example. Let us evaluate the integrals

$$\int_{\partial \mathbb{D}} \frac{\sin^2 z}{z^5}\, dz, \qquad \int_{\partial \mathbb{D}} \frac{\sin^2 z}{(z - \pi/6)^5}\, dz.$$

Since the boundary of the unit circle winds around both 0 and $\pi/6$ once in the positive direction, the Cauchy integral formula gives

$$f^{(4)}(0) = \frac{24}{2\pi i} \int_{\partial \mathbb{D}} \frac{\sin^2 z}{z^5}\, dz, \qquad f^{(4)}\left(\frac{\pi}{6}\right) = \frac{24}{2\pi i} \int_{\partial \mathbb{D}} \frac{\sin^2 z}{(z - \pi/6)^5}\, dz,$$

where $f(z) = \sin^2 z$. We calculate that $f^{(4)}(z) = 8(\sin^2 z - \cos^2 z)$ so that $f^{(4)}(0) = -8$ and $f^{(4)}(\pi/6) = -4$. Therefore the first integral equals $-2\pi i/3$, and the second equals $-\pi i/3$.

4.2.6 Example. We shall use the Cauchy integral formula to evaluate

$$\int_{\partial D(n;1)} \frac{e^z}{z^2 - 1}\, dz, \qquad n = -1, 1.$$

First, for $n = -1$, note that

$$\int_{\partial D(-1;1)} \frac{e^z}{z^2 - 1}\, dz = \int_{\partial D(-1;1)} \frac{f(z)}{z + 1}\, dz,$$

where $f(z) = e^z/(z - 1)$. Now f is analytic for all z except $z = 1$. Therefore

$$\int_{\partial D(-1;1)} \frac{e^z}{z^2 - 1}\, dz = 2\pi i f(-1) = \frac{-\pi i}{e}.$$

Similarly, for $n = 1$, we have

$$\int_{\partial D(1;1)} \frac{e^z}{z^2 - 1}\, dz = \int_{\partial D(1;1)} \frac{g(z)}{z - 1}\, dz,$$

where $g(z) = e^z/(z + 1)$. Since g is analytic except when $z = -1$, we have

$$\int_{\partial D(1;1)} \frac{e^z}{z^2 - 1}\, dz = 2\pi i g(1) = e\pi i.$$

Summary and Notes for Section 4.2.

The Cauchy integral formula is one of the pillars of complex analysis and was fore-shadowed in Section 3.2. Notice that Corollary 3.2.3 is the integral formula on circles. The integral formula for derivatives is a notable generalization of Corollary 3.2.6, which is a version of the formula for circles where the denominator of the integrand must involve the center of the circle.

We now consider a couple of things the integral formula tells us. The first is that for an analytic function f, the value of f or one of its derivatives at a point z at which a contour γ has nonzero index is completely determined by the values of f on γ^*. More simply, if we know the values of f on a simple closed contour, then we know the values of f and all of its derivatives inside the contour!

Secondly, certain integrals can now be calculated by taking derivatives of a related function. This idea seems to run completely against our intuition from calculus, and we've only just begun! In the next chapter, we will see an even more powerful method to evaluate more general integrals.

Augustin-Louis Cauchy presented his integral formula in 1831 for functions that were complex continuously differentiable in disks. (Recall from Section 3.2 that Edouard Jean-Baptiste Goursat is responsible for removing the need for a continuous derivative.) Cauchy was motivated by proving that such functions had convergent power series expansions in such disks. With his integral formula, he expanded the integrand as a series and integrated term by term, as we did in the proof of Theorem 3.2.4. Of course, the rigorous justification for integrating term by term is uniform convergence on compact subsets of the disk, which was observed by Karl Weierstrass years later, as we mentioned in Section 2.2.

Exercises for Section 4.2.

1. Calculate the following integrals over circles.

   (a) $\displaystyle \int_{\partial D} \frac{\cos z}{z^3}\, dz$

   (b) $\displaystyle \int_{\partial D(0;2)} \frac{z^2}{z^2+1}\, dz$

   (c) $\displaystyle \int_{\partial D(1;2)} \frac{\cos^2 \pi z}{(z-i)^3}\, dz$

   (d) $\displaystyle \int_{\partial D(i;\sqrt{2})} \frac{dz}{z^3 \cos z}$

2. Let $\gamma\colon [0, 2\pi] \to \mathbb{C}$ be the contour given by

   $$\gamma(t) = 2\sin t + i\sin 2t.$$

   Calculate the following integrals over γ.

   (a) $\displaystyle \int_{\gamma} \frac{3z^2 - 2iz + 1}{z-1}\, dz$

   (b) $\displaystyle \int_{\gamma} \frac{e^{2z}}{(z+1)^2}\, dz$

   (c) $\displaystyle \int_{\gamma} \frac{\sin \pi z}{(z-1)^4}\, dz$

   (d) $\displaystyle \int_{\gamma} \frac{e^{i\pi z}}{(z-1)^2(z+1)^2}\, dz$

3. Let $r \in [0, 1)$. Calculate the real integral

   $$\int_0^{2\pi} \frac{1 - r\cos t}{1 - 2r\cos t + r^2}\, dt$$

   by using a contour integral over the unit circle.

4. Define $f\colon \mathbb{C} \setminus \partial D(0; 2) \to \mathbb{C}$ by

   $$f(z) = \int_{\partial D(0;2)} \frac{\zeta^3 - 2i\zeta^2 + 4}{(\zeta - z)^3}\, d\zeta.$$

   Find all values of f.

5. Let $\Omega \subseteq \mathbb{C}$ be open and let γ be a closed contour in Ω such that $\mathrm{Ind}_\gamma z = 0$ for all $z \in \mathbb{C} \setminus \Omega$. Show that for all $z \in \Omega \setminus \gamma^*$ and $f \in H(\Omega)$,

   $$\int_\gamma \frac{f'(\zeta)}{\zeta - z}\, d\zeta = \int_\gamma \frac{f(\zeta)}{(\zeta - z)^2}\, d\zeta.$$

6. Show that the function φ in Lemma 4.2.1 is continuous at the point $(z, w) \in \Omega \times \Omega$ if $z \neq w$.

7. Let $r > 0$ and $a \in \mathbb{C}$ be such that $|a| \neq r$. Evaluate

$$\int_{\partial D(0;r)} \frac{|dz|}{|z - a|^2}$$

by converting the integral with respect to arc length into a standard contour integral in order to apply the Cauchy integral formula.

8. Let $\Omega \subseteq \mathbb{C}$ be open and $C \subseteq \mathbb{C}$ be a contour. Suppose that $\varphi \colon \Omega \times C \to \mathbb{C}$ is continuous, and define $f \colon \Omega \to \mathbb{C}$ by

$$f(z) = \int_C \varphi(z, w) \, dw.$$

Prove that f is continuous. How does this apply to the proof of the Cauchy integral formula? (*Hint*: Exercise 16 in Section 2.1 is helpful, as is the proof of Leibniz's rule in Appendix B.)

9. Use the following steps to prove a complex version of Leibniz's rule. Assume the hypotheses of Exercise 8 and that for each $w \in C$, the function $\varphi_w(z) = \varphi(z, w)$ is analytic on Ω and $\partial \varphi / \partial z$ is continuous on $\Omega \times C$.

   (a) Let $z \in \Omega$. Choose $r > 0$ such that $\overline{D}(z; r) \subseteq \Omega$. Show that

   $$f(z) = \frac{1}{2\pi i} \int_{\partial D(z;r)} \frac{f(\zeta)}{\zeta - z} \, d\zeta.$$

   (b) Conclude that $f \in H(\Omega)$ and

   $$f'(z) = \int_C \frac{\partial \varphi}{\partial z}(z, w) \, dw.$$

   (c) Why can we not use this argument to show the function g_1 is analytic in the proof of the Cauchy integral formula?

4.3 Cauchy's Theorem

We now arrive at the point where we can state and prove Cauchy's theorem. Do not underestimate its significance because of the simplicity of our proof. Due to the arrangement of the previous sections, all of the necessary work has already been done. In other developments, the proof of Cauchy's theorem may be much more complicated.

4.3.1 Cauchy's Theorem. *Let $\Omega \subseteq \mathbb{C}$ be open and $f \in H(\Omega)$. If γ is a closed contour in Ω such that $\mathrm{Ind}_\gamma z = 0$ for all $z \in \mathbb{C} \setminus \Omega$, then*

$$\int_\gamma f(z) \, dz = 0. \tag{4.3.1}$$

Proof. Let $a \in \Omega \setminus \gamma^*$ and $g \colon \Omega \to \mathbb{C}$ be given by $g(z) = (z - a)f(z)$. Then by Cauchy's integral formula,

$$\int_\gamma f(z) \, dz = \int_\gamma \frac{g(z)}{z - a} \, dz = 2\pi i g(a) \, \mathrm{Ind}_\gamma \, a = 0,$$

as desired.   □

As with the Cauchy integral formula, we present the commonly used version as a corollary.

4.3.2 Corollary. *Let $\Omega \subseteq \mathbb{C}$ be open and γ be a simple closed contour in Ω. If the inside of γ^* lies in Ω, then*

$$\int_{\gamma} f(z)\, dz = 0$$

for all $f \in H(\Omega)$.

We now present a useful result that is a direct consequence of Cauchy's theorem. The reader may find its proof to be frustratingly less precise than those that have preceded this discussion. A rigorous justification of the constructions used below is beyond the scope of our development. An alternative method of proof is developed in the exercises that, while rigorous, provides less insight than the proof below.

4.3.3 Theorem. *Let $\Omega \subseteq \mathbb{C}$ be open, and let C be a positively oriented simple closed contour in Ω. Suppose that $C_1, \ldots, C_n$ are positively oriented simple closed contours lying inside of C such that every point inside of C lies inside of or on at most one C_k. If Ω contains every point that lies both inside of C and outside of each C_k, then*

$$\int_C f(z)\, dz = \sum_{k=1}^{n} \int_{C_k} f(z)\, dz \qquad (4.3.2)$$

for every $f \in H(\Omega)$.

Proof. Construct a simple contour L_0 starting at a point of C and ending at a point of C_1 such that L_0 intersects none of the contours $C, C_1, \ldots, C_n$ except at its endpoints. Assuming contours $L_0, \ldots, L_{k-1}$ for some $0 < k < n$ have been defined, construct a simple contour L_k starting at a point of C_k and ending at a point of C_{k+1} such that L_k intersects none of the contours $C, C_1, \ldots, C_n$ except at its endpoints and L_k intersects none of the contours $L_0, \ldots, L_{k-1}$ anywhere. Lastly, construct a simple contour L_n starting at a point of C_n and ending at a point of C such that L_n intersects none of the contours $C, C_1, \ldots, C_n$ except at its endpoints and does not intersect the contours $L_0, \ldots, L_{n-1}$ anywhere.

Evidently, the set of points inside of C and outside of each C_k that do not lie on any L_k has two connected components. (See Figure 4.3.) Denote by U_1 the component with the property that if ∂U_1 is traversed in the positive orientation, then each L_k is traversed in its forward direction. Denote the other component by U_2. Set $C^j, C_1^j, \ldots, C_n^j$ to be the parts of the contours $C, C_1, \ldots, C_n$ intersecting ∂U_j for $j = 1, 2$ in their original directions. Recall the notation developed in Theorem 2.9.11. We then have

$$\partial U_1 = C^1 + L_0 - C_1^1 + \cdots + L_{n-1} - C_n^1 + L_n$$
$$\partial U_2 = C^2 - L_n - C_n^2 - \cdots - L_1 - C_1^2 - L_0.$$

Let $f \in H(\Omega)$. Now ∂U_1 and ∂U_2 are simple closed contours in Ω, and all points inside of each lie in Ω, and therefore

$$\int_{\partial U_1} f(z)\,dz = 0 = \int_{\partial U_2} f(z)\,dz$$

by Corollary 4.3.2. Applying this with Theorem 2.9.11, we have

$$\int_{C^1} f(z)\,dz = \sum_{k=1}^{n} \int_{C_k^1} f(z)\,dz - \sum_{k=0}^{n} \int_{L_k} f(z)\,dz$$

$$\int_{C^2} f(z)\,dz = \sum_{k=1}^{n} \int_{C_k^2} f(z)\,dz + \sum_{k=0}^{n} \int_{L_k} f(z)\,dz.$$

Now by adding the above lines, we have

$$\int_{C^1} f(z)\,dz + \int_{C^2} f(z)\,dz = \sum_{k=1}^{n} \left(\int_{C_k^1} f(z)\,dz + \int_{C_k^2} f(z)\,dz \right).$$

This gives (4.3.2). □

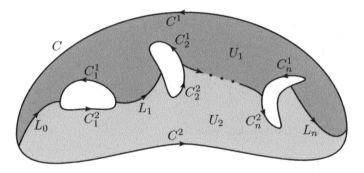

Figure 4.3 Contours used in the proof of Theorem 4.3.3; U_1 and U_2 are the shaded regions

The following is a special case.

4.3.4 Deformation of Contours. *Let $\Omega \subseteq \mathbb{C}$ be open and $f \in H(\Omega)$. If C_1 and C_2 are positively oriented simple closed contours in Ω, C_2 lies inside of C_1, and Ω contains all points lying inside of C_1 and outside of C_2, then*

$$\int_{C_1} f(z)\,dz = \int_{C_2} f(z)\,dz.$$

In the situation described in the above corollary, we often say that C_1 can be *continuously deformed* into C_2 within Ω. Such curves are said to be *homologous* in Ω. See Figure 4.4.

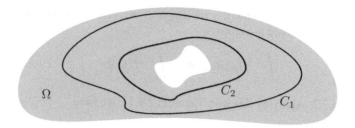

Figure 4.4   The hypotheses of Corollary 4.3.4

Let us consider an example.

4.3.5 Example. Let C be the positively oriented circle of radius 2 centered at 0. We shall calculate

$$\int_C \frac{dz}{z^2 + 1}.$$

If $f(z) = 1/(z^2 + 1)$, then f is analytic in all of $\mathbb{C}$ except where $z = \pm i$. Therefore let C_1 and C_2 be the positively oriented circles of radius $1/2$ centered at i and $-i$, respectively. Then C_1 and C_2 lie inside of C and no points lie on or inside both C_1 and C_2. Thus

$$\int_C \frac{dz}{z^2 + 1} = \int_{C_1} \frac{dz}{z^2 + 1} + \int_{C_2} \frac{dz}{z^2 + 1}$$

by Theorem 4.3.3. Now

$$\int_{C_1} \frac{dz}{z^2 + 1} = \int_{C_1} \frac{1/(z + i)}{z - i} \, dz = \frac{2\pi i}{i + i} = \pi$$

by the Cauchy integral formula. Likewise,

$$\int_{C_2} \frac{dz}{z^2 + 1} = \int_{C_2} \frac{1/(z - i)}{z + i} \, dz = \frac{2\pi i}{-i - i} = -\pi.$$

Therefore the original integral is equal to 0.

Summary and Notes for Section 4.3.

Cauchy's theorem (sometimes called Cauchy's integral theorem) states that the integral of an analytic function over a closed contour is zero provided that the function is defined at all points for which the index of the contour is nonzero. From one point of view, we see that "interesting" contour integrals are those for which the integrand is undefined somewhere "inside" the contour.

The theorem follows quickly from Cauchy's integral formula. In other books, Cauchy's theorem is proved first (with some work) and then the integral formula follows.

The ability to write the integral over a simple closed contour as the sum of integrals over suitable interior contours is a useful tool for calculations. The deformation of contours result hints at the topological notion of homotopy, which can used to obtain some nice results related to Cauchy's theorem. However, this is certainly beyond the scope of this text.

Cauchy proved the integral theorem years before the integral formula. Indeed, Cauchy gained his first serious recognition because of this result. It should be noted that Carl Friedrich Gauss proposed it three years earlier with no proof.

Exercises for Section 4.3.

1. Calculate the following integrals without resorting to partial fractions.

(a) $\displaystyle \int_{\partial D(0;2)} \frac{2z}{2z^2 - z - 1} \, dz$

(b) $\displaystyle \int_{\partial D(1;2)} \frac{e^{\pi z/3}}{z^2 + 1} \, dz$

(c) $\displaystyle \int_{\partial D(0;4\pi)} e^{\cos z} \, dz$

(d) $\displaystyle \int_{\partial D(3i;5)} \frac{z^3}{z^4 - 1} \, dz$

2. Let $f \in H(\mathbb{C} \setminus \{0\})$, and suppose that

$$\int_{\partial \mathbb{D}} f(z) \, dz = 1.$$

Let $C \subseteq \mathbb{C} \setminus \{0\}$ be a simple closed contour such that 0 lies inside of C. Explain why

$$\int_C f(z) \, dz = 1.$$

3. $\triangleright$ Let $\Omega \subseteq \mathbb{C}$ be open. We say that Ω is *simply connected* provided that Ω is connected and $\mathrm{Ind}_\gamma z = 0$ for every closed contour γ in Ω and every $z \in \mathbb{C} \setminus \Omega$. (Loosely, this means Ω has no "holes." This is not the traditional definition.)

   (a) Suppose that $\Omega \subseteq \mathbb{C}$ is a simply connected domain. Show that

   $$\int_{C_1} f(z) \, dz = \int_{C_2} f(z) \, dz$$

   for any $f \in H(\Omega)$ if $C_1, C_2 \subseteq \Omega$ are contours with the same respective initial and terminal points.

   (b) Show that if $\Omega \subseteq \mathbb{C}$ is a simply connected domain and $f \in H(\Omega)$, then f has an antiderivative $F \in H(\Omega)$. (Compare to Theorem 3.1.3.)

4. Suppose that $\Omega \subseteq \mathbb{C} \setminus \{0\}$ is a simply connected domain. Show that a branch of the logarithm can be defined on Ω. (*Hint:* Use part (b) of Exercise 3 to define an antiderivative of $f(z) = 1/z$ on Ω. Adjust by a constant, if necessary.) Compare with Exercise 21 in Section 2.4.

5. $\triangleright$ Let $\Omega \subseteq \mathbb{C}$ be open. Let $\{f_n\}_{n=1}^{\infty}$ be a sequence of functions in $H(\Omega)$. Suppose that $f : \Omega \to \mathbb{C}$ and $f_n \to f$ uniformly on compact subsets of Ω. Show that $f \in H(\Omega)$.

6. Let $\Omega = \{z \in \mathbb{C} : \mathrm{Re}\, z > 1\}$. The *Riemann zeta function* $\zeta : \Omega \to \mathbb{C}$ is given by

$$\zeta(z) = \sum_{n=1}^{\infty} \frac{1}{n^z},$$

where the principal branch of the powers is assumed. Prove that ζ is well defined and analytic on Ω. (*Hint*: Exercise 5 is helpful.)

7. Recall *Green's theorem* from multivariable calculus: Let $R \subseteq \mathbb{R}^2$ be open such that $C = \partial R$ is a simple closed contour (thought of as positively oriented). If $P, Q : \overline{R} \to \mathbb{R}$ are continuous and have continuous first partial derivatives on R, then

$$\int_C P(x, y)\, dx + Q(x, y)\, dy = \iint_R \left(\frac{\partial Q}{\partial x}(x, y) - \frac{\partial P}{\partial y}(x, y) \right) dA(x, y).$$

Show how Green's theorem gives a simple proof of Corollary 4.3.2.

8. Let C be a simple closed contour in $\mathbb{C}$. Use Green's theorem to show that the area of the inside of C is given by

$$A = \frac{1}{2i} \int_C \overline{z}\, dz.$$

9. Show how the Cauchy integral formula follows from Cauchy's theorem. (*Hint*: Apply Cauchy's theorem to the function given in Lemma 4.2.1.)

10. Consider the following, more rigorous, alternative to the proof of Theorem 4.3.3. Let $\Omega \subseteq \mathbb{C}$ be open.

    (a) Suppose that $\gamma_1, \ldots, \gamma_n$ are closed contours in Ω such that $\sum_{k=1}^{n} \mathrm{Ind}_{\gamma_k} z = 0$ for all $z \in \mathbb{C} \setminus \Omega$. Mimic the proof of the Cauchy integral formula to show that

$$f(z) \sum_{k=1}^{n} \mathrm{Ind}_{\gamma_k} z = \frac{1}{2\pi i} \sum_{k=1}^{n} \int_{\gamma_k} \frac{f(\zeta)}{\zeta - z}\, d\zeta$$

    for all $z \in \Omega \setminus \bigcup_{k=1}^{n} \gamma_k^*$ and $f \in H(\Omega)$. (Often, the "formal sum" of closed contours is called a *cycle*.)

    (b) Show that if $\gamma_1, \ldots, \gamma_n$ are as in part (a), then

$$\sum_{k=1}^{n} \int_{\gamma_k} f(z)\, dz = 0$$

    for all $f \in H(\Omega)$.

    (c) Use part (b) to provide an alternate proof to Theorem 4.3.3.

11. Part (a) of Exercise 10 provides a generalization of the Cauchy integral formula. Is there a similar generalization of the Cauchy integral formula for derivatives (Corollary 4.2.3)? If so, prove it.

CHAPTER 5

THE RESIDUE THEOREM

Cauchy's residue theorem represents the pinnacle of complex integration as studied in this text. We will see that integrals of analytic functions over closed contours are determined by the functions' behavior inside of the contours at precisely the points where they are undefined, their singularities. After gaining an understanding of singularities and proving the residue theorem, we will consider what it tells us about real-variable calculus. In particular, we will find that certain integrals of real functions can be evaluated by expansion into the complex plane and that the Laplace transform from differential equations can be inverted (without those pesky tables) for common types of functions.

5.1 Laurent Series

We begin by considering doubly infinite series of both complex numbers and functions.

5.1.1 Definition. Let $\{z_n\}_{n=-\infty}^{\infty}$ be a doubly infinite sequence of complex numbers (a function from $\mathbb{Z}$ into $\mathbb{C}$). If the two series $A = \sum_{n=0}^{\infty} z_n$ and $B = \sum_{n=1}^{\infty} z_{-n}$

Complex Analysis: A Modern First Course in Function Theory, First Edition. Jerry R. Muir, Jr.
© 2015 John Wiley & Sons, Inc. Published 2015 by John Wiley & Sons, Inc.

converge, then the *doubly infinite series* $\sum_{n=-\infty}^{\infty} z_n$ *converges* to their sum. That is,

$$\sum_{n=-\infty}^{\infty} z_n = \sum_{n=0}^{\infty} z_n + \sum_{n=-\infty}^{-1} z_n. \tag{5.1.1}$$

The doubly infinite series *converges absolutely* if both A and B converge absolutely.

5.1.2 Definition. Let $E \subseteq \mathbb{C}$ and suppose $f_n \colon E \to \mathbb{C}$ for all $n \in \mathbb{Z}$. Then the doubly infinite series $\sum_{n=-\infty}^{\infty} f_n$ *converges pointwise* if $\sum_{n=-\infty}^{\infty} f_n(z)$ converges for each $z \in E$. The series *converges uniformly* on E if both $\sum_{n=0}^{\infty} f_n$ and $\sum_{n=1}^{\infty} f_{-n}$ converge uniformly on E.

We now introduce a new type of planar region.

5.1.3 Definition. Let $a \in \mathbb{C}$ and $0 \le r < R \le \infty$. The *annulus* centered at a with *inner radius* r and *outer radius* R is the set

$$A(a; r, R) = \{z \in \mathbb{C} : r < |z - a| < R\}. \tag{5.1.2}$$

See Figure 5.1. The meaning of $R = \infty$ in the above definition should be clear.

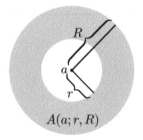

Figure 5.1   The annulus $A(a; r, R)$

Up to this point, we have considered the power series expansion of analytic functions on a disk (of possibly infinite radius). The following theorem allows for a similar representation on an annulus.

5.1.4 Laurent Series Expansion. *Let* $a \in \mathbb{C}$ *and* $0 \le r < R \le \infty$. *If* $f \in H(A(a; r, R))$, *then for all* $z \in A(a; r, R)$,

$$f(z) = \sum_{n=-\infty}^{\infty} c_n(z - a)^n = \sum_{n=0}^{\infty} c_n(z - a)^n + \sum_{n=-\infty}^{-1} c_n(z - a)^n, \tag{5.1.3}$$

where

$$c_n = \frac{1}{2\pi i} \int_C \frac{f(z)}{(z - a)^{n+1}} \, dz \tag{5.1.4}$$

for all $n \in \mathbb{Z}$ *and any positively oriented simple closed contour* $C \subseteq A(a; r, R)$ *surrounding* a. *The series of terms of nonnegative degree converges absolutely for* each $z \in D(a; R)$ and uniformly on compact subsets of $D(a; R)$, the series of terms of negative degree converges absolutely for each $z \in A(a; r, \infty)$ and uniformly on compact subsets of $A(a; r, \infty)$, and hence the doubly infinite series converges absolutely for each $z \in A(a; r, R)$ and uniformly on compact subsets of $A(a; r, R)$. Furthermore the series is unique.

To say that the series (5.1.3) is unique, we mean that if the series $\sum_{n=-\infty}^{\infty} d_n(z - a)^n$ converges to $f(z)$ on $A(a; r, R)$, then $d_n = c_n$ for all $n \in \mathbb{Z}$.

It is, of course, often the case that f is analytic on an open set Ω that contains such an annulus, and we represent f by its Laurent series expansion on that annulus.

Proof. For all $\rho \in (r, R)$, let $C_\rho = \partial D(a; \rho)$, and use Lemma 4.1.1 to define $f_\rho \in H(\mathbb{C} \setminus C_\rho)$ by

$$f_\rho(z) = \frac{1}{2\pi i} \int_{C_\rho} \frac{f(\zeta)}{\zeta - z} \, d\zeta.$$

We begin by finding series representations of f_ρ on each of $D(a; \rho)$ and $A(a; \rho, \infty)$.

Repeated application of Lemma 4.1.1 gives

$$f_\rho^{(n)}(a) = \frac{n!}{2\pi i} \int_{C_\rho} \frac{f(\zeta)}{(\zeta - a)^{n+1}} \, d\zeta$$

for $n = 0, 1, \dots$. Since f_ρ is analytic on $D(a; \rho)$, its power series expansion

$$
\begin{aligned}
f_\rho(z) &= \sum_{n=0}^{\infty} \frac{f_\rho^{(n)}(a)}{n!}(z - a)^n \\
&= \sum_{n=0}^{\infty} \left[\frac{1}{2\pi i} \int_{C_\rho} \frac{f(\zeta)}{(\zeta - a)^{n+1}} \, d\zeta \right] (z - a)^n
\end{aligned}
\tag{5.1.5}
$$

converges absolutely at each point of, and uniformly on compact subsets of, $D(a; \rho)$.

Now for $z \in A(a; \rho, \infty)$,

$$
\begin{aligned}
\frac{f(\zeta)}{\zeta - z} &= \frac{-f(\zeta)}{z - a} \frac{1}{1 - (\zeta - a)/(z - a)} \\
&= \frac{-f(\zeta)}{z - a} \sum_{n=0}^{\infty} \left(\frac{\zeta - a}{z - a} \right)^n \\
&= - \sum_{n=1}^{\infty} \frac{f(\zeta)(\zeta - a)^{n-1}}{(z - a)^n}
\end{aligned}
$$

for all $\zeta \in C_\rho$, since $|\zeta - a|/|z - a| < 1$. There is an upper bound $M > 0$ of $|f|$ on the compact set C_ρ, and thus the Weierstrass M-test gives that convergence is uniform in $\zeta \in C_\rho$ for a given z. Hence

$$f_\rho(z) = \frac{1}{2\pi i} \int_{C_\rho} \frac{f(\zeta)}{\zeta - z} d\zeta$$

$$= -\sum_{n=1}^{\infty} \left[\frac{1}{2\pi i} \int_{C_\rho} \frac{f(\zeta)}{(\zeta - a)^{-n+1}} d\zeta \right] (z - a)^{-n} \qquad (5.1.6)$$

holds for $z \in A(a; \rho, \infty)$. Let $K \subseteq A(a; \rho, \infty)$ be compact and $s = \inf\{|z - a| : z \in K\}$. Then $s > \rho$ by the compactness of K. For $z \in K$ and $n \in \mathbb{N}$,

$$\left| \left[\frac{1}{2\pi i} \int_{C_\rho} \frac{f(\zeta)}{(\zeta - a)^{-n+1}} d\zeta \right] (z - a)^{-n} \right| \leq \frac{1}{2\pi |z - a|^n} \int_{C_\rho} \frac{|f(\zeta)|}{|\zeta - a|^{-n+1}} |d\zeta|$$

$$\leq \frac{1}{2\pi s^n} \frac{M}{\rho^{-n+1}} 2\pi \rho$$

$$= M \left(\frac{\rho}{s} \right)^n,$$

which is the nth term of a convergent geometric series. Thus the series (5.1.6) converges uniformly on K by the Weierstrass M-test. It also converges absolutely for each $z \in K$.

Now consider $z \in A(a; r, R)$. Choose $r < r_1 < r_2 < R$ and $\varepsilon > 0$ such that $\overline{D}(z; \varepsilon) \subseteq A(a; r_1, r_2)$. Set $S = \partial D(z; \varepsilon)$. (See Figure 5.2.) Then S and C_{r_1} are closed contours inside C_{r_2}, no points lie on or inside both S and C_{r_1}, and $f(\zeta)/(\zeta - z)$ is analytic in the variable ζ on the set $A(a; r, R) \setminus \{z\}$, and hence

$$\int_{C_{r_2}} \frac{f(\zeta)}{\zeta - z} d\zeta = \int_S \frac{f(\zeta)}{\zeta - z} d\zeta + \int_{C_{r_1}} \frac{f(\zeta)}{\zeta - z} d\zeta$$

by Theorem 4.3.3. By the Cauchy integral formula,

$$f(z) = \frac{1}{2\pi i} \int_S \frac{f(\zeta)}{\zeta - z} d\zeta = f_{r_2}(z) - f_{r_1}(z). \qquad (5.1.7)$$

We now take advantage of the arbitrary choice of ρ. If C is a positively oriented simple closed contour in $A(a; r, R)$ surrounding a and ρ is chosen large enough so that $C \subseteq D(a; \rho)$ (see Exercise 13 in Section 1.6), then the integrals in (5.1.5) can be taken over C by Corollary 4.3.4. Thus $\sum_{n=0}^{\infty} c_n (z - a)^n$ converges on $D(a; \rho)$ for all such ρ and hence on $D(a; R)$. Likewise, if ρ is chosen small enough so that $C \subseteq A(a; \rho, \infty)$ (see the exercises), then the integrals in (5.1.6) can be taken over C, showing $\sum_{n=-\infty}^{-1} c_n (z - a)^n$ converges on $A(a; \rho, \infty)$ for all such ρ and hence on $A(a; r, \infty)$. Hence for any $z \in A(a; r, R)$, r_1 and r_2 can be chosen small enough and large enough in (5.1.7), respectively, so that

$$f(z) = \sum_{n=0}^{\infty} c_n (z - a)^n + \sum_{n=-\infty}^{-1} c_n (z - a)^n = \sum_{n=-\infty}^{\infty} c_n (z - a)^n,$$

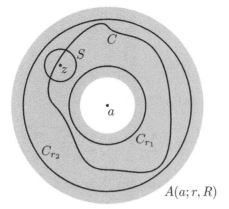

Figure 5.2 Constructions used in the proof of Theorem 5.1.4

with convergence uniform on compact sets and absolute at each point.

The proof of uniqueness is left as an exercise. □

The reader should notice how the development of the series for f_ρ on $A(a; \rho, \infty)$ in the above proof mirrors the development of the power series for a continuously differentiable function on a disk in the proof of Theorem 3.2.4. It should also be apparent that if f is analytic in two different annuli with the same center and nonempty intersection, then the Laurent series of f is the same in both annuli. Therefore if $f \in H(\Omega)$ for some open set $\Omega \subseteq \mathbb{C}$ and $z \in \Omega$ and $a \in \mathbb{C}$ are such that Ω contains an annulus centered at a containing z, then f has a Laurent series representation converging in the largest annulus in Ω centered at a containing z. (Compare with Corollary 3.2.6.)

5.1.5 Definition. Assume the notation in Theorem 5.1.4. The series of terms of negative degree in (5.1.3) is called the *principal part* of the Laurent series of f at a.

Now that we are aware of the existence of the Laurent series expansion, let us consider some examples.

5.1.6 Example. The function $f(z) = e^{1/z}$ is analytic on $A(0; 0, \infty)$. For all such z, we have

$$e^{1/z} = \sum_{n=0}^{\infty} \frac{(1/z)^n}{n!} = \sum_{n=-\infty}^{0} \frac{z^n}{(-n)!},$$

where the last series is in the standard form given in (5.1.3).

5.1.7 Example. The function

$$f(z) = \frac{1}{(z-1)(z-2)}$$

is analytic on the domains $\mathbb{D}$, $A(0; 1, 2)$, and $A(0; 2, \infty)$. Therefore it has three series expansions centered at 0. First, we use partial fractions to write

$$f(z) = \frac{-1}{z-1} + \frac{1}{z-2}.$$

For $z \in \mathbb{D}$, we have $|z| < 1$ and $|z/2| < 1$, and thus

$$f(z) = \frac{1}{1-z} - \frac{1}{2}\frac{1}{1-z/2} = \sum_{n=0}^{\infty} z^n - \frac{1}{2}\sum_{n=0}^{\infty}\frac{z^n}{2^n} = \sum_{n=0}^{\infty}\left(1 - \frac{1}{2^{n+1}}\right)z^n.$$

This is, of course, a power series in the disk. Now for $z \in A(0; 1, 2)$, $|1/z| < 1$ and $|z/2| < 1$, and hence

$$\begin{aligned}
f(z) &= -\frac{1}{z}\frac{1}{1-1/z} - \frac{1}{2}\frac{1}{1-z/2} \\
&= -\frac{1}{z}\sum_{n=0}^{\infty}\frac{1}{z^n} - \frac{1}{2}\sum_{n=0}^{\infty}\frac{z^n}{2^n} \\
&= -\sum_{n=-\infty}^{-1} z^n - \sum_{n=0}^{\infty}\frac{z^n}{2^{n+1}}.
\end{aligned}$$

Lastly for $z \in A(0; 2, \infty)$, we have $|1/z| < 1$ and $|2/z| < 1$. Therefore

$$\begin{aligned}
f(z) &= -\frac{1}{z}\frac{1}{1-1/z} + \frac{1}{z}\frac{1}{1-2/z} \\
&= -\frac{1}{z}\sum_{n=0}^{\infty}\frac{1}{z^n} + \frac{1}{z}\sum_{n=0}^{\infty}\frac{2^n}{z^n} \\
&= \sum_{n=-\infty}^{-1}\left(\frac{1}{2^{n+1}} - 1\right)z^n.
\end{aligned}$$

Summary and Notes for Section 5.1.

We have seen that power series converge naturally on disks. Adding to that the idea that series of negative power terms converge naturally on the outside of circles, results in the ability to expand an analytic function defined on an annulus as a series containing both positive and negative power terms, a Laurent series. Like with power series, these series converge uniformly on compact subsets of the largest annulus in the domain of the function with prescribed center and interior point.

Pierre Alphonse Laurent's work was first reported by Augustin-Louis Cauchy in 1843. Although Cauchy was supportive of it, he also remarked that it was a special case of an earlier result of his and provided a proof.

Exercises for Section 5.1.

1. Find a Laurent series based at 0 for each of the following functions f. Where does the series converge?

   (a) $f(z) = z^2 \sin \dfrac{1}{z}$

   (b) $f(z) = \dfrac{e^{2z}}{z^4}$

   (c) $f(z) = \dfrac{\text{Log}(z+1)}{z^2}$

2. In each of the following, a function f is given together with a point $a \in \mathbb{C}$. Determine all possible series expansions (power or Laurent) of f based at a and describe the set of convergence of each.

   (a) $f(z) = \dfrac{1}{(z+1)(z-3i)}, \qquad a = 0$

   (b) $f(z) = \dfrac{z+1}{(z-1)(z^2+1)}, \qquad a = 1$

   (c) $f(z) = \dfrac{z}{(z-i)(z-2)}, \qquad a = -i$

3. Let $a \in \mathbb{C}$ and $\{c_n\}_{n=-\infty}^{-1} \subseteq \mathbb{C}$, and suppose that the set E of all $z \in \mathbb{C} \setminus \{a\}$ such that

$$f(z) = \sum_{n=-\infty}^{-1} c_n (z-a)^n$$

   converges is nonempty. For appropriate z, define

$$g(z) = f\left(\frac{1}{z} + a\right).$$

   (a) Show that $g(z)$ is equal to a power series based at 0 for all z in its domain. Use the nature of power series convergence to show that $E^\circ = A(a; r, \infty)$ for some $r \geq 0$.

   (b) Write $f(z)$ in terms of g to prove that f can be differentiated term by term in $A(a; r, \infty)$, showing $f \in H(A(a; r, \infty))$.

4. Prove the uniqueness of the Laurent series in Theorem 5.1.4.

5. Let $a \in \mathbb{C}$ and $0 < r < R < \infty$. Suppose that $f \in H(A(a; r, R))$ and $|f(z)| \leq M$ for all $z \in A(a; r, R)$ and some $M \geq 0$. If $\sum_{n=-\infty}^{\infty} c_n(z-a)^n$ is the Laurent series of f on $A(a; r, R)$, show that $|c_n| \leq M/R^n$ for all $n \geq 0$ and $|c_n| \leq M/r^n$ for all $n \leq 0$.

6. Let $0 \leq r < R \leq \infty$ and $f \in H(A(0; r, R))$.

   (a) Show that if f is even, then the Laurent series of f on $A(0; r, R)$ has nonzero terms only of even degree.

   (b) Show that if f is odd, then the Laurent series of f on $A(0; r, R)$ has nonzero terms only of odd degree.

7. Let $f: A(0; 0, \infty) \to \mathbb{C}$ be given by

$$f(z) = \exp\left(\frac{1}{2}\left[z + \frac{1}{z}\right]\right).$$

Write $f(z) = \sum_{n=-\infty}^{\infty} c_n z^n$. Show that for all $n \in \mathbb{Z}$,

$$c_{|n|} = \frac{1}{\pi} \int_0^\pi e^{\cos t} \cos nt \, dt.$$

8. Let $\alpha \in (-1, 1)$. Derive the real series

$$\sum_{n=1}^{\infty} \alpha^n \cos n\theta = \frac{\alpha \cos \theta - \alpha^2}{1 - 2\alpha \cos \theta + \alpha^2}$$

for all $\theta \in \mathbb{R}$. (*Hint:* Consider a Laurent series for $\alpha/(z - \alpha)$ for $|z| > |\alpha|$. Substitute $z = e^{i\theta}$.)

9. Verify the following claim in proof of Theorem 5.1.4: If $a \in \mathbb{C}$, $0 \le r < R \le \infty$, and C is a simple closed contour in $A(a; r, R)$ surrounding a, then there exists $\rho > r$ such that $C \subseteq A(a; \rho, \infty)$.

5.2 Classification of Singularities

The main goal of this chapter is to study an analytic function by utilizing the behavior of the function near points at which the function *is not defined*. Here, we define the points in question.

5.2.1 Definition. Let $\Omega \subseteq \mathbb{C}$ be open and $E \subseteq \Omega$ be a set with no limit points in Ω. If $f \in H(\Omega \setminus E)$, then the points in E are called *isolated singularities* of f.

It is helpful to compare this to the notion of isolated zeros discussed following Theorem 3.4.5. For each isolated singularity a of f, there is $r > 0$ such that f is analytic on $A(a; 0, r)$. (Such an annulus is called a *punctured disk*.) For this reason, we will consider the simpler hypothesis that $f \in H(A(a; 0, r))$ in most of this section's results. Careful examination of the above definition reveals that there is nothing preventing f from being analytic on $D(a; r)$.

From the previous section, we know that such an f has a Laurent series expansion on $A(a; 0, r)$,

$$f(z) = \sum_{n=-\infty}^{\infty} c_n (z - a)^n. \tag{5.2.1}$$

Using this series, we classify the three types of isolated singularities.

5.2.2 Definition. Let $a \in \mathbb{C}$, $r > 0$, and $f \in H(A(a; 0, r))$. Express f by the Laurent series expansion (5.2.1) on $A(a; 0, r)$.

(a) If $c_n = 0$ for all $n < 0$, then f has a *removable singularity* at a.

(b) If $c_{-m} \ne 0$ for some $m \in \mathbb{N}$ and $c_n = 0$ for all $n < -m$, then f has a *pole* of order m at a. If $m = 1$, the pole is *simple*.

(c) If $c_n \ne 0$ for infinitely many $n < 0$, then f has an *essential singularity* at a.

We now explain the "removable" in "removable singularity."

5.2.3 Theorem. *Let $f \in H(A(a; 0, r))$ for some $a \in \mathbb{C}$ and $r > 0$. Then f has a removable singularity at a if and only if there exists $g \in H(D(a; r))$ such that $f(z) = g(z)$ for all $z \in A(a; 0, r)$.*

Proof. If f has a removable singularity at a, then f has the Laurent series expansion

$$f(z) = \sum_{n=0}^{\infty} c_n (z - a)^n$$

for $z \in A(a; 0, r)$. Define $g \colon D(a; r) \to \mathbb{C}$ by the same series.
The converse is immediate. $\square$

5.2.4 Theorem. *Suppose $f \in H(A(a; 0, r))$ for some $a \in \mathbb{C}$ and $r > 0$. Then f has a removable singularity at a if and only if there is some $\varepsilon \in (0, r)$ such that f is bounded on $A(a; 0, \varepsilon)$.*

Proof. If f has a removable singularity at a, then define g as in Theorem 5.2.3. Since g is continuous on $D(a; r)$, it is bounded on $\overline{D}(a; \varepsilon)$ for any $\varepsilon \in (0, r)$.

Conversely, suppose $|f(z)| \leq M$ for some $M \geq 0$ and all $z \in A(a; 0, \varepsilon)$, where $\varepsilon \in (0, r)$. Assume that f has the Laurent series expansion (5.2.1). Then by Theorem 5.1.4, we have for a given $n \in \mathbb{N}$,

$$
\begin{aligned}
|c_{-n}| &= \left| \frac{1}{2\pi i} \int_{\partial D(a; \rho)} \frac{f(z)}{(z - a)^{-n+1}} \, dz \right| \\
&\leq \frac{1}{2\pi} \int_{\partial D(a; \rho)} \frac{|f(z)|}{|z - a|^{-n+1}} \, |dz| \\
&\leq \frac{1}{2\pi} \frac{M}{\rho^{-n+1}} 2\pi \rho \\
&= M\rho^n
\end{aligned}
$$

for all $\rho \in (0, \varepsilon)$. Hence $c_{-n} = 0$. Therefore f has a removable singularity at a. $\square$

5.2.5 Corollary. *If $f \in H(A(a; 0, r))$ for some $a \in \mathbb{C}$ and $r > 0$, then f has a removable singularity at a if and only if $\lim_{z \to a} f(z)$ exists (and is finite).*

We now provide a similar analysis for poles. The first theorem is a straightforward application of Laurent series.

5.2.6 Theorem. *Let $a \in \mathbb{C}$, $r > 0$, and $f \in H(A(a; 0, r))$. Then f has a pole of order $m \in \mathbb{N}$ at a if and only if there is a function $g \in H(D(a; r))$ such that $g(a) \neq 0$ and*

$$f(z) = \frac{g(z)}{(z - a)^m}, \qquad z \in A(a; 0, r). \tag{5.2.2}$$

Proof. If f has a pole of order m at a, then let $f(z) = \sum_{n=-m}^{\infty} c_n(z-a)^n$ be the Laurent series expansion of f. Now $c_{-m} \neq 0$ and the function $g \in H(D(a;r))$ given by

$$g(z) = \sum_{n=0}^{\infty} c_{n-m}(z-a)^n$$

satisfies $g(a) = c_{-m}$ and $g(z) = (z-a)^m f(z)$ for all $z \in A(a;0,r)$.

Conversely, suppose $g(z) = \sum_{n=0}^{\infty} c_n(z-a)^n$ for $z \in D(a;r)$ and $g(a) = c_0 \neq 0$. Then $f(z) = g(z)/(z-a)^m$ implies

$$f(z) = \sum_{n=-m}^{\infty} c_{n+m}(z-a)^n, \qquad z \in A(a;0,r).$$

Therefore f has a pole of order m at a.   □

It is left as an exercise to show that $A(a;0,r)$ may be replaced by some open set Ω, where f is analytic on Ω except at the point $a \in \Omega$. This is a direct parallel of Theorem 3.4.4, a result about zeros of order m. The reader should review Section 3.4 to see the symmetry between zeros and poles.

5.2.7 Theorem. *Let $f \in H(A(a;0,r))$ for some $a \in \mathbb{C}$ and $r > 0$. Then f has a pole at a if and only if $\lim_{z \to a} f(z) = \infty$.*

Proof. If f has a pole at a of order $m \in \mathbb{N}$, then let g be as in Theorem 5.2.6. We see that $\lim_{z \to a} f(z) = \infty$, using (5.2.2).

Conversely, suppose that $\lim_{z \to a} f(z) = \infty$. Choose $\rho \in (0,r)$ such that $f(z) \neq 0$ for all $z \in A(a;0,\rho)$. Then set $g(z) = 1/f(z)$ for $z \in A(a;0,\rho)$. Clearly, g is analytic on its domain and $\lim_{z \to a} g(z) = 0$. Accordingly, g has a removable singularity at a by Corollary 5.2.5. Let $g_1 \in H(D(a;\rho))$ be such that $g_1(z) = g(z) \neq 0$ for all $z \in A(a;0,\rho)$. Now $g_1(a) = 0$, and we let m be the order of the zero. Then by Theorem 3.4.4, there is some $h \in H(D(a;\rho))$ such that $h(a) \neq 0$ and $g_1(z) = (z-a)^m h(z)$. Then $h(z) \neq 0$ for all $z \in D(a;\rho)$, and so $1/h$ is analytic on $D(a;\rho)$. Now

$$f(z) = \frac{1/h(z)}{(z-a)^m}, \qquad z \in A(a;0,\rho).$$

Theorem 5.2.6 implies that f has a pole of order m at a.   □

By Corollary 5.2.5 and Theorem 5.2.7, we now know that if $f \in H(A(a;0,r))$ for some $a \in \mathbb{C}$ and $r > 0$, then f has an essential singularity at a if and only if $\lim_{z \to a} f(z)$ fails to either converge or diverge to ∞. How bad can this failure be? In every case, it is quite spectacular. The following theorem hints at this, but it merely scratches the surface.

5.2.8 Casorati–Weierstrass Theorem. *Let $a \in \mathbb{C}$, $r > 0$, and $f \in H(A(a;0,r))$. If f has an essential singularity at a, then*

$$\overline{f(A(a;0,\rho))} = \mathbb{C}$$

for every $\rho \in (0, r)$.

Proof. For eventual contradiction, suppose that for some $\rho \in (0, r)$, there exists $b \in \mathbb{C}$ such that $b \notin \overline{f(A(a; 0, \rho))}$. Then b is an exterior point of $f(A(a; 0, \rho))$, and hence there is some $\varepsilon > 0$ such that $D(b; \varepsilon) \cap f(A(a; 0, \rho)) = \varnothing$. Define $g \colon A(a; 0, \rho) \to \mathbb{C}$ by

$$g(z) = \frac{1}{f(z) - b}.$$

Then g is analytic, and for all $z \in A(a; 0, \rho)$,

$$|g(z)| = \frac{1}{|f(z) - b|} < \frac{1}{\varepsilon}.$$

Thus g has a removable singularity at a by Theorem 5.2.4. Accordingly, $\lim_{z \to a} g(z)$ exists.

Now $f(z) = b + 1/g(z)$. If $\lim_{z \to a} g(z) \neq 0$, then $\lim_{z \to a} f(z)$ exists, and hence f has a removable singularity at a by Corollary 5.2.5. If $\lim_{z \to a} g(z) = 0$, then $\lim_{z \to a} f(z) = \infty$, and hence f has a pole at a by Theorem 5.2.7. In either situation, we have the desired contradiction. $\qquad \square$

5.2.9 Example. Let $f(z) = e^{1/z}$ for $z \in A(0; 0, \infty)$. The Laurent series expansion of f about 0 is calculated in Example 5.1.6. From it, we conclude that f has an essential singularity at 0. Let $b \in \mathbb{C} \setminus \{0\}$. For $n \in \mathbb{N}$, define

$$z_n = \frac{1}{\log b + 2n\pi i},$$

where the principal value of $\log b$ is used. Then

$$f(z_n) = e^{\log b + 2n\pi i} = b.$$

Clearly $z_n \to 0$, and therefore given any $\rho > 0$, $z_n \in A(0; 0, \rho)$ for some n. Hence $b \in f(A(0; 0, \rho))$ for all $\rho > 0$. Since $f(z) \neq 0$ for all z, this proves

$$f(A(0; 0, \rho)) = \mathbb{C} \setminus \{0\}$$

for all $\rho > 0$. Clearly, we then have

$$\overline{f(A(0; 0, \rho))} = \mathbb{C}$$

for all $\rho > 0$.

We end this section with the definition of an important class of functions.

5.2.10 Definition. Let $\Omega \subseteq \mathbb{C}$ be open and $E \subseteq \Omega$ have no limit points in Ω. If $f \in H(\Omega \setminus E)$ and f has a pole or removable singularity at each $a \in E$, then f is said to be *meromorphic* on Ω.

It should be noted that if f is as described in this definition, then $\lim_{z \to a} f(z)$ exists in $\mathbb{C}_\infty$ for all $a \in E$. By defining $f(a)$ to be this value for all $a \in E$, we have a function f continuous from Ω into $\mathbb{C}_\infty$. This allows for us to think of meromorphic functions on Ω as functions with domain all of Ω.

Suppose that $\Omega \subseteq \mathbb{C}$ is a domain and that $g, h \in H(\Omega)$ are not equivalently 0. Now $f = g/h$ is analytic on Ω except at the points in $Z(h)$. Since $Z(h)$ has no limit points in Ω by Theorem 3.4.5, these points are isolated singularities of f. Let $a \in Z(h)$ be a zero of order $m \in \mathbb{N}$. We may then write

$$h(z) = (z - a)^m H(z),$$

where $H \in H(\Omega)$ and $H(a) \neq 0$. We may likewise write

$$g(z) = (z - a)^n G(z),$$

where $n \in \mathbb{N} \cup \{0\}$ and $G \in H(\Omega)$ is such that $G(a) \neq 0$. (Either $a \in Z(g)$, in which case $n \in \mathbb{N}$ is the order of the zero, or $n = 0$ and $G = g$.) Now for $z \in \Omega \setminus Z(h)$,

$$f(z) = (z - a)^{n-m} \frac{G(z)}{H(z)}.$$

Since $G(a)/H(a) \neq 0$, it is apparent that f has a pole of order $m - n$ at a if $n < m$ or a removable singularity if $n \geq m$. Further, if $n = m$, then f can be defined to have a nonzero value at a, and if $n > m$, then f can be defined to be 0 at a, and the zero will be of order $n - m$. We summarize these observations in the following theorem.

5.2.11 Theorem. *Let $\Omega \subseteq \mathbb{C}$ be a domain, and suppose that $g, h \in H(\Omega)$ are not equivalently 0. Then the function $f = g/h$ is meromorphic on Ω. Moreover, if $a \in Z(h)$ is a zero of order m, then one of the following occurs.*

(a) *If $g(a) \neq 0$, then f has a pole of order m at a.*

(b) *If $a \in Z(g)$ has order $n < m$, then f has a pole of order $m - n$ at a.*

(c) *If $a \in Z(g)$ has order $n = m$, then f has a removable singularity at a, and f can be defined to have a nonzero value there.*

(d) *If $a \in Z(g)$ has order $n > m$, then f has a removable singularity at a, and f can be defined to have a zero of order $n - m$ there.*

5.2.12 Example. Consider the meromorphic function

$$f(z) = \frac{\cos z - 1}{z^3 - 2iz^2 - z}.$$

The factorization of the denominator is $z(z - i)^2$, and hence the denominator has a simple zero at 0 and a zero of order 2 at i. Examination of the numerator reveals it has zeros of order 2 at $2n\pi$, $n \in \mathbb{Z}$. Therefore f has a removable singularity at 0, which can be removed by defining $f(0) = 0$. The resulting zero of f at 0 is simple. In addition, f has a pole of order 2 at i.

Summary and Notes for Section 5.2.

There is an important congruence between the analysis of zeros and that of singularities. Recall that the order of a zero of an analytic function is the degree of the first nonzero term of the power series expansion of the function about the zero. The classification of isolated singularities works in a similar manner with the Laurent series expanded about the singularity. If the degree of the nonzero term of lowest power is at least 0, the singularity is removable. If it is negative, then the singularity is a pole of order -1 times that degree. If there is no such term, the singularity is essential. Furthermore, there is a factorization theorem for poles that mirrors the factorization theorem for zeros.

The behavior of an analytic function f about an isolated singularity a exposes the nature of the singularity. If $\lim_{z \to a} f(z)$ exists, the singularity is removable, and if $\lim_{z \to a} f(z) = \infty$, then the singularity is a pole. Although that means all other possibilities indicate an essential singularity, we actually see from the Casorati–Weierstrass theorem that the behavior there must be wild.

Felice Casorati and Karl Weierstrass discussed this theorem while meeting in Berlin. Casorati published it in 1868 and Weierstrass several years later. It was discovered independently by Yulian Vasil'evich Sokhotskii at the same time. More can be said about the behavior of analytic functions near essential singularities, as Charles Émile Picard discovered, results that are covered in more advanced texts.

Exercises for Section 5.2.

1. For each of the following functions f, find the isolated singularities of f and classify each as a removable singularity, a pole, or an essential singularity. If a singularity is a pole, determine its order. If a singularity is removable, assign a value to f at that point to make f analytic there.

   (a) $f(z) = \dfrac{z^2 + 2z - 3}{z^4 - z^3}$

   (b) $f(z) = \cos \dfrac{1}{z}$

   (c) $f(z) = \cot z$

   (d) $f(z) = \dfrac{z}{\sin^2 z}$

   (e) $f(z) = \dfrac{1}{1 - e^z}$

   (f) $f(z) = \dfrac{\tan^2 z}{z^2}$

2. Let $a \in \mathbb{C}$ and $r > 0$. Suppose that $f, g \in H(A(a; 0, r))$ have poles of orders n and m, respectively, at a.

   (a) What can be said about the singularity of $f + g$ at a? Give examples.

   (b) What can be said about the singularity of fg at a? Give examples.

3. Let $\Omega \subseteq \mathbb{C}$ be open, and suppose f is analytic on Ω except for a pole of order n at the point $a \in \Omega$. What can be said about the singularity of f' at a?

4. Let $f \colon \mathbb{C} \setminus \{0\} \to \mathbb{C}$ be given by

$$f(z) = \sin \frac{1}{z},$$

and let $\alpha \in \mathbb{C}$. In the spirit of Example 5.2.9, construct a sequence $\{z_n\}_{n=1}^{\infty} \subseteq \mathbb{C} \setminus \{0\}$ such that $z_n \to 0$ and $f(z_n) \to \alpha$ as $n \to \infty$.

5. Let $\Omega \subseteq \mathbb{C}$ be open and $a \in \Omega$. Suppose that f is analytic on Ω except for an isolated singularity at a. Extend Theorem 5.2.6 by showing that f has a pole of order $m \in \mathbb{N}$ at a if and only if there is a function $g \in H(\Omega)$ such that $g(a) \neq 0$ and

$$f(z) = \frac{g(z)}{(z-a)^m}$$

for all $z \in \Omega \setminus \{a\}$.

6. $\triangleright$ Let $\Omega \subseteq \mathbb{C}$ be open, and suppose that $f, g \in H(\Omega)$ each have an isolated zero at the point $a \in \Omega$. Prove the complex version of l'Hôpital's rule:

$$\lim_{z \to a} \frac{f(z)}{g(z)} = \lim_{z \to a} \frac{f'(z)}{g'(z)}.$$

(This includes infinite limits.) Specifically, if f and g both have zeros of order $m \in \mathbb{N}$ at a, then

$$\lim_{z \to a} \frac{f(z)}{g(z)} = \frac{f^{(m)}(a)}{g^{(m)}(a)}.$$

7. Let $\Omega \subseteq \mathbb{C}$ be open. Suppose that f is analytic on Ω except for an isolated singularity at the point $a \in \Omega$. Show that a is a removable singularity of f if and only if

$$\lim_{z \to a} (z - a) f(z) = 0.$$

8. Let f and g be entire functions such that $|f(z)| \leq |g(z)|$ for all $z \in \mathbb{C}$. Prove that there is some constant $c \in \overline{\mathbb{D}}$ such that $f = cg$.

9. Let $\Omega \subseteq \mathbb{C}$ be open and $a \in \Omega$. Suppose f is analytic on Ω except for a pole at a. Set $g(z) = e^{f(z)}$ for $z \in \Omega \setminus \{a\}$. Show that g has an essential singularity at a. (*Hint:* Consider g'/g.)

10. Let $\Omega \subseteq \mathbb{C}$ be open and such that $A(0; r, \infty) \subseteq \Omega$. Suppose that $f \in H(\Omega)$. We say that ∞ is a removable singularity, pole of order $m \in \mathbb{N}$, or essential singularity of f if 0 is the same for the function $g(z) = f(1/z)$ defined on $\Omega' = \{z \in \mathbb{C} \setminus \{0\} : 1/z \in \Omega\}$. If f is entire, classify the singularity of f at ∞ by giving conditions on the power series of f based at 0. What are the only entire functions f such that $\lim_{z \to \infty} f(z) = \infty$?

5.3 Residues

Consider an open set $\Omega \subseteq \mathbb{C}$ and a function f analytic on Ω except for isolated singularities in Ω. For such a singularity a, we can write f as its Laurent series expansion

$$f(z) = \sum_{n=-\infty}^{\infty} c_n (z-a)^n \tag{5.3.1}$$

for all $z \in A(a; 0, r) \subseteq \Omega$ for some $r > 0$. Let $\rho \in (0, r)$. Using uniform convergence on compact sets, we calculate the familiar integral

$$\frac{1}{2\pi i} \int_{\partial D(a;\rho)} f(z)\, dz = \sum_{n=-\infty}^{\infty} \frac{c_n}{2\pi i} \int_{\partial D(a;\rho)} (z-a)^n \, dz = c_{-1},$$

using Example 2.9.15. That c_{-1} is all that remains from the integration motivates the following definition.

5.3.1 Definition. If $f \in H(A(a; 0, r))$ for some $a \in \mathbb{C}$ and $r > 0$, then the *residue* of f at a is given by

$$\operatorname*{Res}_{z=a} f(z) = c_{-1}, \tag{5.3.2}$$

where c_{-1} is the coefficient of $(z-a)^{-1}$ in the Laurent series expansion of f given in (5.3.1).

Observe that if f has a removable singularity at a, then $\operatorname{Res}_{z=a} f(z) = 0$. In the case that the singularity is a pole, we can say the following.

5.3.2 Theorem. *Let $a \in \mathbb{C}$ and $r > 0$, and suppose that $f \in H(A(a; 0, r))$ has a pole of order $m \in \mathbb{N}$ at a. Then*

$$\operatorname*{Res}_{z=a} f(z) = \frac{1}{(m-1)!} \lim_{z \to a} \frac{d^{m-1}}{dz^{m-1}} (z-a)^m f(z). \tag{5.3.3}$$

Of course, in the case that the pole is simple (of order 1), then there is no differentiation.

Proof. This is a direct series manipulation. By hypothesis, we can write

$$f(z) = \sum_{n=-m}^{\infty} c_n (z-a)^n$$

for all $z \in A(a; 0, r)$. For such z,

$$(z-a)^m f(z) = \sum_{n=0}^{\infty} c_{n-m} (z-a)^n,$$

and hence

$$\frac{d^{m-1}}{dz^{m-1}} (z-a)^m f(z) = \sum_{n=m-1}^{\infty} \frac{n!}{(n-m+1)!} c_{n-m} (z-a)^{n-m+1}.$$

This expression has a removable singularity at a, and the result follows by taking the limit. $\qquad \square$

Theorem 5.3.2 requires prior knowledge of the order of the pole at a, which can often come from Theorem 5.2.11. If the singularity of f at a is essential, we do not have such a nice tool.

The primary use of residues is given in the following theorem, a generalization of Cauchy's theorem and our final integral formula for analytic functions.

5.3.3 Residue Theorem. *Let $\Omega \subseteq \mathbb{C}$ be open, and suppose that f is analytic on Ω except for a finite number of distinct isolated singularities $a_1, \ldots, a_n \in \Omega$. If γ is a closed contour in $\Omega \setminus \{a_1, \ldots, a_n\}$ such that $\mathrm{Ind}_\gamma z = 0$ for all $z \in \mathbb{C} \setminus \Omega$, then*

$$\frac{1}{2\pi i} \int_\gamma f(z)\, dz = \sum_{j=1}^{n} \operatorname*{Res}_{z=a_j} f(z)\, \mathrm{Ind}_\gamma a_j. \tag{5.3.4}$$

Proof. For a given $j \in \{1, \ldots, n\}$, there is $r_j > 0$ such that f has the Laurent series

$$f(z) = \sum_{k=-\infty}^{\infty} c_k (z - a_j)^k$$

for all $z \in A(a_j; 0, r_j)$. Theorem 5.1.4 implies that the function

$$g_j(z) = \sum_{k=-\infty}^{-1} c_k (z - a_j)^k,$$

the principal part of the Laurent series, is an analytic function on $\mathbb{C} \setminus \{a_j\}$. Furthermore, the function $f - g_j$ has a removable singularity at a_j. For $k \leq -2$, we know that $\int_\gamma (z - a_j)^k\, dz = 0$ by Corollary 2.9.14, as the integrand is antidifferentiable on $\mathbb{C} \setminus \{a_j\}$. We therefore conclude, by uniform convergence on compact sets, that

$$\frac{1}{2\pi i} \int_\gamma g_j(z)\, dz = \sum_{k=-\infty}^{-1} \frac{c_k}{2\pi i} \int_\gamma (z - a_j)^k\, dz$$

$$= \frac{c_{-1}}{2\pi i} \int_\gamma \frac{dz}{z - a_j}$$

$$= \operatorname*{Res}_{z=a_j} f(z)\, \mathrm{Ind}_\gamma a_j.$$

For $z \in \Omega \setminus \{a_1, \ldots, a_n\}$, define

$$F(z) = f(z) - \sum_{j=1}^{n} g_j(z).$$

For a given $j_0 \in \{1, \ldots, n\}$, $f - g_{j_0}$ has a removable singularity at a_{j_0} and g_j is defined and analytic at a_{j_0} for all $j \neq j_0$. Therefore F has a removable singularity at a_{j_0}. As this can be done for all j_0, F can be extended to be analytic on Ω.

It now follows by Cauchy's theorem (Theorem 4.3.1) that

$$0 = \int_\gamma F(z)\, dz = \int_\gamma f(z)\, dz - \sum_{j=1}^{n} \int_\gamma g_j(z)\, dz.$$

It is immediately seen that

$$\frac{1}{2\pi i} \int_\gamma f(z) \, dz = \sum_{j=1}^n \frac{1}{2\pi i} \int_\gamma g_j(z) \, dz = \sum_{j=1}^n \operatorname*{Res}_{z=a_j} f(z) \operatorname{Ind}_\gamma a_j,$$

as desired. $\qquad\qquad\qquad\qquad\qquad\qquad\qquad\qquad\qquad\qquad\qquad\square$

As was the case with both Cauchy's theorem and Cauchy's integral formula, there is an easier form that occurs when the contour is simple and closed.

5.3.4 Corollary. *Let $\Omega \subseteq \mathbb{C}$ be open, and let γ be a simple closed contour in Ω such that the inside of γ^* lies in Ω. If f is analytic on Ω except for isolated singularities in $\Omega \setminus \gamma^*$, of which $a_1, \ldots, a_n$ lie inside of γ^*, then*

$$\frac{1}{2\pi i} \int_\gamma f(z) \, dz = \sum_{j=1}^n \operatorname*{Res}_{z=a_j} f(z).$$

This follows immediately from the residue theorem, but it can also be proved using Theorem 4.3.3. The latter is left as an exercise.

We now consider some examples of the residue theorem in action. As is common for these types of calculations, we make use of l'Hôpital's rule. This works as in calculus, and its proof is requested in Exercise 6 of Section 5.2.

5.3.5 Example. We will calculate the integral

$$\int_{\partial D(0;\pi)} \tan z \, dz.$$

Since $\tan z = \sin z / \cos z$, we apply Theorem 5.2.11 to see that f has simple poles at $\pm\pi/2$. Theorem 5.3.2 and l'Hôpital's rule allow us to calculate

$$\operatorname*{Res}_{z=\pi/2} \tan z = \lim_{z\to\pi/2} \left(z - \frac{\pi}{2}\right) \frac{\sin z}{\cos z} = -1.$$

Similarly, $\operatorname{Res}_{z=-\pi/2} \tan z = -1$. The conditions of the residue theorem are met, and hence

$$\int_{\partial D(0;\pi)} \tan z \, dz = 2\pi i \left(\operatorname*{Res}_{z=\pi/2} \tan z + \operatorname*{Res}_{z=-\pi/2} \tan z \right) = -4\pi i.$$

5.3.6 Example. We will calculate the integral

$$\int_{\partial D(0;2)} \frac{e^{\pi z} - 1}{z^5 + z^3} \, dz.$$

First, observe that the integrand

$$\frac{e^{\pi z} - 1}{z^5 + z^3} = \frac{e^{\pi z} - 1}{z^3(z + i)(z - i)}$$

has a pole of order 2 at 0 (the numerator has a zero of order 1 and the denominator has a zero of order 3) and simple poles at $\pm i$. We calculate the residue at 0 using Theorem 5.3.2. Note first that

$$\frac{d}{dz} z^2 \frac{e^{\pi z} - 1}{z^5 + z^3} = \frac{\pi e^{\pi z}(z^3 + z) - (e^{\pi z} - 1)(3z^2 + 1)}{(z^3 + z)^2}.$$

We take the limit using l'Hôpital's rule to see

$$\operatorname*{Res}_{z=0} \frac{e^{\pi z} - 1}{z^5 + z^3} = \lim_{z \to 0} \frac{\pi^2 e^{\pi z}(z^3 + z) - 6z(e^{\pi z} - 1)}{2(z^3 + z)(3z^2 + 1)} = \frac{\pi^2}{2}.$$

The last step follows after factoring out z from the numerator and denominator. Now we calculate

$$\operatorname*{Res}_{z=i} \frac{e^{\pi z} - 1}{z^5 + z^3} = \lim_{z \to i} \frac{e^{\pi z} - 1}{z^3(z + i)} = -1.$$

Likewise, the residue at $-i$ is also -1. Therefore by the residue theorem,

$$\int_{\partial D(0;2)} \frac{e^{\pi z} - 1}{z^5 + z^3} \, dz = 2\pi i \left(\frac{\pi^2}{2} - 1 - 1 \right) = \pi i (\pi^2 - 4).$$

It may be interesting to compare this calculation with the method that could have been used to evaluate the integral following Theorem 4.3.3.

Summary and Notes for Section 5.3.

The residue of an analytic function at an isolated singularity is the coefficient of the -1 power term in its Laurent series expansion about the singularity. Residues are not always easy to calculate, especially when the singularity is essential. For poles, Theorem 5.3.2 is quite helpful.

The residue theorem gives a formula for calculating the integral of an analytic function about a closed contour in its domain in terms of the residues of the function at its (finitely many) isolated singularities. This generalization of Cauchy's theorem and integral formula is the culmination of our work on complex integration and is the most computationally useful integral formula in this text. Going all the way back to Example 2.9.6, we have been considering integrals over closed contours within which the integrands have singularities, and now we see the final state of that theory.

It should come as no surprise at this point that Cauchy developed the residue theorem, and it often bears his name. This was done during the same time that he developed his integral theorem.

Exercises for Section 5.3.

1. For each of the following functions f, find the residue of f at each of the isolated singularities of f.

   (a) $f(z) = \dfrac{z^2 + 2z + 1}{z^4 + 3z^3}$

(b) $f(z) = \dfrac{\sin z}{1 + z^2}$

(c) $f(z) = \csc z$

(d) $f(z) = \dfrac{e^{2/z}}{z}$

(e) $f(z) = \dfrac{1 - e^{z/2}}{1 - e^z}$

(f) $f(z) = \dfrac{\operatorname{Log} z}{(z-1)^2(z-2)}$

2. Evaluate the following integrals using the residue theorem.

(a) $\displaystyle\int_{\partial D(0;4)} \dfrac{z^2 + 1}{z^3 + 2z^2 - 3z}\, dz$

(b) $\displaystyle\int_{\partial D(\pi;\pi/2)} \csc z\, dz$

(c) $\displaystyle\int_{\partial D(0;\pi)} z^2 \sin \dfrac{1}{z}\, dz$

(d) $\displaystyle\int_{\partial D(1+i;2)} \dfrac{z+1}{z^4 + z^2}\, dz$

3. Evaluate the integrals in Exercise 1 of Section 4.3 using the residue theorem.

4. Let $\gamma\colon [0, 2\pi] \to \mathbb{C}$ be the contour given by

$$\gamma(t) = (1 + 2\cos t)e^{it}.$$

(Note γ^* is the limaçon described by the polar equation $r = 1 + 2\cos\theta$.) Use the residue theorem to calculate the following integrals over γ.

(a) $\displaystyle\int_\gamma \tan \pi z\, dz$

(b) $\displaystyle\int_\gamma \dfrac{\operatorname{Log}(z+1)}{2z^2 - 5z + 2}\, dz$

(c) $\displaystyle\int_\gamma \dfrac{e^{iz}}{(z^2 - 2z + 2)^2}\, dz$

5. Prove Corollary 5.3.4 directly from Theorem 4.3.3. Do not appeal to the residue theorem.

6. Let $\Omega \subseteq \mathbb{C}$ be open such that $-z \in \Omega$ whenever $z \in \Omega$. Suppose f is even and analytic on Ω except for isolated singularities in Ω. Prove that if $a \in \Omega$ is a pole of order $m \in \mathbb{N}$ or an essential singularity of f, then $-a$ is a singularity of the same type and

$$\operatorname*{Res}_{z=-a} f(z) = -\operatorname*{Res}_{z=a} f(z).$$

Under what conditions on a contour γ in Ω can we guarantee that $\int_\gamma f(z)\, dz = 0$?

7. Let Ω be open, $E \subseteq \Omega$ be a set with no limit points in Ω, and $C \subseteq \Omega \setminus E$ be a simple closed contour such that the inside of C lies in Ω. (The set E could be the set of isolated singularities of an analytic function.) Why can only finitely many points of E lie inside of C?

8. Suppose that $C \subseteq \mathbb{C}$ is a positively oriented simple closed contour. Show that if f is analytic on $\mathbb{C}$ except for a finite number of isolated singularities inside of C, then

$$\frac{1}{2\pi i} \int_C f(z)\,dz = \operatorname*{Res}_{z=0} \frac{1}{z^2} f\left(\frac{1}{z}\right).$$

(*Hint*: Use a "large" circle centered at 0.)

9. Show that

$$\int_{\partial \mathbb{D}} e^{z+1/z}\,dz = 2\pi i \sum_{n=0}^{\infty} \frac{1}{n!(n+1)!}.$$

10. Let $\Omega \subseteq \mathbb{C}$ be open, and let f be analytic in Ω except for distinct poles $p_1, \ldots, p_N \in \Omega$ of orders $n_1, \ldots, n_N$, respectively. In addition, suppose that f has distinct zeros $a_1, \ldots, a_M \in \Omega$ of orders $m_1, \ldots, m_M$, respectively. Choose a closed contour γ with trace in $\Omega \setminus \{a_1, \ldots, a_M, p_1, \ldots, p_N\}$ such that $\operatorname{Ind}_\gamma z = 0$ for all $z \in \mathbb{C} \setminus \Omega$.

   (a) Show that

   $$\frac{1}{2\pi i} \int_\gamma \frac{f'(z)}{f(z)}\,dz = \sum_{j=1}^{M} m_j \operatorname{Ind}_\gamma a_j - \sum_{k=1}^{N} n_k \operatorname{Ind}_\gamma p_k.$$

   (b) This result is called the *argument principle*. Give an explanation related to that following the proof of Theorem 4.1.3 of why this name is appropriate.

11. In this exercise, we explore how residues may be used to sum certain series. To begin, let P and Q be polynomials such that $\deg Q \geq 2 + \deg P$, $g = P/Q$, and $f(z) = g(z) \cot \pi z$. Note that f is defined on $\mathbb{C}$ except for the isolated singularities $\mathbb{Z} \cup Z(Q)$.

   (a) Show that there exist $M > 0$ and $R > 0$ such that $|g(z)| \leq M/|z|^2$ for all $z \in \mathbb{C}$ such that $|z| > R$.

   (b) For $N \in \mathbb{N}$ with $N > R$, let C_N denote the rectangular contour with sides lying on the lines $\operatorname{Re} z = \pm(N + 1/2)$ and $\operatorname{Im} z = \pm N$. Show that

   $$\lim_{N \to \infty} \int_{C_N} f(z)\,dz = 0$$

   by parameterizing the four separate sides of C_N and showing their moduli each go to 0.

   (c) Verify that for all $n \in \mathbb{Z} \setminus Z(Q)$, $\operatorname{Res}_{z=n} f(z) = g(n)/\pi$.

   (d) Conclude that

   $$\sum_{n=-\infty}^{\infty} \operatorname*{Res}_{z=n} f(z) = - \sum_{a \in Z(Q) \setminus \mathbb{Z}} \operatorname*{Res}_{z=a} f(z).$$

   (This includes arguing that the series on the left-hand side converges. The right-hand side is understood to be 0 if $Z(Q) \subseteq \mathbb{Z}$.)

   (e) Use parts (c) and (d) and the given g to verify the summation identity.

   i. $g(z) = \dfrac{1}{z^2}$, $\quad \displaystyle\sum_{n=1}^{\infty} \frac{1}{n^2} = \frac{\pi^2}{6}$

   ii. $g(z) = \dfrac{1}{z^2 + c^2}$, $c > 0$, $\quad \displaystyle\sum_{n=1}^{\infty} \frac{1}{n^2 + c^2} = \frac{\pi}{2c} \coth \pi c - \frac{1}{2c^2}$

5.4   Evaluation of Real Integrals

The power of the residue theorem has already been seen in the calculation of several contour integrals in Section 5.3. What is especially noteworthy is how the residue theorem can also be used to calculate *real* integrals that are either difficult or impossible to evaluate using the methods of calculus, another significant instance of complex methods facilitating the solution of real problems. The examples in this section illustrate a variety of techniques for evaluating real integrals.

We begin with a trigonometric integral over a finite interval that is not easy to calculate using calculus.

5.4.1 Example. We shall calculate the integral

$$I = \int_0^{2\pi} \frac{dx}{2 + \cos x}.$$

First observe that

$$2 + \cos t = 2 + \frac{e^{it} + e^{-it}}{2} = \frac{e^{2it} + 4e^{it} + 1}{2e^{it}}.$$

Since $z = e^{it}, 0 \leq t \leq 2\pi$, parameterizes $\partial \mathbb{D}$, we recognize our integral as a contour integral by writing

$$\int_{\partial \mathbb{D}} \frac{2}{z^2 + 4z + 1}\, dz = \int_0^{2\pi} \frac{2ie^{it}}{e^{2it} + 4e^{it} + 1}\, dt = i \int_0^{2\pi} \frac{dt}{2 + \cos t}.$$

The integrand on the left-hand side has simple poles at $-2 \pm \sqrt{3}$. Only $-2 + \sqrt{3}$ lies in $\mathbb{D}$, and thus

$$\int_{\partial \mathbb{D}} \frac{2}{z^2 + 4z + 1}\, dz = 2\pi i \operatorname*{Res}_{z = -2+\sqrt{3}} \frac{2}{z^2 + 4z + 1}$$

$$= 2\pi i \lim_{z \to -2+\sqrt{3}} \frac{2}{z + 2 + \sqrt{3}} = \frac{2\pi i}{\sqrt{3}}.$$

Therefore $I = 2\pi/\sqrt{3}$.

More common applications of the residue theorem involve the evaluation of improper real integrals. It is sometimes necessary, before evaluating such an integral using residues, to verify that it is convergent. To wit, if $\int_{-\infty}^{\infty} f(x)\, dx$ is known to exist, then it must equal $\lim_{R \to \infty} \int_{-R}^{R} f(x)\, dx$. However, the converse is not true, as seen in calculus. The following lemma from real analysis, a parallel to Theorem 1.7.9, is helpful in this regard. Because of its use in the next section, we state it for a *piecewise continuous* function f on $[0, \infty)$, meaning f is piecewise continuous on $[0, R]$ for all $R > 0$. (See Definition 2.8.2.) Similar to our work in Section 2.8, a range of $\mathbb{C}$ instead of $\mathbb{R}$ does not change the proof, which we leave as an exercise.

5.4.2 Lemma. *Let the complex-valued f and nonnegative-valued g be piecewise continuous on $[0, \infty)$. If*

$$\int_0^\infty g(t)\, dt$$

is convergent and $|f(t)| \le g(t)$ for all t, then

$$\left| \int_0^\infty f(t)\, dt \right| \le \int_0^\infty |f(t)|\, dt \le \int_0^\infty g(t)\, dt,$$

implying the convergence of the left-hand and middle integrals.

Note that this lemma can immediately be extended to integrals over $(-\infty, 0]$ or $(-\infty, \infty)$. We now consider the following.

5.4.3 Example. Let us calculate the value of

$$I = \int_{-\infty}^\infty \frac{\cos x}{1 + x^2}\, dx.$$

Lemma 5.4.2 gives that I converges, since it is bounded by the convergent integral $\int_{-\infty}^\infty dx/(1 + x^2)$. Therefore $I = \lim_{R \to \infty} I_R$, where

$$I_R = \int_{-R}^R \frac{\cos x}{1 + x^2}\, dx = \text{Re} \int_{-R}^R \frac{e^{ix}}{1 + x^2}\, dx$$

for $R > 1$.

For such R, let C_R^1 be the interval $[-R, R]$ and C_R^2 be the upper half of $\partial D(0; R)$ parameterized in the counterclockwise direction. Define C_R to be the closed contour $C_R^1 + C_R^2$. (See Figure 5.3.) Inside C_R, the expression $e^{iz}/(1 + z^2)$ has a simple pole at i. Therefore by the residue theorem,

$$\int_{C_R} \frac{e^{iz}}{1 + z^2}\, dz = 2\pi i \operatorname*{Res}_{z=i} \frac{e^{iz}}{1 + z^2} = 2\pi i \lim_{z \to i} \frac{e^{iz}}{z + i} = \frac{\pi}{e}.$$

For $t \in [0, \pi]$, observe that

$$|e^{iRe^{it}}| = |e^{iR\cos t} e^{-R\sin t}| = e^{-R\sin t} \le 1.$$

Thus

$$\left| \int_{C_R^2} \frac{e^{iz}}{1 + z^2}\, dz \right| = \left| \int_0^\pi \frac{iRe^{it} e^{iRe^{it}}}{1 + R^2 e^{2it}}\, dt \right| \le \int_0^\pi \frac{R}{R^2 - 1}\, dt \to 0$$

as $R \to \infty$.

We now have

$$\int_{-R}^R \frac{e^{ix}}{1 + x^2}\, dx = \int_{C_R^1} \frac{e^{iz}}{1 + z^2}\, dz = \int_{C_R} \frac{e^{iz}}{1 + z^2}\, dz - \int_{C_R^2} \frac{e^{iz}}{1 + z^2}\, dz \to \frac{\pi}{e}$$

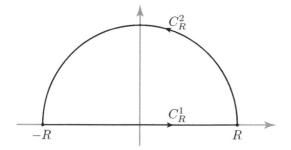

Figure 5.3   Contours in Example 5.4.3

as $R \to \infty$. Therefore $I = \pi/e$.

In the next more challenging example, the integral is improper at both 0 and ∞, causing us to use an *indented contour*. The following lemma will be helpful.

5.4.4 Lemma. *Let $a \in \mathbb{C}$, $R > 0$, and $f \in H(A(a; 0, R))$ have a simple pole at a. For $r \in (0, R)$, $\theta \in \mathbb{R}$, and $\alpha \in (0, 2\pi]$, define $\gamma_r : [\theta, \theta + \alpha] \to \mathbb{C}$ by $\gamma_r(t) = a + re^{it}$. Then*

$$\lim_{r \to 0^+} \int_{\gamma_r} f(z)\, dz = i\alpha \operatorname*{Res}_{z=a} f(z).$$

Proof. The Laurent series expansion of f about a can be written

$$f(z) = \sum_{n=-1}^{\infty} c_n (z-a)^n = \frac{c_{-1}}{z-a} + g(z),$$

where $g \in H(D(a; R))$ contains the terms of nonnegative degree of the series. Using the parameterization of γ_r, we have

$$\int_{\gamma_r} \frac{c_{-1}}{z-a}\, dz = \int_{\theta}^{\theta+\alpha} ic_{-1}\, dt = i\alpha \operatorname*{Res}_{z=a} f(z)$$

for all $r \in (0, R)$. The analytic function g is bounded by some $M > 0$ on the compact set $\overline{D}(a; R/2)$, and hence for all $r \in (0, R/2)$,

$$\left| \int_{\gamma_r} g(z)\, dz \right| \le \int_{\gamma_r} |g(z)|\, |dz| \le \alpha r M.$$

Hence $\int_{\gamma_r} g(z)\, dz \to 0$ as $r \to 0^+$, giving the result. $\qquad\square$

5.4.5 Example. Consider the integral

$$I = \int_0^\infty \frac{\sin x}{x}\, dx.$$

Observe first, through substitution, that for $0 < r < R$,

$$\int_r^R \frac{\sin x}{x} dx = \frac{1}{2i} \left(\int_r^R \frac{e^{ix}}{x} dx - \int_r^R \frac{e^{-ix}}{x} dx \right)$$

$$= \frac{1}{2i} \left(\int_r^R \frac{e^{ix}}{x} dx + \int_{-R}^{-r} \frac{e^{ix}}{x} dx \right).$$

Let C_r and C_R be the top halves of the circles $\partial D(0; r)$ and $\partial D(0; R)$, respectively, parameterized in the counterclockwise direction. Define $C_{r,R}$ to be the closed positively oriented contour formed by $[r, R]$, C_R, $[-R, -r]$, and $-C_r$. (See Figure 5.4.) Then Cauchy's theorem implies

$$\int_{C_{r,R}} \frac{e^{iz}}{z} dz = 0.$$

Our above observations then provide

$$2i \int_r^R \frac{\sin x}{x} dx = \int_{C_r} \frac{e^{iz}}{z} dz - \int_{C_R} \frac{e^{iz}}{z} dz.$$

The function $f(z) = e^{iz}/z$ is analytic on $\mathbb{C} \setminus \{0\}$ with a simple pole at 0. It follows from Lemma 5.4.4 that

$$\lim_{r \to 0+} \int_{C_r} \frac{e^{iz}}{z} dz = \pi i \operatorname*{Res}_{z=0} \frac{e^{iz}}{z} = \pi i \lim_{z \to 0} e^{iz} = \pi i.$$

Using the concavity of $\sin t$, we see that $\sin t \geq 2t/\pi$ for $t \in [0, \pi/2]$. With this observation, we have

$$|e^{iRe^{it}}| = |e^{iR\cos t} e^{-R\sin t}| = e^{-R\sin t} \leq e^{-2Rt/\pi}$$

for $t \in [0, \pi/2]$. We use the symmetry of $\sin t$ about $t = \pi/2$ to calculate

$$\left| \int_{C_R} \frac{e^{iz}}{z} dz \right| = \left| \int_0^\pi \frac{iRe^{it} e^{iRe^{it}}}{Re^{it}} dt \right|$$

$$\leq \int_0^\pi e^{-R\sin t} dt$$

$$= 2 \int_0^{\pi/2} e^{-R\sin t} dt$$

$$\leq 2 \int_0^{\pi/2} e^{-2Rt/\pi} dt$$

$$= \frac{\pi}{R}(1 - e^{-R}) \to 0$$

as $R \to \infty$.

It is now evident that

$$2i \int_r^R \frac{\sin x}{x} \, dx \to \pi i$$

as $r \to 0^+$ and $R \to \infty$, and hence $I = \pi/2$.

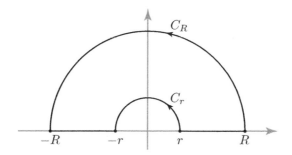

Figure 5.4   Contours used in Example 5.4.5

We now break form by considering an improper integral that, while accessible using calculus methods, illustrates another useful technique.

5.4.6 Example. We will evaluate the integral

$$I = \int_0^\infty \frac{dx}{\sqrt{x}\,(1+x)}.$$

Let $\varepsilon > 0$ and $0 < r < 1 < R$. Define C to be the closed contour formed by the segment $[re^{i\varepsilon}, Re^{i\varepsilon}]$, the arc of the circle $\partial D(0; R)$ from $Re^{i\varepsilon}$ to $Re^{i(2\pi-\varepsilon)}$ oriented counterclockwise, the segment $[Re^{i(2\pi-\varepsilon)}, re^{i(2\pi-\varepsilon)}]$ and the arc of the circle $\partial D(0;r)$ from $re^{i(2\pi-\varepsilon)}$ to $re^{i\varepsilon}$ oriented clockwise. (See Figure 5.5.) We consider $f(z) = 1/[z^{1/2}(1+z)]$, where the branch of $z^{1/2}$ is such that $0 < \arg z < 2\pi$, so that f is defined and analytic on $\mathbb{C} \setminus [0, \infty)$ except for a simple pole at -1. It follows that

$$\int_C f(z)\,dz = 2\pi i \operatorname*{Res}_{z=-1} f(z) = 2\pi i \lim_{z \to -1} \frac{1}{z^{1/2}} = 2\pi.$$

Parameterizing C then gives

$$2\pi = \int_r^R \frac{e^{i\varepsilon}}{t^{1/2}e^{i\varepsilon/2}(1+te^{i\varepsilon})}\,dt + \int_\varepsilon^{2\pi-\varepsilon} \frac{iRe^{it}}{R^{1/2}e^{it/2}(1+Re^{it})}\,dt$$
$$+ \int_r^R \frac{e^{-i\varepsilon}}{t^{1/2}e^{-i\varepsilon/2}(1+te^{-i\varepsilon})}\,dt - \int_\varepsilon^{2\pi-\varepsilon} \frac{ire^{it}}{r^{1/2}e^{it/2}(1+re^{it})}\,dt,$$

where in the third integral, we use that $e^{i(2\pi-\varepsilon)} = e^{-i\varepsilon}$ and $e^{i(2\pi-\varepsilon)/2} = -e^{-i\varepsilon/2}$. The integrands of the first and third integrals are jointly continuous for $(\varepsilon, t) \in [0, \pi/2] \times [r, R]$, and hence the integrals are continuous for $\varepsilon \in [0, \pi/2]$ by Leibniz's rule (see Appendix B), and the other two integrals are continuous for $\varepsilon \in [0, \pi/2]$ by the fundamental theorem of calculus. Hence we may take the limit as $\varepsilon \to 0^+$ to obtain

$$2\pi = 2 \int_r^R \frac{dx}{\sqrt{x}\,(1+x)} + \int_0^{2\pi} \frac{iR^{1/2}e^{it/2}}{1 + Re^{it}}\, dt - \int_0^{2\pi} \frac{ir^{1/2}e^{it/2}}{1 + re^{it}}\, dt. \qquad (5.4.1)$$

Now

$$\left| \int_0^{2\pi} \frac{iR^{1/2}e^{it/2}}{1 + Re^{it}}\, dt \right| \leq \int_0^{2\pi} \frac{\sqrt{R}}{R - 1}\, dt \to 0$$

as $R \to \infty$ and

$$\left| \int_0^{2\pi} \frac{ir^{1/2}e^{it/2}}{1 + re^{it}}\, dt \right| \leq \int_0^{2\pi} \frac{\sqrt{r}}{1 - r}\, dt \to 0$$

as $r \to 0^+$. Therefore taking $R \to \infty$ and $r \to 0^+$ in (5.4.1) shows $I = \pi$.

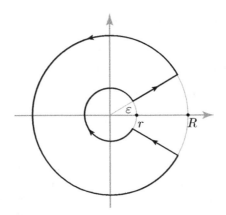

Figure 5.5   Contours used in Example 5.4.6

Our final example uses a rectangular contour, whose top is carefully chosen for algebraic simplification.

5.4.7 Example. We will calculate the integral

$$I = \int_{-\infty}^{\infty} \frac{e^x}{1 + e^{3x}}\, dx.$$

Lemma 5.4.2 gives the bounds

$$\int_0^{\infty} \frac{e^x}{1 + e^{3x}}\, dx \leq \int_0^{\infty} e^{-2x}\, dx, \qquad \int_{-\infty}^0 \frac{e^x}{1 + e^{3x}}\, dx \leq \int_{-\infty}^0 e^x\, dx,$$

showing I is convergent.

For $R > 0$, let C_R be the closed rectangle obtained by summing the directed segments $C_R^1 = [-R, R]$, $C_R^2 = [R, R + 2\pi i/3]$, $C_R^3 = [R + 2\pi i/3, -R + 2\pi i/3]$, and $C_R^4 = [-R + 2\pi i/3, -R]$. (See Figure 5.6.) Observe that

$$\int_{C_R^3} \frac{e^z}{1 + e^{3z}} \, dz = -\int_{-R}^{R} \frac{e^{t+2\pi i/3}}{1 + e^{3t+2\pi i}} \, dt = -e^{2\pi i/3} \int_{-R}^{R} \frac{e^t}{1 + e^{3t}} \, dt.$$

Therefore

$$\int_{C_R^1} \frac{e^z}{1 + e^{3z}} \, dz + \int_{C_R^3} \frac{e^z}{1 + e^{3z}} \, dz = (1 - e^{2\pi i/3}) \int_{-R}^{R} \frac{e^x}{1 + e^{3x}} \, dx.$$

Now we address the vertical sides. First,

$$\left| \int_{C_R^2} \frac{e^z}{1 + e^{3z}} \, dz \right| = \left| \int_0^{2\pi/3} \frac{ie^{R+it}}{1 + e^{3R+3it}} \, dt \right| \leq \int_0^{2\pi/3} \frac{e^R}{e^{3R} - 1} \, dt \to 0$$

as $R \to \infty$. Similarly,

$$\left| \int_{C_R^4} \frac{e^z}{1 + e^{3z}} \, dz \right| = \left| \int_0^{2\pi/3} \frac{-ie^{-R+it}}{1 + e^{-3R+3it}} \, dt \right| \leq \int_0^{2\pi/3} \frac{e^{-R}}{1 - e^{-3R}} \, dt \to 0$$

as $R \to \infty$.

We now see that

$$\lim_{R \to \infty} \int_{C_R} \frac{e^z}{1 + e^{3z}} \, dz = (1 - e^{2\pi i/3}) \lim_{R \to \infty} \int_{-R}^{R} \frac{e^x}{1 + e^{3x}} \, dx = (1 - e^{2\pi i/3})I.$$

The integrand on the left-hand side has singularities when $z = (2k + 1)\pi i/3$ for $k \in \mathbb{Z}$. Therefore the only singularity inside of C_R is at $\pi i/3$. Since the denominator has a simple zero at $\pi i/3$, there is a simple pole there. Using l'Hôpital's rule, we calculate the residue

$$\operatorname*{Res}_{z=\pi i/3} \frac{e^z}{1 + e^{3z}} = \lim_{z \to \pi i/3} \frac{(z - \pi i/3)e^z}{1 + e^{3z}} = \lim_{z \to \pi i/3} \frac{(z - \pi i/3)e^z + e^z}{3e^{3z}} = -\frac{e^{\pi i/3}}{3}.$$

By the residue theorem,

$$\int_{C_R} \frac{e^z}{1 + e^{3z}} \, dz = \frac{-2\pi i e^{\pi i/3}}{3}$$

for all $R > 0$. We can then solve to see that

$$I = -\frac{2\pi i}{3} \frac{e^{\pi i/3}}{1 - e^{2\pi i/3}} = \frac{\pi}{3} \frac{2i}{e^{\pi i/3} - e^{-\pi i/3}} = \frac{\pi}{3} \csc \frac{\pi}{3} = \frac{2\pi}{3\sqrt{3}}.$$

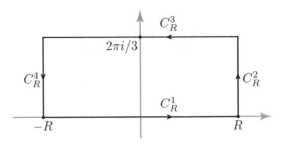

Figure 5.6   Contours used in Example 5.4.7

Summary and Notes for Section 5.4.

We have seen examples in previous sections of real-variable problems being solved more easily by moving into the complex plane, but none are as impressive as the methods by which the residue theorem and contour integration can be used to evaluate real integrals. Our work in this section does not form a complete inventory of the techniques that can be used to apply complex integration theory to calculate real integrals, but should give the reader a good feel for the topic.

The desire to evaluate real integrals led many prominent mathematicians to explore complex numbers. Depending on the era of their work, they had to deal with varying degrees of distrust of these "impossible" numbers, as Leonhard Euler called them. Euler, Pierre-Simon Laplace, Siméon-Denis Poisson, and Adrien-Marie Legendre all achieved success using complex numbers in this way. Cauchy, who, as we have seen, deserves much credit for the development of modern complex analysis, was drawn to the subject by the observation that his predecessors had solved these integrals by "passage from the real to the imaginary." Once he proved the residue theorem, he was quick to show off its application to the evaluation of real integrals, using methods very much as in this section.

Exercises for Section 5.4.

1. Calculate the following integrals using a method similar to Example 5.4.1.

   (a) $\displaystyle\int_0^{2\pi} \frac{dx}{4\sin x + 5}$

   (b) $\displaystyle\int_0^{2\pi} \frac{\sin^2 x}{5 - 3\cos x}\, dx$

   (c) $\displaystyle\int_0^{2\pi} \frac{dx}{(3 - \cos x)^2}$

2. Evaluate the following integrals using contours as in Figure 5.3.

   (a) $\displaystyle\int_{-\infty}^{\infty} \frac{dx}{x^4 + 4}$

   (b) $\displaystyle\int_{-\infty}^{\infty} \frac{x^4}{x^6 + 1}\, dx$

(c) $\displaystyle\int_{-\infty}^{\infty} \frac{\cos x}{x^2 - 4x + 5}\, dx$

3. Calculate the following integrals using contours as in Figure 5.4.

(a) $\displaystyle\int_{0}^{\infty} \frac{\sin 2x}{x(x^2 + 1)^2}\, dx$

(b) $\displaystyle\int_{0}^{\infty} \frac{\cos x - 1}{x^2}\, dx$

(c) $\displaystyle\int_{0}^{\infty} \frac{\ln x}{(x^2 + 1)^2}\, dx$

4. Evaluate the following integrals using contours as in Figure 5.6.

(a) $\displaystyle\int_{-\infty}^{\infty} \frac{e^{x/4}}{1 + e^x}\, dx$

(b) $\displaystyle\int_{-\infty}^{\infty} \frac{e^x}{1 + e^{6x}}\, dx$

5. Generalize the technique in Example 5.4.6 to evaluate the integral

$$\int_{0}^{\infty} \frac{dx}{x^a(1 + x)}$$

for any $a \in (0, 1)$.

6. Evaluate the integral

$$\int_{-\infty}^{\infty} \frac{\cos bx}{x^2 + a^2}\, dx,$$

for any $a, b > 0$.

7. Let $a, b > 0$. Evaluate the integral

$$\int_{-\infty}^{\infty} \frac{\sin ax}{x(x^2 + b^2)}\, dx.$$

8. Let $n \in \mathbb{N}$. For $R > 1$, let C_R be the wedge contour formed by summing the segment $[0, R]$, the circular arc $\{Re^{i\theta} : 0 \leq \theta \leq \pi/n\}$, and the segment $[Re^{i\pi/n}, 0]$. Use these contours to evaluate the integral

$$\int_{-\infty}^{\infty} \frac{dx}{1 + x^{2n}}.$$

9. Use an indented version of the contour in Exercise 8 to evaluate

$$\int_{0}^{\infty} \frac{\sin^2(x^n)}{x^{n+1}}\, dx$$

for all $n \in \mathbb{N}$. (*Hint:* It is useful to note that

$$\sin^2 \theta = \frac{1 - e^{2i\theta}}{4} + \frac{1 - e^{-2i\theta}}{4}$$

for all $\theta \in \mathbb{R}$.)

10. Let $n \in \mathbb{N}$ with $n \geq 2$ and $a > 0$. Use contours as in Figure 5.4 to evaluate

$$\int_0^\infty \frac{\sqrt[n]{x}}{x^2 + a^2}\, dx.$$

(*Hint*: Wisely choose a branch of the power function $f(z) = z^{1/n}$.)

11. Answer the following.

(a) Let f be analytic on $\mathbb{C}$ except for the isolated singularities $a_1, \ldots, a_n \in \mathbb{C} \setminus [0, \infty)$. Use the technique of Example 5.4.6 to show that if

$$\lim_{R \to \infty} R(\ln R) \sup\{|f(z)| : |z| = R\} = 0,$$

then

$$\int_0^\infty f(x)\, dx = -\sum_{k=1}^n \operatorname*{Res}_{z=a_k} f(z) \log z,$$

where the branch of the logarithm corresponds to $0 < \arg z < 2\pi$.

(b) Prove that if $f = P/Q$, where P and Q are polynomials, $\deg Q \geq 2 + \deg P$, and $Z(Q) \cap [0, \infty) = \varnothing$, then f satisfies the conditions in part (a).

(c) Evaluate the following integrals. How would these integrals have to be considered using calculus methods? (*Hint*: The last one is tricky. Recognize the sum of residues as the derivative of a certain telescoping sum.)

i. $\displaystyle\int_0^\infty \frac{dx}{x^2 + 3x + 2}$

ii. $\displaystyle\int_0^\infty \frac{x^2}{(x+1)^2(x^2+1)}\, dx$

iii. $\displaystyle\int_0^\infty \frac{dx}{1 + x^n}$, $n \in \mathbb{N}, n \geq 2$ (Compare with Exercise 8.)

12. Prove Lemma 5.4.2. (*Hint*: First consider the series $\sum_{n=1}^\infty \int_{n-1}^n f(t)\, dt$.)

5.5   The Laplace Transform

It is common in a first course in differential equations to encounter the Laplace transform of a real function. Due to the algebraic relationship between the transform of a function and the transform of its derivatives, applying the Laplace transform to an ordinary differential equation results in an algebraic equation more easily solved.

Our purpose is not to study the Laplace transform for the sake of solving differential equations per se. Instead, we shall see how the theory of analytic functions impacts the theory of the Laplace transform. The only method available to find the inverse transform of a function when one first encounters it, and hence to complete the process of solving a differential equation as described above, is to manipulate the function until it is recognizable as a combination of known transforms, usually consulting a table. We will use residues to find a formula for the inverse of the Laplace transform in certain (commonly occurring) cases.

5.5.1 Definition. The *Laplace transform* of the complex-valued piecewise continuous function f on $[0, \infty)$ is the function

$$F(z) = \int_0^\infty e^{-zt} f(t)\, dt, \tag{5.5.1}$$

defined for all $z \in \mathbb{C}$ for which the integral exists. This relationship is often denoted $F = \mathcal{L}f$.

The proof of the following linearity property is left as an exercise.

5.5.2 Theorem. *Let f and g be complex-valued piecewise continuous functions on $[0, \infty)$ whose Laplace transforms are each defined for certain common values $z \in \mathbb{C}$. Then $[\mathcal{L}(f + g)](z)$ exists for all z such that both $[\mathcal{L}f](z)$ and $[\mathcal{L}g](z)$ exist and*

$$[\mathcal{L}(af + bg)](z) = a[\mathcal{L}f](z) + b[\mathcal{L}g](z)$$

for all such z and constants $a, b \in \mathbb{C}$.

Initially, we should find some conditions on a function f to guarantee the existence of its Laplace transform. We start with the following definition.

5.5.3 Definition. Let f be complex valued and piecewise continuous on $[0, \infty)$, and let $\alpha \in \mathbb{R}$. Then f is of *exponential order* α if there exists $M \geq 0$ such that

$$|f(t)| \leq M e^{\alpha t} \tag{5.5.2}$$

for all $t \in [0, \infty)$ at which f is defined.

For $\alpha \in \mathbb{R}$, let us denote right half-planes by

$$H_\alpha = \{z \in \mathbb{C} : \operatorname{Re} z > \alpha\}$$

for the remainder of this section.

5.5.4 Theorem. *Let f be a complex-valued piecewise continuous function on $[0, \infty)$ of exponential order $\alpha \in \mathbb{R}$. Then $F = \mathcal{L}f$ is defined and analytic on H_α.*

Proof. Let $z \in H_\alpha$, and set $x = \operatorname{Re} z$. Suppose $M \geq 0$ is such that $|f(t)| \leq M e^{\alpha t}$ for all $t \in [0, \infty)$ at which f is defined. Then

$$\int_0^\infty |e^{-zt} f(t)|\, dt \leq \int_0^\infty M e^{(\alpha - x)t}\, dt,$$

and the convergence of the right-hand side implies the convergence of the left-hand side. Therefore F is defined on H_α by Lemma 5.4.2.

Let $n \in \mathbb{N}$. As in Definition 2.8.2, partition $[0, n]$ by $0 = t_0 < \cdots < t_m = n$ such that f can be extended continuously to $[t_{j-1}, t_j]$ for each $j = 1, \ldots, m$. Then for every j, $|f|$ is bounded by some $C_j > 0$. Following (2.8.4), we define $F_n \colon \mathbb{C} \to \mathbb{C}$ by

$$F_n(z) = \int_0^n e^{-zt} f(t)\, dt = \sum_{j=1}^m \int_{t_{j-1}}^{t_j} \sum_{k=0}^\infty \frac{(-1)^k z^k t^k}{k!} f(t)\, dt. \tag{5.5.3}$$

For a given $z \in \mathbb{C}$ and $j = 1, \ldots, m$,

$$\left| \frac{(-1)^k z^k t^k}{k!} f(t) \right| \leq \frac{|z|^k n^k}{k!} C_j$$

is satisfied by the extension of f to $[t_{j-1}, t_j]$. The right-hand side contains terms of a convergent (exponential) series (with index k), and therefore the series in the integrands on the right-hand side of (5.5.3) converge uniformly for $t \in [t_{j-1}, t_j]$ for each j and z by the Weierstrass M-test. Theorem 2.9.19 allows us to write

$$F_n(z) = \sum_{j=1}^{m} \left(\sum_{k=0}^{\infty} \left[\frac{(-1)^k}{k!} \int_{t_{j-1}}^{t_j} t^k f(t) \, dt \right] z^k \right), \qquad z \in \mathbb{C}.$$

We have expressed F_n as a finite sum of entire functions, hence F_n is entire.

Let $K \subseteq H_\alpha$ be compact, and set $\beta = \inf\{\operatorname{Re} z : z \in K\}$. Then $\beta > \alpha$ because K is compact. If $z \in K$, then $|e^{-zt}| \leq e^{-\beta t}$. For a given $\varepsilon > 0$, there is $N \in \mathbb{N}$ such that for all $n \geq N$, $e^{(\alpha - \beta)n} < \varepsilon(\beta - \alpha)/M$. Thus for all $z \in K$ and $n \geq N$,

$$|F(z) - F_n(z)| \leq \int_n^\infty |e^{-zt} f(t)| \, dt \leq \int_n^\infty M e^{(\alpha - \beta)t} \, dt = \frac{M e^{(\alpha - \beta)n}}{\beta - \alpha} < \varepsilon.$$

Therefore $F_n \to F$ uniformly on K. We conclude that F is analytic on H_α by Exercise 5 in Section 4.3. □

Let us consider a couple of fundamental examples.

5.5.5 Example. Let $a \in \mathbb{C}$, and set $\alpha = \operatorname{Re} a$. Define $f \colon [0, \infty) \to \mathbb{C}$ by $f(t) = e^{at}$. For $z \in H_\alpha$, let $x = \operatorname{Re} z$. Then

$$|e^{(a-z)t}| = e^{(\alpha - x)t} \to 0$$

as $t \to \infty$. Therefore $F = \mathcal{L}f$ exists on H_α and is given by

$$F(z) = \int_0^\infty e^{-zt} e^{at} \, dt = \int_0^\infty e^{(a-z)t} \, dt = \frac{1}{z - a}.$$

5.5.6 Example. For $n = 0, 1, \ldots$, let $f_n \colon [0, \infty) \to \mathbb{C}$ be given by $f_n(t) = t^n$. If $z \in H_0$ and $x = \operatorname{Re} z$, then successive applications of l'Hôpital's rule yield

$$\left| \frac{t^n e^{-zt}}{z} \right| = \frac{t^n e^{-xt}}{|z|} \to 0$$

as $t \to \infty$. Therefore, we may integrate by parts to see that

$$[\mathcal{L} f_n](z) = \int_0^\infty e^{-zt} t^n \, dt = \frac{n}{z} \int_0^\infty e^{-zt} t^{n-1} \, dt = \frac{n}{z} [\mathcal{L} f_{n-1}](z)$$

for all $n \in \mathbb{N}$ and $z \in H_0$ where $\mathcal{L} f_{n-1}$ exists. By Example 5.5.5, $[\mathcal{L} f_0](z) = 1/z$ for all $z \in H_0$. Therefore induction gives

$$[\mathcal{L} f_n](z) = \frac{n!}{z^{n+1}}.$$

for all $z \in H_0$.

If we combine Examples 5.5.5 and 5.5.6 with Theorem 5.5.2, we see that the Laplace transform F of any linear combination of exponential functions and polynomials (and thus sines, cosines, and hyperbolic sines and cosines) exists on a half-plane and is equal to a rational function with denominator of greater degree than numerator. Any function F of this type can be defined on all of $\mathbb{C}$ except possibly for a finite number of isolated singularities. (We think of this as an *analytic continuation* from the half-plane.) Further, $\lim_{z \to \infty} F(z) = 0$. As an example of the power of the residue calculus, we shall show how to calculate the inverse transform of functions with these characteristics.

We should remark that we have given no reason to assume that an inverse is well defined. Indeed, the following theorem guarantees that for a given analytic F with certain properties, there is a function f whose Laplace transform is equal to F on a half-plane. For f to be considered *the* inverse Laplace transform of F, it should be known that there is no other function g (up to a certain point) that has Laplace transform F. It is the case that g would have to equal f wherever they are both continuous. Specifically, f is unique if continuity is required. A proof of this is difficult and is developed in one of the exercises.

5.5.7 Theorem. *Let F be analytic on $\mathbb{C}$ except for a finite number of distinct isolated singularities $a_1, \ldots, a_n \in \mathbb{C}$, and suppose $\lim_{z \to \infty} F(z) = 0$. Let $\alpha = \max\{\operatorname{Re} a_k : k = 1, \ldots, n\}$. Then the function $f \colon [0, \infty) \to \mathbb{C}$ given by*

$$f(t) = \sum_{k=1}^{n} \operatorname*{Res}_{z=a_k} e^{zt} F(z)$$

is continuous and satisfies $\mathcal{L}f = F$ on H_α.

Proof. Fix $z \in H_\alpha$, $x = \operatorname{Re} z$, and $\beta \in (\alpha, x)$. Choose $R > 0$ arbitrarily so that $z, a_1, \ldots, a_n \in D(\beta; R)$. Set $C_R = \partial D(\beta; R)$. Further, let C_R^1 be the left side of C_R together with the segment $L_R = [\beta - iR, \beta + iR]$ and C_R^2 be $-L_R$ together with the right side of C_R, both parameterized positively. Then $C_R = C_R^1 + C_R^2$. Moreover, the singularities $a_1, \ldots, a_n$ lie inside of C_R^1 and z lies inside of C_R^2. (See Figure 5.7.)

By the residue theorem,

$$f(t) = \frac{1}{2\pi i} \int_{C_R^1} e^{\zeta t} F(\zeta) \, d\zeta.$$

It follows from Leibniz's rule (see Appendix B) that f is continuous. For $T > 0$, calculate

$$\int_0^T e^{-zt} f(t) \, dt = \frac{1}{2\pi i} \int_0^T e^{-zt} \int_{C_R^1} e^{\zeta t} F(\zeta) \, d\zeta \, dt$$

$$= \frac{1}{2\pi i} \int_{C_R^1} F(\zeta) \int_0^T e^{(\zeta - z)t} \, dt \, d\zeta$$

$$= \frac{1}{2\pi i} \int_{C_R^1} \frac{(e^{(\zeta - z)T} - 1)F(\zeta)}{\zeta - z} \, d\zeta$$

using Fubini's theorem. (See Exercise 16 in Section 2.9.) For all $\zeta \in C_R^1$, $\operatorname{Re}\zeta \leq \beta < x$, and hence

$$\left| \frac{e^{(\zeta - z)T}}{\zeta - z} \right| \leq \frac{e^{(\beta - x)T}}{x - \beta} \to 0$$

as $T \to \infty$. There is some $M \geq 0$ such that $|F(\zeta)| \leq M$ for all $\zeta \in C_R^1$. Thus

$$\left| \int_{C_R^1} \frac{e^{(\zeta - z)T} F(\zeta)}{\zeta - z} \, d\zeta \right| \leq M \int_{C_R^1} \left| \frac{e^{(\zeta - z)T}}{\zeta - z} \right| \, |d\zeta| \to 0$$

as $T \to \infty$. It is therefore evident that

$$\int_0^\infty e^{-zt} f(t) \, dt = \lim_{T \to \infty} \frac{1}{2\pi i} \int_{C_R^1} \frac{(e^{(\zeta - z)T} - 1)F(\zeta)}{\zeta - z} \, d\zeta$$

$$= -\frac{1}{2\pi i} \int_{C_R^1} \frac{F(\zeta)}{\zeta - z} \, d\zeta.$$

Rewriting using the Cauchy integral formula, we have

$$2\pi i [\mathcal{L}f](z) = \int_{C_R^2} \frac{F(\zeta)}{\zeta - z} \, d\zeta - \int_{C_R} \frac{F(\zeta)}{\zeta - z} \, d\zeta = 2\pi i F(z) - \int_{C_R} \frac{F(\zeta)}{\zeta - z} \, d\zeta.$$

The proof will then be complete once it is shown that

$$\lim_{R \to \infty} \int_{C_R} \frac{F(\zeta)}{\zeta - z} \, d\zeta = 0.$$

To that end, let $M_R = \sup\{|F(\zeta)| : \zeta \in C_R\}$. Since $F(\zeta) \to 0$ as $\zeta \to \infty$, we have $M_R \to 0$ as $R \to \infty$. Note that for all $\zeta \in C_R$,

$$|\zeta - z| \geq |\zeta| - |z| \geq R - |\beta| - |z|.$$

Then for $R > |\beta| + |z|$,

$$\left| \int_{C_R} \frac{F(\zeta)}{\zeta - z} \, d\zeta \right| \leq \int_{C_R} \frac{|F(\zeta)|}{|\zeta - z|} \, |d\zeta| \leq \left(\frac{2\pi R}{R - |\beta| - |z|} \right) M_R.$$

The expression in parentheses in the above line tends to 2π as $R \to \infty$, giving the result. $\qquad \square$

The following is a corollary to our reasoning in the above proof.

5.5.8 Corollary. *Assume the hypotheses and notation developed in the statement and proof of Theorem 5.5.7. For all $t > 0$,*

$$f(t) = \frac{1}{2\pi i} \lim_{R \to \infty} \int_{L_R} e^{zt} F(z) \, dz = \frac{1}{2\pi i} \operatorname{PV} \int_{\beta - i\infty}^{\beta + i\infty} e^{zt} F(z) \, dz. \qquad (5.5.4)$$

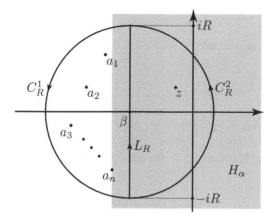

Figure 5.7   Constructions used the the proof of Theorem 5.5.7

The unfortunate notation for the final expression in (5.5.4) is common for the *Bromwich integral.* The PV stands for *principal value* and is used to denote values of improper (often divergent) integrals by taking the limit of $\int_{-R}^{R} f(t)\,dt$ as $R \to \infty$. Consider the expression to be defined by the one preceding it.

Proof. Let $t > 0$. For R as in the proof of Theorem 5.5.7, define S_R to be the left side of $\partial D(\beta; R)$ parameterized in the counterclockwise direction. Then

$$2\pi i f(t) = \int_{C_R^1} e^{zt} F(z)\,dz = \int_{L_R} e^{zt} F(z)\,dz + \int_{S_R} e^{zt} F(z)\,dz.$$

We must then show that

$$\lim_{R \to \infty} \int_{S_R} e^{zt} F(z)\,dz = 0.$$

Using a technique similar to that in Example 5.4.5, we calculate

$$\left| \int_{S_R} e^{zt} F(z)\,dz \right| = \left| \int_{\pi/2}^{3\pi/2} iRe^{i\theta} e^{(\beta + Re^{i\theta})t} F(\beta + Re^{i\theta})\,d\theta \right|$$

$$\leq M_R Re^{\beta t} \int_{\pi/2}^{3\pi/2} e^{Rt \cos\theta}\,d\theta$$

$$= 2M_R Re^{\beta t} \int_{0}^{\pi/2} e^{-Rt \sin\theta}\,d\theta$$

$$\leq 2M_R Re^{\beta t} \int_{0}^{\pi/2} e^{-2Rt\theta/\pi}\,d\theta$$

$$= \frac{M_R \pi e^{\beta t}}{t}(1 - e^{-Rt}).$$

This tends to 0 as $R \to \infty$. $\square$

We conclude with an example.

5.5.9 Example. Let

$$F(z) = \frac{2z - 1}{z^3 + z^2 + z + 1}.$$

The denominator factors to $(z + i)(z - i)(z + 1)$, and hence F has simple poles at i, $-i$, and -1. Further, $\lim_{z \to \infty} F(z) = 0$, and hence we can calculate f such that $\mathcal{L}f = F$ on H_0 using Theorem 5.5.7. We must compute the residues of $e^{zt} F(z)$ at the singularities.

$$\operatorname*{Res}_{z=i} e^{zt} F(z) = \lim_{z \to i} \frac{e^{zt}(2z - 1)}{(z + i)(z + 1)} = \frac{e^{it}(2i - 1)}{2i(i + 1)} = \frac{e^{it}(3 - i)}{4}$$

$$\operatorname*{Res}_{z=-i} e^{zt} F(z) = \lim_{z \to -i} \frac{e^{zt}(2z - 1)}{(z - i)(z + 1)} = \frac{e^{-it}(-2i - 1)}{-2i(-i + 1)} = \frac{e^{-it}(3 + i)}{4}$$

$$\operatorname*{Res}_{z=-1} e^{zt} F(z) = \lim_{z \to -1} \frac{e^{zt}(2z - 1)}{z^2 + 1} = \frac{-3e^{-t}}{2}$$

Hence

$$f(t) = \frac{e^{it}(3 - i)}{4} + \frac{e^{-it}(3 + i)}{4} - \frac{3e^{-t}}{2} = \frac{3}{2} \cos t + \frac{1}{2} \sin t - \frac{3}{2} e^{-t}.$$

Summary and Notes for Section 5.5.

The Laplace transform is an example of an *integral transform*. By applying it to both sides of an ordinary differential equation, one obtains an algebraic equation that is likely much easier to solve. The solution of the differential equation is then the inverse Laplace transform of the solution to the algebraic equation. In typical differential equations texts, tables are made of the Laplace transforms of many common functions so that the inverse transform can be found through rearrangement and recognition. However, we now see that the inverse transform can be directly calculated under certain conditions using the residue theorem, a much more satisfying approach.

Laplace developed his transform in the late 18th century, motivated by his work in probability and inspired by earlier work of Euler and Joseph Louis Lagrange.

Exercises for Section 5.5.

1. In each of the following, a function $f \colon [0, \infty) \to \mathbb{C}$ is given. Calculate its Laplace transform. Include the transform's domain.

   (a) $f(t) = \cos t$

   (b) $f(t) = t - ie^{it} + \sin 2it.$

   (c) $f(t) = te^{it}$

(d) $f(t) = \begin{cases} e^{it} & \text{if } 0 \le t \le 2\pi, \\ 0 & \text{if } t > 2\pi \end{cases}$

2. In each of the following, a function F is given that is analytic on $\mathbb{C}$ except for a finite number of isolated singularities and $\lim_{z \to \infty} F(z) = 0$. Find a function $f : [0, \infty) \to \mathbb{C}$ such that $F = \mathcal{L}f$ on a right half-plane.

   (a) $F(z) = \dfrac{z^3}{z^4 - 1}$

   (b) $F(z) = \dfrac{z^2 - 4}{(z^2 + 4)^2}$

   (c) $F(z) = \dfrac{z + 2}{z^3 + 2iz^2 + 3z}$

   (d) $F(z) = \dfrac{z^2}{(z^2 + 1)^3}$

3. Suppose that $f : [0, \infty) \to \mathbb{C}$ is continuous, piecewise continuously differentiable, and of exponential order $\alpha \in \mathbb{R}$. Show that the Laplace transform of f' exists on H_α and is given by

$$[\mathcal{L}(f')](z) = z[\mathcal{L}f](z) - f(0).$$

4. Prove Theorem 5.5.2.

5. Solve the given initial value problems using the following strategy: Take the Laplace transform of both sides of the equation, use Theorem 5.5.2, Exercise 3, and the initial conditions to solve for the Laplace transform of y, and then find y using Theorem 5.5.7. Express the answer in a complex-free form.

   (a) $y'' + 4y' + 3y = \cos t$, $y(0) = -1$, $y'(0) = 4$

   (b) $y'' + 9y = t$, $y(0) = 1$, $y'(0) = -2$

6. Let f be a complex-valued piecewise continuous function of exponential order $\alpha \in \mathbb{R}$ with Laplace transform F on H_α. Show that

$$F'(z) = -\int_0^\infty te^{-zt} f(t) \, dt.$$

   (*Hint*: Use the sequence $\{F_n\}$ from the proof of Theorem 5.5.4, the Cauchy integral formula, and Fubini's theorem.)

7. Let $p, q : \mathbb{C} \to \mathbb{C}$ be polynomials of degrees m and n, respectively, where $m < n$. Further, suppose that q can be written

$$q(z) = c \prod_{k=1}^{n} (z - a_k),$$

   for all $z \in \mathbb{C}$, where $c \in \mathbb{C} \setminus \{0\}$ and $a_1, \ldots, a_n \in \mathbb{C}$ are distinct. If $F = p/q$, show that the function $f : [0, \infty) \to \mathbb{C}$ given by

$$f(t) = \sum_{k=1}^{n} \frac{e^{a_k t} p(a_k)}{q'(a_k)}$$

   satisfies $\mathcal{L}f = F$ on a right half-plane.

8. Use the following steps to show that if $f, g \colon [0, \infty) \to \mathbb{C}$ are continuous and of exponential order $\alpha \in \mathbb{R}$, and if $\mathcal{L}f = \mathcal{L}g$, then $f = g$.

(a) The *Weierstrass approximation theorem* from real analysis states (with $\mathbb{R}$ in place of $\mathbb{C}$): If $\varphi \colon [a, b] \to \mathbb{C}$ is continuous for some closed interval $[a, b] \subseteq \mathbb{R}$, then there is a sequence of polynomials $p_n \colon [a, b] \to \mathbb{C}$ such that $p_n \to \varphi$ uniformly on $[a, b]$. (The proof of this theorem is the topic of Exercise 12 in Section 6.4.) Use this to prove that if $f \colon [0, 1] \to \mathbb{C}$ is continuous and

$$\int_0^1 x^n f(x) \, dx = 0$$

for all $n = 0, 1, \ldots$, then $f(x) = 0$ for all $x \in [0, 1]$.

(b) Let $f \colon [0, \infty) \to \mathbb{C}$ be continuous and suppose that $F = \mathcal{L}f$ is defined in a half-plane H_α for some $\alpha \in \mathbb{R}$. Fix $x > \alpha + 1$, and for $n = 0, 1, \ldots$, let $z_n = x + n$. Use the substitution $s = e^{-t}$ to show that

$$F(z_n) = \int_0^1 s^n h(s) \, ds,$$

where $h(s) = s^{x-1} f(-\ln s)$ for $s \in (0, 1]$.

(c) Let $f \colon [0, \infty) \to \mathbb{C}$ be continuous and of exponential order α. Prove that if $\mathcal{L}f$ is equal to zero on H_α, then f is 0.

(d) Conclude the desired result.

CHAPTER 6

HARMONIC FUNCTIONS AND FOURIER SERIES

Harmonic functions are abundant in certain areas of applied mathematics and are worthy objects of study for their own sake. We will see that in the plane, these functions are intimately related to analytic functions, a fact we will repeatedly exploit to develop properties of each type of function using characteristics of the other. Finally, through harmonic functions, we will consider Fourier series as motivated from a complex-analytic point of view.

6.1 Harmonic Functions

In this chapter, we are mainly concerned with *real-valued* functions defined on domains in $\mathbb{C}$ (or equivalently $\mathbb{R}^2$). While the reader may wonder what purpose this has in a course on complex analysis, we recall that we are considering functions of a complex *variable* and note that many properties of analytic functions will be applied here to study this important new family of functions.

As was done in Sections 3.6 and 3.7, we will freely move between the forms $z = x + iy$ and $z = (x, y)$ of a complex number $z \in \mathbb{C}$ and accordingly think of subsets of $\mathbb{C}$ as also lying in $\mathbb{R}^2$.

Complex Analysis: A Modern First Course in Function Theory, First Edition. Jerry R. Muir, Jr.
© 2015 John Wiley & Sons, Inc. Published 2015 by John Wiley & Sons, Inc.

We begin by defining harmonic functions and follow with a theorem showing their connection to analytic functions.

6.1.1 Definition. Let $\Omega \subseteq \mathbb{C}$ be open. A function $u \colon \Omega \to \mathbb{R}$ whose second partial derivatives exist and are continuous on Ω is *harmonic* if it satisfies *Laplace's equation*

$$\frac{\partial^2 u}{\partial x^2} + \frac{\partial^2 u}{\partial y^2} = 0 \tag{6.1.1}$$

everywhere on Ω.

We note that it is common to denote Laplace's equation using a differential operator Δ called the *Laplacian*. This is defined for functions $u \colon \Omega \to \mathbb{R}$ with continuous second partial derivatives by

$$\Delta u = \frac{\partial^2 u}{\partial x^2} + \frac{\partial^2 u}{\partial y^2}, \tag{6.1.2}$$

and Laplace's equation becomes $\Delta u = 0$. We see that the set of harmonic functions is the kernel (nullspace) of the linear transformation Δ. (Compare this to Exercise 6 in Section 3.6.)

6.1.2 Theorem. *Let $\Omega \subseteq \mathbb{C}$ be open and $f \in H(\Omega)$. Then $u = \operatorname{Re} f$ and $v = \operatorname{Im} f$ are harmonic functions on Ω.*

Proof. If $f = u + iv$, then (3.6.1) and (3.6.2) allow us to write

$$f' = \frac{\partial u}{\partial x} + i\frac{\partial v}{\partial x} = \frac{\partial v}{\partial y} - i\frac{\partial u}{\partial y}. \tag{6.1.3}$$

Expressing f'' in terms of the partial derivatives of both representations of f' in the same manner shows that all second partial derivatives of u and v exist. Furthermore, continuity of f'' implies the continuity of these second partial derivatives.

Differentiating the Cauchy–Riemann equations (3.6.3) with respect to x and y, respectively, yields

$$\frac{\partial^2 u}{\partial x^2} = \frac{\partial^2 v}{\partial x\,\partial y}, \qquad \frac{\partial^2 u}{\partial y^2} = -\frac{\partial^2 v}{\partial y\,\partial x}.$$

It is well known from multivariable calculus that mixed partial derivatives are equal when they are continuous, and hence we can add these equations to obtain (6.1.1), showing that u is harmonic.

Since $v = \operatorname{Re}(-if)$, we see that it is also harmonic since $-if \in H(\Omega)$. □

An immediate question is whether the converse holds. That is, (when) is a harmonic function equal to the real part of an analytic function? (As we see from the above proof, the question of the imaginary part is redundant.) Before giving an answer, we consider the following example.

6.1.3 Example. Define $u\colon \mathbb{C} \setminus \{0\} \to \mathbb{R}$ by $u(z) = \ln|z|$. That u is harmonic follows from a direct verification of Laplace's equation. An argument could also be made using Theorem 6.1.2 with a bit of care. If $\Omega \subseteq \mathbb{C} \setminus \{0\}$ is the domain of a branch of the logarithm $f(z) = \log z$, then $u(z) = \operatorname{Re} f(z)$ for all $z \in \Omega$. This shows that u is harmonic on Ω, and, since Ω was chosen arbitrarily, u is harmonic on all of $\mathbb{C} \setminus \{0\}$.

Closer inspection shows that u is *not* equal to the real part of a function $g \in H(\mathbb{C} \setminus \{0\})$ for if it were, then with f and Ω chosen as above, we would have $\operatorname{Re} h(z) = 0$ for all $z \in \Omega$, where $h = f - g$. Thus $h(\Omega)$ is a subset of the imaginary axis, and by the open mapping theorem (Theorem 3.5.1), we see that h must be constant. It would follow that, since f is equal to g plus a constant, f could be extended to be analytic on all of $\mathbb{C} \setminus \{0\}$, a contradiction.

We will now see that the desired result does hold, provided that we restrict the type of domain in question. See Exercise 3 in Section 4.3 for the necessary background on simply connected domains.

6.1.4 Theorem. *Let $\Omega \subseteq \mathbb{C}$ be a simply connected domain and $u\colon \Omega \to \mathbb{R}$ be harmonic. Then u is the real part of a function in $H(\Omega)$.*

Proof. Define $g\colon \Omega \to \mathbb{C}$ by

$$g = \frac{\partial u}{\partial x} - i\frac{\partial u}{\partial y}.$$

By Laplace's equation, we have

$$\frac{\partial}{\partial x}\left(\frac{\partial u}{\partial x}\right) = \frac{\partial^2 u}{\partial x^2} = -\frac{\partial^2 u}{\partial y^2} = \frac{\partial}{\partial y}\left(-\frac{\partial u}{\partial y}\right).$$

Using continuity of second mixed partial derivatives, we have

$$\frac{\partial}{\partial y}\left(\frac{\partial u}{\partial x}\right) = \frac{\partial^2 u}{\partial y\, \partial x} = \frac{\partial^2 u}{\partial x\, \partial y} = -\frac{\partial}{\partial x}\left(-\frac{\partial u}{\partial y}\right).$$

We therefore see that g satisfies the Cauchy–Riemann equations, and hence $g \in H(\Omega)$.

It follows from Exercise 3 in Section 4.3 that there is an $f \in H(\Omega)$ such that $f' = g$. Setting $U = \operatorname{Re} f$ and using (6.1.3), we see that

$$f' = \frac{\partial U}{\partial x} - i\frac{\partial U}{\partial y}.$$

Setting real and imaginary parts of f' and g equal gives that u and U are each differentiable functions of two real variables whose partial derivatives are equal. It follows that u and U must differ by a constant $c \in \mathbb{R}$, and hence $u = \operatorname{Re}(f + c)$. $\qquad\square$

It is important to realize that all is not lost if the domain $\Omega \subseteq \mathbb{C}$ of a harmonic function u is not simply connected, for it is still the case that u is the real part of an analytic function on any given disk in Ω.

A key to the above proof is that two differentiable functions of two real variables whose first partial derivatives are equal on a domain differ by a real constant on that domain. This is certainly a well-known fact from calculus for functions of one variable, and a proof using the one-variable result is outlined in the exercises.

The proof of the following corollary is left as an exercise.

6.1.5 Corollary. *Let $\Omega \subseteq \mathbb{C}$ be open and $u \colon \Omega \to \mathbb{R}$ be harmonic. Then all partial derivatives of u of all orders exist and are continuous.*

Theorem 6.1.4 prompts the following definition.

6.1.6 Definition. Let $\Omega \subseteq \mathbb{C}$ be open and $u \colon \Omega \to \mathbb{R}$ be harmonic. Any harmonic function $v \colon \Omega \to \mathbb{R}$ such that $u + iv \in H(\Omega)$ is called a *harmonic conjugate* of u.

Theorem 6.1.4 gives that any harmonic function on a simply connected domain has a harmonic conjugate, but the proof does not give a convenient recipe for constructing one. Here, we do so when the domain is a disk.

6.1.7 Theorem. *Let $a = x_0 + iy_0 \in \mathbb{C}$, $r > 0$, and $u \colon D(a; r) \to \mathbb{R}$ be harmonic. Then the function $v \colon D(a; r) \to \mathbb{R}$ given by*

$$v(x, y) = \int_{y_0}^{y} \frac{\partial u}{\partial x}(x, t) \, dt - \int_{x_0}^{x} \frac{\partial u}{\partial y}(s, y_0) \, ds$$

is a harmonic conjugate of u.

Proof. Due to the geometry of $D(a; r)$, the function v is well defined. We may differentiate v with respect to x using Leibniz's rule (Theorem B.16) for the first integral (because $\partial^2 u / \partial x^2$ is continuous) and the fundamental theorem of calculus for the second. As a result, we obtain

$$\begin{aligned}
\frac{\partial v}{\partial x}(x, y) &= \int_{y_0}^{y} \frac{\partial^2 u}{\partial x^2}(x, t) \, dt - \frac{\partial u}{\partial y}(x, y_0) \\
&= -\int_{y_0}^{y} \frac{\partial^2 u}{\partial y^2}(x, t) \, dt - \frac{\partial u}{\partial y}(x, y_0) \\
&= -\frac{\partial u}{\partial y}(x, t) \Big]_{t=y_0}^{t=y} - \frac{\partial u}{\partial y}(x, y_0) \\
&= -\frac{\partial u}{\partial y}(x, y)
\end{aligned}$$

for all $(x, y) \in D(a; r)$, using Laplace's equation. Because the second integral in the definition of v is constant in y, we see that

$$\frac{\partial v}{\partial y}(x, y) = \frac{\partial u}{\partial x}(x, y).$$

We have therefore verified the Cauchy–Riemann equations for the function $u + iv$, as needed. $\square$

6.1.8 Example. Let $u: \mathbb{C} \to \mathbb{R}$ be given by $u(x, y) = x^3 - 3xy^2 + 2y$. A direct use of Laplace's equation verifies that u is harmonic. Since u is harmonic on any disk $D(0; r)$, $r > 0$, we can use Theorem 6.1.7 to find a harmonic conjugate of u valid on each $D(0; r)$, and hence on $\mathbb{C}$. To that end, noting

$$\frac{\partial u}{\partial x}(x, y) = 3x^2 - 3y^2, \qquad \frac{\partial u}{\partial y}(x, y) = 2 - 6xy,$$

we have the harmonic conjugate

$$v(x, y) = \int_0^y (3x^2 - 3t^2)\, dt - \int_0^x 2\, ds = 3x^2 y - y^3 - 2x.$$

Now that we understand the relationship between harmonic functions and analytic functions, we present our first theorem that gives a result about harmonic functions using analytic functions as a tool. Note that parts (a) and (b) imply that the set of harmonic functions on a domain is a *vector space* over $\mathbb{R}$.

6.1.9 Theorem. *Let $\Omega \subseteq \mathbb{C}$ be open.*

(a) *If $u: \Omega \to \mathbb{R}$ is harmonic and $c \in \mathbb{R}$, then cu is harmonic.*

(b) *If $u, v: \Omega \to \mathbb{R}$ are harmonic, then $u + v$ is harmonic.*

(c) *If $f \in H(\Omega)$, $f(\Omega)$ is open, and $u: f(\Omega) \to \mathbb{R}$ is harmonic, then $u \circ f$ is harmonic.*

Proof. Parts (a) and (b) can be directly verified from Laplace's equation.

Let $a \in \Omega$. There is $r > 0$ such that $D(f(a); r) \subseteq f(\Omega)$. By Theorem 6.1.4, there is $g \in H(D(f(a); r))$ such that $u(w) = \operatorname{Re} g(w)$ for all $w \in D(f(a); r)$. By continuity (see Theorem 2.1.9), $V = f^{-1}(D(f(a); r)) \subseteq \Omega$ is open, $g \circ f$ is analytic on V, and

$$(u \circ f)(z) = u(f(z)) = \operatorname{Re} g(f(z)) = \operatorname{Re}(g \circ f)(z), \qquad z \in V.$$

By Theorem 6.1.2, $u \circ f$ is harmonic on V, an open set containing a. Since a was arbitrarily chosen, we see that $u \circ f$ is harmonic on Ω, giving (c). $\qquad\square$

6.1.10 Remark. There is a bit of a subtlety worth acknowledging in the preceding proof. In part (c), if we knew that $f(\Omega)$ was simply connected, then we could directly write $u = \operatorname{Re} g$ for $g \in H(f(\Omega))$ by Theorem 6.1.4. That $f(\Omega)$ is known only to be open requires the more complicated proof. This same issue is at play in the next theorem, a more impressive use of the relationship between harmonic and analytic functions. In this case, we need a strategy similar to that used in the proof of Theorem 3.4.5.

6.1.11 Maximum/Minimum Principle. *Let $\Omega \subseteq \mathbb{C}$ be a domain and $u: \Omega \to \mathbb{R}$ be harmonic.*

(a) *If there exists $a \in \Omega$ such that $u(a) \geq u(z)$ for all $z \in \Omega$, then u is constant.*

(b) *If there exists $b \in \Omega$ such that $u(b) \leq u(z)$ for all $z \in \Omega$, then u is constant.*

Proof. Since (b) follows from applying (a) to the function $-u$, we need only prove part (a). Let $\alpha = u(a)$, and set $E = \{z \in \Omega : u(z) = \alpha\}$. Since $a \in E$, $E \neq \varnothing$. The proof will be completed by showing $E = \Omega$.

Let $z \in E$ and $r > 0$ such that $D(z;r) \subseteq \Omega$. By Theorem 6.1.4, $u = \mathrm{Re}\, f$ on $D(z;r)$ for some $f \in H(D(z;r))$. If u is nonconstant on $D(z;r)$ then so is f, and hence $f(D(z;r))$ is open by the open mapping theorem. Thus we may choose $\varepsilon > 0$ such that $D(f(z);\varepsilon) \subseteq f(D(z;r))$. There is then some $w \in D(z;r)$ such that $f(w) = f(z)+\varepsilon/2$. Taking the real part of both sides yields $u(w) = \alpha+\varepsilon/2 > u(a)$, a contradiction. Hence u is constant on $D(z;r)$, showing $D(z;r) \subseteq E$. We conclude that E is open.

By continuity, the set $\Omega \setminus E = u^{-1}((-\infty, \alpha))$ is open. Theorem 1.4.16 then gives $\Omega \setminus E = \varnothing$. $\qquad\square$

We finish the section with a direct application of the Cauchy integral formula. Notice how the proof once again must accommodate that a harmonic function is the real part of an analytic function only on certain domains.

6.1.12 Definition. Let $\Omega \subseteq \mathbb{C}$ be open. A continuous function $\varphi \colon \Omega \to \mathbb{R}$ has the *mean value property* if for every $a \in \Omega$ and $r > 0$ such that $\overline{D}(a;r) \subseteq \Omega$,

$$\varphi(a) = \frac{1}{2\pi} \int_0^{2\pi} \varphi(a + re^{i\theta})\, d\theta. \tag{6.1.4}$$

6.1.13 Theorem. *Every harmonic function on an open set $\Omega \subseteq \mathbb{C}$ has the mean value property.*

Proof. Let $u \colon \Omega \to \mathbb{R}$ be harmonic, $a \in \Omega$, and $R > 0$ such that $\overline{D}(a;R) \subseteq \Omega$. By Theorem 6.1.4, there is $f \in H(D(a;R))$ such that $u(z) = \mathrm{Re}\, f(z)$ for $z \in D(a;R)$. For all $r \in (0, R)$, the Cauchy integral formula gives

$$f(a) = \frac{1}{2\pi i} \int_{\partial D(a;r)} \frac{f(z)}{z - a}\, dz = \frac{1}{2\pi} \int_0^{2\pi} f(a + re^{i\theta})\, d\theta,$$

by letting $z = a + re^{i\theta}$. Taking the real part of both sides gives

$$u(a) = \frac{1}{2\pi} \int_0^{2\pi} u(a + re^{i\theta})\, d\theta.$$

The result follows by taking the limit as $r \to R^-$. (See Exercise 2.) $\qquad\square$

Summary and Notes for Section 6.1.

By way of the Cauchy–Riemann equations, we see that the solutions to Laplace's equation (6.1.1) are, on simply connected domains (and hence locally), exactly those functions that are real parts of analytic functions. It would therefore be tempting, given our prior work on analytic functions, to simply define harmonic functions to be the real parts of analytic functions and proceed to study their properties. This approach, however, would shortchange the substantial theory of harmonic functions, of which we will only scratch the surface.

Indeed, the concept of a harmonic function as a solution to Laplace's equation can be defined on open sets in any Euclidean space $\mathbb{R}^n$. The close relationship with analytic functions when $n = 2$ explains our restriction to this case. Many results of this and later sections, including the maximum/minimum principle, the satisfaction of the mean value property, and the harmonic Liouville's theorem (Exercise 7) hold in the higher-dimensional setting, but must be proved by different methods since our proofs rely strongly on the analytic connection.

Laplace's equation itself has its origins in mathematical physics. Recall from multivariable calculus that a vector field (for example a gravitational or electrostatic field) on a region is called conservative if it is the gradient of a scalar-valued function on that region. (The negative of this scalar function is called a potential, and hence pursuit of these ideas leads one into the mathematical area known as potential theory.) When such a vector field is divergence free (such as when a gravitational field is free from attracting masses or an electrostatic field is free of charges), the corresponding potential satisfies Laplace's equation. (The Laplacian is the divergence of the gradient, and hence $\nabla^2 = \nabla \cdot \nabla$ is sometimes used in place of Δ to denote it.) A steady-state temperature function in a region also satisfies Laplace's equation.

Pierre Simon de Laplace, for whom Laplace's equation is named, showed the equation held for gravitational potentials related to celestial bodies in the late 1700s. Earlier in that century, in his work on hydrodynamics, Jean le Rond d'Alembert showed that the solutions to (what are now called) the Cauchy–Riemann equations satisfy Laplace's equation.

Laplace's equation is the canonical example of a homogeneous elliptic partial differential equation, and much time is spent on it and its nonhomogeneous counterpart, Poisson's equation, in an introductory study of partial differential equations.

We could easily have chosen $\mathbb{C}$ as the range of our harmonic functions (see Exercise 12), and the theory of complex-valued harmonic functions eventually diverges as a separate area of study. But for the results we give, either such functions are not appropriate (the maximum/minimum principle, for instance) or the result could be easily extended by passing through real and imaginary parts (the mean value property, for instance), and hence we chose the simpler range.

Exercises for Section 6.1.

1. For each of the following functions $u \colon \Omega \to \mathbb{R}, \Omega \subseteq \mathbb{C}$, use Laplace's equation to verify that u is harmonic and use Theorem 6.1.7 to find a harmonic conjugate of v on Ω.

   (a) $u(x, y) = y^2 - x^2 + 4xy^3 - 4x^3 y$, $\quad \Omega = \mathbb{C}$

   (b) $u(x, y) = \cos x \sinh y$, $\quad \Omega = \mathbb{C}$

   (c) $u(x, y) = \dfrac{y}{x^2 + y^2}$, $\quad \Omega = D(1; 1)$

2. ▷ Let $a \in \mathbb{C}, 0 \le r < R < \infty$, and $f: \overline{A}(a; r, R) \to \mathbb{C}$ be continuous. Prove that

$$\lim_{\rho \to R^-} \int_0^{2\pi} f(a + \rho e^{i\theta}) \, d\theta = \int_0^{2\pi} f(a + R e^{i\theta}) \, d\theta.$$

(*Hint*: Use uniform continuity and Theorem B.13.)

3. ▷ Let $\Omega \subseteq \mathbb{C}$ be open and $u: \Omega \to \mathbb{R}$ be harmonic. Show that if v is a harmonic conjugate of u on Ω, then $-u$ is a harmonic conjugate of v on Ω.

4. ▷ Let $\Omega \subseteq \mathbb{C}$ be a domain and $u: \Omega \to \mathbb{R}$ be harmonic with harmonic conjugate v. Prove that $V: \Omega \to \mathbb{R}$ is a harmonic conjugate of u if and only if $V = v + c$ for a constant $c \in \mathbb{R}$.

5. Note that, unlike for analytic functions in Theorem 3.3.1, products and quotients of harmonic functions are not addressed in Theorem 6.1.9. Is the product or quotient of harmonic functions necessarily harmonic? Give an affirmative proof or a counterexample for each.

6. Prove Corollary 6.1.5.

7. Prove *Liouville's theorem* for harmonic functions: If the harmonic function $u: \mathbb{C} \to \mathbb{R}$ is bounded above or below by a constant, then u is a constant function. (*Hint*: Apply the analytic Liouville's theorem to an entire function of the form $g(z) = e^{f(z)}$.)

8. To derive the formula given in Theorem 6.1.7, we can begin with the Cauchy–Riemann equation

$$\frac{\partial v}{\partial y}(x, t) = \frac{\partial u}{\partial x}(x, t)$$

and integrate both sides with respect to t to obtain

$$v(x, y) = \int_{y_0}^y \frac{\partial u}{\partial x}(x, t) \, dt + \varphi(x),$$

where φ is a differentiable function dependent only on x. Differentiation of both sides with respect to x and using the other Cauchy–Riemann equation allows us to solve for φ. Come up with an alternate formula for the harmonic conjugate by beginning with the other Cauchy–Riemann equation, justifying all steps.

9. Let $\Omega \subseteq \mathbb{C} \backslash \{0\}$ be open, and consider the function $u: \Omega \to \mathbb{R}$ in polar coordinates (r, θ) instead of rectangular coordinates (x, y). Show that Laplace's equation is equivalent to

$$\frac{\partial^2 u}{\partial r^2} + \frac{1}{r} \frac{\partial u}{\partial r} + \frac{1}{r^2} \frac{\partial^2 u}{\partial \theta^2} = 0.$$

10. Let $\Omega \subseteq \mathbb{C}$ be open. Show that if the continuous function $\varphi: \Omega \to \mathbb{R}$ has the mean value property, then for every $a \in \Omega$ and $r > 0$ such that $\overline{D}(a; r) \subseteq \Omega$,

$$\varphi(a) = \frac{1}{\pi r^2} \iint_{D(a;r)} \varphi(x, y) \, dA(x, y).$$

11. Recall the definition of the gradient from Exercise 5 of Section 3.6. Suppose $\Omega \subseteq \mathbb{C}$ is a simply connected domain and $u: \Omega \to \mathbb{R}$ is harmonic. Let

$$E = \{(x, y) \in \Omega : \nabla u(x, y) = (0, 0)\}.$$

Prove that if E has a limit point in Ω, then u is constant. Does the result hold if the domain is not assumed to be simply connected?

12. Let $\Omega \subseteq \mathbb{C}$ be open. A function $f \colon \Omega \to \mathbb{C}$ is *harmonic* provided that the functions $\operatorname{Re} f$ and $\operatorname{Im} f$ are harmonic. Show that if Ω is simply connected, then f is harmonic if and only if there exist $g, h \in H(\Omega)$ such that $f = h + \bar{g}$. (Observe the connection to Exercise 6 in Section 3.6.)

13. Let $\Omega \subseteq \mathbb{C}$ be a domain and $u, v \colon \Omega \to \mathbb{R}$ be differentiable functions of two real variables such that the respective first partial derivatives of u and v are equal on Ω. Let $\alpha, \beta \in \Omega$. By Exercise 18 in Section 2.9, there is a contour $\gamma \colon [a, b] \to \Omega$ such that $\gamma(a) = \alpha$ and $\gamma(b) = \beta$. Show that $u(\alpha) - v(\alpha) = u(\beta) - v(\beta)$ by applying the chain rule from multivariable calculus to the functions $u \circ \gamma$ and $v \circ \gamma$. Conclude that u and v must differ by a constant.

6.2 The Poisson Integral Formula

Our initial study of harmonic functions in the last section benefitted from the strong connection between harmonic and analytic functions. Notably, the objects most central to our study of analytic functions, series, were not mentioned. Here, we cash in that chip and get a huge payoff.

We desire to derive a useful formula in an uncomplicated manner, so we begin by making the assumption, upon which we will eventually improve, that $R > 1$ and $u \colon D(0; R) \to \mathbb{R}$ is harmonic. By Theorem 6.1.4, we know that $u = \operatorname{Re} f$ for some $f \in H(D(0; R))$. As usual, we write

$$f(z) = \sum_{n=0}^{\infty} c_n z^n,$$

noting that the series converges uniformly on compact subsets of $D(0; R)$, namely on $\overline{\mathbb{D}}$. We then may expand u as

$$u(z) = \frac{f(z) + \overline{f(z)}}{2} = \frac{1}{2} \left(\sum_{n=0}^{\infty} c_n z^n + \sum_{n=0}^{\infty} \bar{c}_n \bar{z}^n \right),$$

with each series converging uniformly on $\overline{\mathbb{D}}$. If we let $a_n = c_n/2$ and $a_{-n} = \bar{c}_n/2$ for $n \in \mathbb{N}$ and $a_0 = (c_0 + \bar{c}_0)/2 = \operatorname{Re} c_0$, then we have the improved polar form

$$u(re^{i\theta}) = \frac{1}{2} \left(\sum_{n=0}^{\infty} c_n r^n e^{in\theta} + \sum_{n=0}^{\infty} \bar{c}_n r^n e^{-in\theta} \right) = \sum_{n=-\infty}^{\infty} a_n r^{|n|} e^{in\theta}.$$

Again, we note that the doubly infinite series converges uniformly on $\overline{\mathbb{D}}$. (See Definition 5.1.2.)

We now look to recover a formula for the coefficients. One way would be to use the formulas for the coefficients c_n given in (3.2.7). A more beneficial (and simpler) strategy is to take advantage of the identity (2.8.5). Using uniform convergence to move the integral inside the sum, we obtain

$$\int_0^{2\pi} e^{-im\theta} u(e^{i\theta}) \, d\theta = \sum_{n=-\infty}^{\infty} a_n \int_0^{2\pi} e^{i(n-m)\theta} \, d\theta = 2\pi a_m$$

for $m \in \mathbb{Z}$. Hence

$$a_n = \frac{1}{2\pi} \int_0^{2\pi} e^{-in\theta} u(e^{i\theta}) \, d\theta, \qquad n \in \mathbb{Z}. \tag{6.2.1}$$

This gives

$$u(re^{i\theta}) = \sum_{n=-\infty}^{\infty} \left(\frac{1}{2\pi} \int_0^{2\pi} e^{-int} u(e^{it}) \, dt \right) r^{|n|} e^{in\theta}$$

$$= \frac{1}{2\pi} \int_0^{2\pi} \left[\sum_{n=-\infty}^{\infty} r^{|n|} e^{in(\theta-t)} \right] u(e^{it}) \, dt, \tag{6.2.2}$$

provided we can move the integral through the series. We can do so for $r \in [0, 1)$.

To justify (6.2.2), note that $|u(\zeta)| \leq M$ for all ζ in the compact set $\partial \mathbb{D}$ and some constant $M > 0$ by continuity. Therefore the series

$$\sum_{n=-\infty}^{\infty} r^{|n|} e^{in(\theta-t)} u(e^{it})$$

has terms bounded by $Mr^{|n|}$ and hence converges uniformly in t for fixed r by the Weierstrass M-test.

This motivates the following.

6.2.1 Definition. The family of 2π-periodic functions $P_r : \mathbb{R} \to \mathbb{R}$, $r \in [0, 1)$, given by

$$P_r(t) = \sum_{n=-\infty}^{\infty} r^{|n|} e^{int} \tag{6.2.3}$$

is called the *Poisson kernel*.

We are now one step removed from having proved the following theorem, which gives that (6.2.2) holds with a notable upgrade of hypotheses.

6.2.2 Poisson Integral Formula. *Let* $u : \overline{\mathbb{D}} \to \mathbb{R}$ *be continuous and such that* u *is harmonic on* $\mathbb{D}$. *Then*

$$u(re^{i\theta}) = \frac{1}{2\pi} \int_0^{2\pi} P_r(\theta - t) u(e^{it}) \, dt, \qquad re^{i\theta} \in \mathbb{D}. \tag{6.2.4}$$

In other words, u is represented on $\mathbb{D}$ by integrating its values on $\partial \mathbb{D}$ "against" the Poisson kernel.

Proof. Fix $re^{i\theta} \in \mathbb{D}$. For $s \in (0, 1)$, define $u_s \colon D(0; 1/s) \to \mathbb{R}$ by $u_s(z) = u(sz)$. Then each u_s is harmonic by Theorem 6.1.9, and the above work gives

$$u_s(re^{i\theta}) = \frac{1}{2\pi} \int_0^{2\pi} P_r(\theta - t) u_s(e^{it}) \, dt = \frac{1}{2\pi} \int_0^{2\pi} \varphi(se^{it}) \, dt,$$

where $\varphi \colon \overline{\mathbb{D}} \setminus \{0\} \to \mathbb{R}$ given by

$$\varphi(z) = u(z) \sum_{n=-\infty}^{\infty} r^{|n|} e^{in\theta} \frac{|z|^n}{z^n}$$

is continuous by a typical application of the Weierstrass M-test. By Exercise 2 in Section 6.1, we may take the limit inside of the integral in the calculation

$$u(re^{i\theta}) = \lim_{s \to 1^-} u(sre^{i\theta}) = \lim_{s \to 1^-} u_s(re^{i\theta}) = \frac{1}{2\pi} \int_0^{2\pi} \varphi(e^{it}) \, dt,$$

giving the result. □

6.2.3 Remark. Note that this representation of u on $\mathbb{D}$ as an integral dependent only on its values on $\partial \mathbb{D}$ is similar to the representation of an analytic function f inside of a disk (or more general domain) as the integral of its values on the boundary times a fixed function. In the case of $\mathbb{D}$, this function is $1/[2\pi i(\zeta - z)]$ for $z \in \mathbb{D}$ and $\zeta \in \partial \mathbb{D}$ and is sometimes referred to as the *Cauchy kernel*.

The Poisson integral formula states that a function harmonic on $\mathbb{D}$ and continuous on $\overline{\mathbb{D}}$ is such that the values of the function on $\mathbb{D}$ are determined by its values on $\partial \mathbb{D}$. We now consider a type of converse. Specifically, (when) is a function continuous on $\partial \mathbb{D}$ continuously extendable to a function harmonic on $\mathbb{D}$? This question has its own name.

6.2.4 Definition. Let Ω be an open proper subset of $\mathbb{C}$. The *Dirichlet problem* on Ω is: Given a continuous function $\varphi \colon \partial \Omega \to \mathbb{R}$, find a continuous function $u \colon \overline{\Omega} \to \mathbb{R}$ such that u is harmonic on Ω and $u(z) = \varphi(z)$ for all $z \in \partial \Omega$.

We begin our pursuit of a solution to the Dirichlet problem on $\mathbb{D}$ by considering some basic, yet useful, characteristics of the Poisson kernel. See Figure 6.1.

6.2.5 Lemma. *The Poisson kernel has the following properties.*

(a) *For all $r \in [0, 1)$ and $t \in \mathbb{R}$, $P_r(t)$ can be expressed in either of the following alternate forms:*

$$P_r(t) = \text{Re} \left(\frac{1 + re^{it}}{1 - re^{it}} \right) = \frac{1 - r^2}{1 - 2r \cos t + r^2}.$$

(b) *For all $r \in [0, 1)$ and $\theta \in \mathbb{R}$,*

$$\frac{1}{2\pi} \int_0^{2\pi} P_r(\theta - t) \, dt = 1.$$

(c) *For all $r \in [0, 1)$ and $t \in \mathbb{R}$, $P_r(t) > 0$ and $P_r(-t) = P_r(t)$.*

(d) *For any $\varepsilon > 0$ and $\delta \in (0, \pi)$, there is $\rho \in (0, 1)$ such that $P_r(t) < \varepsilon$ for all $r \in (\rho, 1)$ and $t \in [\delta, 2\pi - \delta]$.*

Proof. To prove (a), we first calculate

$$P_r(t) = 1 + \sum_{n=1}^{\infty} r^n e^{int} + \sum_{n=1}^{\infty} r^n e^{-int}$$

$$= \mathrm{Re} \left(1 + 2 \sum_{n=1}^{\infty} r^n e^{int} \right)$$

$$= \mathrm{Re} \left(\frac{1 + re^{it}}{1 - re^{it}} \right)$$

by summing the geometric series. By multiplying the top and bottom of the last fraction by the conjugate of the denominator, we obtain

$$P_r(t) = \mathrm{Re} \left(\frac{1 + 2ir \sin t - r^2}{1 - 2r \cos t + r^2} \right) = \frac{1 - r^2}{1 - 2r \cos t + r^2}.$$

For (b), we can use uniform convergence of the series in t for a given r to move the integral inside the series in the calculation

$$\frac{1}{2\pi} \int_0^{2\pi} P_r(\theta - t) \, dt = \frac{1}{2\pi} \sum_{n=-\infty}^{\infty} r^{|n|} e^{in\theta} \int_0^{2\pi} e^{-int} \, dt = 1,$$

once again using (2.8.5).

Part (c) follows immediately from the second form given in part (a).

To prove (d), we first note that periodicity and part (c) show we only need consider $t \in [\delta, \pi]$. For $r \in (0, 1)$, we calculate

$$P_r'(t) = -\frac{2r(1 - r^2) \sin t}{(1 - 2r \cos t + r^2)^2} < 0$$

for $t \in (0, \pi)$, using part (a). Therefore P_r is decreasing on this interval, and hence $P_r(t) < \varepsilon$ for $t \in [\delta, \pi]$ if $P_r(\delta) < \varepsilon$. Since $\lim_{r \to 1^-} P_r(\delta) = 0$, a ρ satisfying the conclusion of (d) exists by the definition of the limit. $\square$

By allowing for piecewise continuous boundary functions, the following theorem *does better* than solve the Dirichlet problem on $\mathbb{D}$. We note that a function φ is *piecewise continuous* on $\partial \mathbb{D}$ if $\varphi(e^{it})$ is piecewise continuous in the variable t on $[0, 2\pi]$. (See Definition 2.8.2.)

6.2.6 Theorem. *Let φ be real valued and piecewise continuous on $\partial \mathbb{D}$, and let $E \subseteq \partial \mathbb{D}$ be its set of discontinuities. Then there is a function $u \colon \overline{\mathbb{D}} \setminus E \to \mathbb{R}$ such that u*

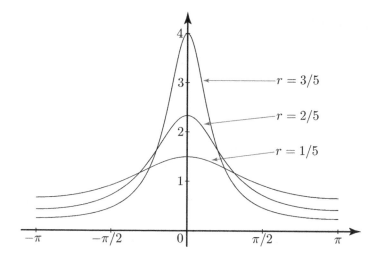

Figure 6.1 Graphs of $P_r(t)$ over $[-\pi, \pi]$ for $r = 1/5, 2/5, 3/5$

is harmonic on $\mathbb{D}$, $u(z) = \varphi(z)$ for all $z \in \partial\mathbb{D} \setminus E$, and u is continuous on $\overline{\mathbb{D}} \setminus E$. Specifically, the function u is given on $\mathbb{D}$ by

$$u(re^{i\theta}) = \frac{1}{2\pi} \int_0^{2\pi} P_r(\theta - t)\varphi(e^{it})\, dt. \tag{6.2.5}$$

On $\mathbb{D}$, u is the real part of the function $f \in H(\mathbb{D})$ given by

$$f(z) = \frac{1}{2\pi} \int_0^{2\pi} \frac{e^{it} + z}{e^{it} - z}\varphi(e^{it})\, dt. \tag{6.2.6}$$

We note that such a solution is *unique*, as seen in Exercise 6.

Proof. We first show that $f \in H(\mathbb{D})$ and then use it to obtain u. Partition $[0, 2\pi]$ by $0 = t_0 < \cdots < t_n = 2\pi$ so that $\varphi(e^{it})$ can be extended continuously to $t \in [t_{k-1}, t_k]$ for each $k = 1, \ldots, n$, and define the contours $\gamma_k : [t_{k-1}, t_k] \to \mathbb{C}$ by $\gamma_k(t) = e^{it}$. Then for each k, let $f_k : \mathbb{D} \to \mathbb{C}$ be given by

$$\begin{aligned}
f_k(z) &= \frac{1}{2\pi i} \int_{\gamma_k} \frac{\zeta + z}{\zeta - z} \frac{\varphi(\zeta)}{\zeta}\, d\zeta \\
&= \frac{1}{2\pi i} \left(\int_{\gamma_k} \frac{\varphi(\zeta)}{\zeta - z}\, d\zeta + z \int_{\gamma_k} \frac{\varphi(\zeta)/\zeta}{\zeta - z}\, d\zeta \right),
\end{aligned}$$

where the continuous extensions of φ are used at the endpoints of the contours. Since their integrands have continuous numerators, each of the last two integrals are of the form in Lemma 4.1.1 and hence define analytic functions in z. Thus $f_k \in H(\mathbb{D})$ for all k. Parameterizing and using (2.8.4) yields

$$f(z) = \frac{1}{2\pi} \sum_{k=1}^{n} \int_{t_{k-1}}^{t_k} \frac{e^{it} + z}{e^{it} - z} \varphi(e^{it}) \, dt = \sum_{k=1}^{n} f_k(z),$$

showing $f \in H(\mathbb{D})$.

If $u \colon \overline{\mathbb{D}} \setminus E \to \mathbb{R}$ is given by

$$u(z) = \begin{cases} \operatorname{Re} f(z) & \text{if } z \in \mathbb{D}, \\ \varphi(z) & \text{if } z \in \partial\mathbb{D} \setminus E, \end{cases}$$

then u is harmonic on $\mathbb{D}$ by Theorem 6.1.2, and for $re^{i\theta} \in \mathbb{D}$, we have

$$u(re^{i\theta}) = \frac{1}{2\pi} \int_0^{2\pi} \operatorname{Re}\left(\frac{1 + re^{i(\theta-t)}}{1 - re^{i(\theta-t)}}\right) \varphi(e^{it}) \, dt = \frac{1}{2\pi} \int_0^{2\pi} P_r(\theta - t)\varphi(e^{it}) \, dt,$$

using part (a) of Lemma 6.2.5. It remains to show that u is continuous on $\overline{\mathbb{D}} \setminus E$. We know u is continuous on $\mathbb{D}$ because it is harmonic, and hence the proof will be complete once it is shown that u is continuous at $e^{i\alpha} \in \partial\mathbb{D} \setminus E$.

Let $\varepsilon > 0$. Since φ is continuous at $e^{i\alpha}$, there is $\beta \in (0, \pi)$ such that $|\varphi(e^{it}) - \varphi(e^{i\alpha})| < \varepsilon/3$ when $|t - \alpha| < \beta$. For each k, the continuous extension of φ to the compact set γ_k^* is bounded on γ_k^*. Since there are finitely many k, we can fix $M > 0$ such that $|\varphi(e^{it})| \le M$ for all t such that $\varphi(e^{it})$ is defined. By part (d) of Lemma 6.2.5, there is $\rho \in (0, 1)$ such that $P_r(t) < \varepsilon/(3M)$ for all $r \in (\rho, 1)$ and $t \in [\beta/2, 2\pi - \beta/2]$.

Let $r \in (\rho, 1)$ and $\theta \in \mathbb{R}$ such that $|\theta - \alpha| < \beta/2$. We use part (b) of Lemma 6.2.5 and periodicity (see Exercise 1) to write

$$u(re^{i\theta}) - u(e^{i\alpha}) = \frac{1}{2\pi} \int_0^{2\pi} P_r(\theta - t)\varphi(e^{it}) \, dt - \frac{\varphi(e^{i\alpha})}{2\pi} \int_0^{2\pi} P_r(\theta - t) \, dt$$

$$= \frac{1}{2\pi} \int_0^{2\pi} P_r(\theta - t)[\varphi(e^{it}) - \varphi(e^{i\alpha})] \, dt$$

$$= \frac{1}{2\pi} \int_{\alpha-\beta}^{\alpha+\beta} P_r(\theta - t)[\varphi(e^{it}) - \varphi(e^{i\alpha})] \, dt$$

$$\quad + \frac{1}{2\pi} \int_{\alpha+\beta}^{\alpha-\beta+2\pi} P_r(\theta - t)[\varphi(e^{it}) - \varphi(e^{i\alpha})] \, dt$$

$$= I_1 + I_2.$$

Now observe that

$$|I_1| \le \frac{1}{2\pi} \int_{\alpha-\beta}^{\alpha+\beta} P_r(\theta - t)|\varphi(e^{it}) - \varphi(e^{i\alpha})| \, dt$$

$$< \frac{1}{2\pi} \int_{\alpha-\beta}^{\alpha+\beta} P_r(\theta - t)\frac{\varepsilon}{3} \, dt$$

$$< \frac{\varepsilon}{3} \frac{1}{2\pi} \int_0^{2\pi} P_r(\theta - t)\, dt$$

$$= \frac{\varepsilon}{3}.$$

When considering I_2, note that we have t such that $\alpha + \beta \le t \le \alpha - \beta + 2\pi$, and our choice of θ satisfies $-\alpha - \beta/2 < -\theta < -\alpha + \beta/2$. Adding these gives $\beta/2 < t - \theta < 2\pi - \beta/2$. Part (c) of Lemma 6.2.5 and our choice of r give $P_r(\theta - t) < \varepsilon/(3M)$ for these t. Hence

$$|I_2| \le \frac{1}{2\pi} \int_{\alpha+\beta}^{\alpha-\beta+2\pi} P_r(\theta - t)(|\varphi(e^{it})| + |\varphi(e^{i\alpha})|)\, dt$$

$$\le \frac{1}{2\pi} \int_{\alpha+\beta}^{\alpha-\beta+2\pi} \frac{\varepsilon}{3M}(2M)\, dt$$

$$< \frac{2\varepsilon}{3}.$$

Therefore $|u(re^{i\theta}) - u(e^{i\alpha})| \le |I_1| + |I_2| < \varepsilon$.

Lastly, it is not hard to see that the set $U = \{re^{i\theta} : r > \rho,\ \alpha - \beta/2 < \theta < \alpha + \beta/2\}$ is open. Hence there is a $\delta > 0$ such that $D(e^{i\alpha}; \delta) \subseteq U$. If $re^{i\theta} \in \overline{\mathbb{D}}$ is such that $|re^{i\theta} - e^{i\alpha}| < \delta$, then either $r = 1$, in which case $|\theta - \alpha| < \beta$ gives $|u(e^{i\theta}) - u(e^{i\alpha})| = |\varphi(e^{i\theta}) - \varphi(e^{i\alpha})| < \varepsilon$, or $r \in (\rho, 1)$ and $|u(re^{i\theta}) - u(e^{i\alpha})| < \varepsilon$ from the above. Hence u is continuous at $e^{i\alpha}$. ∎

6.2.7 Remark. It is certainly not the case that the Poisson integral formula can only be used for functions defined on $\mathbb{D}$. Consider a general disk $D(a; R)$ and continuous $u : \overline{D}(a; R) \to \mathbb{R}$ that is harmonic on $D(a; R)$. The mapping $f \in H(\mathbb{C})$ given by $f(z) = a + Rz$ is such that $f(\overline{\mathbb{D}}) = \overline{D}(a; R)$, and $u \circ f$ is harmonic on $\mathbb{D}$ by Theorem 6.1.9. Hence using the form in Lemma 6.2.5, part (a), we have

$$(u \circ f)(\rho e^{i\theta}) = \frac{1}{2\pi} \int_0^{2\pi} \frac{1 - \rho^2}{1 - 2\rho\cos(\theta - t) + \rho^2}(u \circ f)(e^{it})\, dt, \qquad \rho e^{i\theta} \in \mathbb{D}.$$

Set $r = \rho R$ to get

$$u(a + re^{i\theta}) = \frac{1}{2\pi} \int_0^{2\pi} \frac{1 - r^2/R^2}{1 - 2(r/R)\cos(\theta - t) + r^2/R^2} u(a + Re^{it})\, dt$$

$$= \frac{1}{2\pi} \int_0^{2\pi} \frac{R^2 - r^2}{R^2 - 2rR\cos(\theta - t) + r^2} u(a + Re^{it})\, dt,$$

valid for all $a + re^{i\theta} \in D(a; R)$. This is the Poisson integral formula for $D(a; R)$.

Likewise, if φ is real valued and piecewise continuous on $\partial D(a; R)$, then

$$u(a + re^{i\theta}) = \frac{1}{2\pi} \int_0^{2\pi} \frac{R^2 - r^2}{R^2 - 2rR\cos(\theta - t) + r^2} \varphi(a + Re^{it})\, dt$$

solves the Dirichlet problem on $D(a; R)$.

We conclude by applying our work with the Poisson integral formula, in particular Theorem 6.2.6, to our study of the mean value property. We now show that the property, already known to be satisfied by harmonic functions, actually completely characterizes the set of harmonic functions! In fact, we prove a slightly stronger result that will be of use later.

6.2.8 Theorem. *Let $\Omega \subseteq \mathbb{C}$ be open and $\varphi \colon \Omega \to \mathbb{R}$ be a continuous function. If for every $z_0 \in \Omega$, there exists $\varepsilon > 0$ such that $D(z_0; \varepsilon) \subseteq \Omega$ and*

$$\varphi(z_0) = \frac{1}{2\pi} \int_0^{2\pi} \varphi(z_0 + re^{i\theta})\, d\theta \tag{6.2.7}$$

for all $r \in (0, \varepsilon)$, then φ is harmonic on Ω.

Proof. Let $a \in \Omega$ and $R > 0$ such that $\overline{D}(a; R) \subseteq \Omega$. It suffices to show that φ is harmonic on $D(a; R)$. Let u be the solution to the Dirichlet problem on $D(a; R)$ with boundary function φ (see Remark 6.2.7). Then u is continuous on $\overline{D}(a; R)$, u is harmonic on $D(a; R)$, and $u(z) = \varphi(z)$ for all $z \in \partial D(a; R)$. Let $\psi = \varphi - u$. Then ψ is continuous on $\overline{D}(a; R)$ and $\psi(z) = 0$ for all $z \in \partial D(a; R)$. The proof will be completed by showing $\psi = 0$, for then $\varphi = u$ is harmonic.

Since $\overline{D}(a; R)$ is compact,

$$\alpha = \max\{\psi(z) : z \in \overline{D}(a; R)\} \geq 0$$

is attained for at least one $z \in \overline{D}(a; R)$. In order to uncover a contradiction, suppose that $\alpha > 0$. We observe that $K = \{z \in \overline{D}(a; R) : \psi(z) = \alpha\}$ is a nonempty compact subset of $D(a; R)$. By continuity of the modulus, there exists $z_0 \in K$ such that

$$|z_0 - a| = \max\{|z - a| : z \in K\}.$$

By hypothesis, there is $\varepsilon > 0$ such that (6.2.7) holds for all $r \in (0, \varepsilon)$.

Let $0 < r < \min\{\varepsilon, R - |z_0 - a|\}$ and $\omega = \arg(z_0 - a)$. Then

$$R > |(z_0 + re^{i\omega}) - a| = |z_0 - a| + r > |z_0 - a|,$$

and hence $z_0 + re^{i\omega} \in D(a; R) \setminus K$. (See Figure 6.2.) Thus $\psi(z_0 + re^{i\omega}) = \beta < \alpha$. Then there is $\delta \in (0, \pi)$ such that $|\theta - \omega| < \delta$ implies $\psi(z_0 + re^{i\theta}) < (\alpha + \beta)/2$. Since u is harmonic on $D(a; R)$, it has the mean value property

$$u(z_0) = \frac{1}{2\pi} \int_0^{2\pi} u(z_0 + re^{i\theta})\, d\theta.$$

Subtracting this from (6.2.7) gives

$$\alpha = \psi(z_0) = \frac{1}{2\pi} \int_0^{2\pi} \psi(z_0 + re^{i\theta})\, d\theta$$

$$= \frac{1}{2\pi} \left(\int_{\omega - \delta}^{\omega + \delta} \psi(z_0 + re^{i\theta})\, d\theta + \int_{\omega + \delta}^{\omega - \delta + 2\pi} \psi(z_0 + re^{i\theta})\, d\theta \right)$$

$$< \frac{1}{2\pi} \left(2\delta \frac{\alpha + \beta}{2} + (2\pi - 2\delta)\alpha \right)$$

$$= \alpha - \frac{(\alpha - \beta)\delta}{2\pi}$$

$$< \alpha,$$

a contradiction. Hence $\alpha = 0$.

An analogous argument shows that the minimum of ψ over $\overline{D}(a; R)$ is also 0. Hence $\psi(z) = 0$ for all $z \in D(a; R)$. $\qquad\qquad\square$

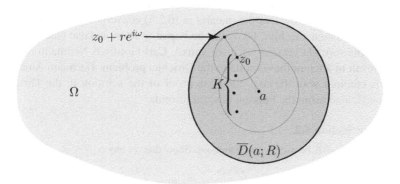

Figure 6.2   Constructions used the the proof of Theorem 6.2.8

We can now combine Theorems 6.1.13 and 6.2.8 to obtain the following classification of harmonic functions.

6.2.9 Corollary. *Let $\Omega \subseteq \mathbb{C}$ be open and $u \colon \Omega \to \mathbb{R}$ be continuous. Then u is harmonic if and only if u has the mean value property.*

Summary and Notes for Section 6.2.

Like the Cauchy integral formula, the Poisson integral formula allows us to recover the values of a function in a region by knowing its values on the region's boundary; in this case, the function is harmonic inside the unit disk and continuous to the boundary. As noted in Section 6.1, many properties of harmonic functions in the plane extend to higher dimensions. This certainly holds for the Poisson integral formula, making it extremely useful for the study of harmonic functions.

The solution to the Dirichlet problem on $\mathbb{D}$ using the Poisson kernel, while closely related to the Poisson integral formula, must be singled out for its significance. Any piecewise continuous function on $\partial\mathbb{D}$ can be continued inside of $\mathbb{D}$ as a harmonic function, continuous to the boundary at points of boundary continuity. This generates a related analytic function, showing that there is an analytic function on $\mathbb{D}$ for every real-valued piecewise continuous boundary function. More advanced study in this area reveals that piecewise continuity of the boundary function can be replaced by even more general criteria, resulting in an extremely useful tool to generate analytic functions.

The Dirichlet problem is an example of a boundary value problem, that is, a problem of finding a function that satisfies a partial differential equation (in this case, Laplace's equation) inside a region while equaling a prescribed function defined on the region's boundary. As mentioned in the conclusion of Section 6.1, solutions of Laplace's equation have myriad physical interpretations. For example, we can view the solution of the Dirichlet problem on $\mathbb{D}$ with boundary function φ as the steady-state temperature inside $\mathbb{D}$ when φ is the temperature on $\partial\mathbb{D}$.

Motivated to prove the convergence of a Fourier series (which we will consider in Section 6.4) to a continuous function φ on $\partial\mathbb{D}$, Siméon Denis Poisson tried to use his integral formula to show that the integral in (6.2.5) converges to φ pointwise as $r \to 1^-$. Unbeknownst to him, the result he hoped to show is false (see Example 6.4.11), but his formula remains of great value! Carl Gottfried Neumann utilized Poisson's result to attempt the solution to the Dirichlet problem. Hermann Amandus Schwarz is credited with the first complete proof of the solution to the Dirichlet problem on $\mathbb{D}$, also using the Poisson integral formula.

Exercises for Section 6.2.

1. ▷ Let $\varphi\colon \partial\mathbb{D} \to \mathbb{R}$ be piecewise continuous. Show that for any $\alpha, \beta \in \mathbb{R}$,

$$\int_0^{2\pi} \varphi(e^{i\theta})\, d\theta = \int_\alpha^{\alpha+2\pi} \varphi(e^{i\theta})\, d\theta = \int_0^{2\pi} \varphi(e^{i(\beta-\theta)})\, d\theta.$$

The first equality is used several times in this section.

2. By choosing an appropriate continuous function $u\colon \overline{\mathbb{D}} \to \mathbb{R}$ that is harmonic on $\mathbb{D}$, use the Poisson integral formula to verify the integral identity

$$\int_0^{2\pi} \frac{(1-r^2)\cos 2t}{1 - 2r\cos(\theta - t) + r^2}\, dt = 2\pi r^2 \cos 2\theta, \qquad 0 \le r < 1,\ \theta \in \mathbb{R}.$$

3. Define $\varphi\colon \partial\mathbb{D} \to \mathbb{R}$ by

$$\varphi(z) = \begin{cases} 1 & \text{if } \operatorname{Im} z \ge 0, \\ 0 & \text{if } \operatorname{Im} z < 0. \end{cases}$$

Find the solution to the Dirichlet problem on $\mathbb{D}$ for the boundary function φ by directly calculating (6.2.6) and taking the real part. (*Hint*: Separate into two integrals and be careful to account for branches of the logarithm.)

4. Provide an alternative proof of the Poisson integral formula using Theorem 6.2.6 (which does not assume the formula) and the following strategy. Let v be the solution to the Dirichlet problem on $\mathbb{D}$ using u on $\partial\mathbb{D}$. Then argue that the harmonic function $u - v$ must be equal to 0 using the maximum/minimum principle.

5. ▷ Let $\Omega \subseteq \mathbb{C}$ be open and $\{u_n\}_{n=1}^\infty$ be a sequence of harmonic functions on Ω converging uniformly on compact subsets of Ω to a function $u\colon \Omega \to \mathbb{R}$. Prove that u is harmonic. (*Hint*: Use the mean value property.)

6. Let $\Omega \subseteq \mathbb{C}$ be a bounded domain. Prove that if there is a solution to the Dirichlet problem on Ω for continuous boundary function $\varphi\colon \partial\Omega \to \mathbb{R}$, then the solution is unique.

7. Let $\Omega \subseteq \mathbb{C}$ be open, $a < b$, and $u: \Omega \times [a, b] \to \mathbb{R}$ be continuous and such that u_t is harmonic for all $t \in [a, b]$, where $u_t(z) = u(z, t)$. Define $U: \Omega \to \mathbb{R}$ by

$$U(z) = \int_a^b u(z, t)\, dt.$$

Show that U is harmonic. Why can't we use Leibniz's rule?

8. Let φ be complex valued and piecewise continuous on $\partial \mathbb{D}$, and define $f: \mathbb{D} \to \mathbb{C}$ by

$$f(re^{i\theta}) = \frac{1}{2\pi} \int_0^{2\pi} P_r(\theta - t)\varphi(e^{it})\, dt.$$

(a) Prove that the functions $\operatorname{Re} f$ and $\operatorname{Im} f$ are harmonic.

(b) Show that $f \in H(\mathbb{D})$ if and only if

$$\int_0^{2\pi} e^{int}\varphi(e^{it})\, dt = 0, \qquad n \in \mathbb{N}.$$

9. Consider the statement of Theorem 6.2.8 with (6.2.7) replaced by

$$\varphi(z_0) = \frac{1}{\pi r^2} \iint_{D(z_0;r)} \varphi(x, y)\, dA(x, y).$$

Use Exercise 10 of Section 6.1 to modify the proof of Theorem 6.2.8 to show the result holds with this alternative hypothesis.

10. Here, we consider an improvement to Cauchy's integral formula on disks. Prove that if $f: \overline{D}(a; r) \to \mathbb{C}$ is continuous and is analytic on $D(a; r)$, then

$$f(z) = \frac{1}{2\pi i} \int_{\partial D(a;r)} \frac{f(\zeta)}{\zeta - z}\, d\zeta$$

for all $z \in D(a; r)$ by mimicking the technique used in the proof of the Poisson integral formula.

6.3 Further Connections to Analytic Functions

With the Poisson integral formula and the characterization of harmonic functions by the mean value property in hand, we consider some other ways that our knowledge of analytic functions informs us about harmonic functions and vice versa.

We know (Theorem 6.2.6) that the Dirichlet problem has a solution on $\mathbb{D}$ for a given piecewise continuous function on $\partial \mathbb{D}$ by way of the Poisson kernel. However, just because the solution is expressible by an integral does not mean it is easy to find in a closed form! In certain simple cases, we may turn to the mapping properties of analytic functions and Theorem 6.1.9 for assistance. Consider the following example.

6.3.1 Example. Let $\alpha, \beta \in \mathbb{R}$ such that $0 < \beta - \alpha < 2\pi$. We consider the solution of the Dirichlet problem on $\mathbb{D}$ for the function $\varphi \colon \partial\mathbb{D} \to \mathbb{R}$ given by

$$\varphi(e^{i\theta}) = \begin{cases} 1 & \text{if } \alpha \leq \theta < \beta, \\ 0 & \text{if } \beta \leq \theta < \alpha + 2\pi. \end{cases}$$

Define the linear fractional transformation

$$T(z) = \frac{1 - e^{-i\alpha}z}{1 - e^{-i\beta}z}.$$

We see that $T(e^{i\alpha}) = 0$ and $T(e^{i\beta}) = \infty$, and hence $T(\partial\mathbb{D})$ is a line through the origin. Furthermore, $T(0) = 1$ and $T(\infty) = e^{i(\beta-\alpha)}$ are symmetric points with respect to $T(\partial\mathbb{D})$, meaning the line passes through $e^{i(\beta-\alpha)/2}$. We conclude that $T(\mathbb{D})$ is the half-plane H bounded by this line and containing 1. Since

$$T(e^{i(\alpha+\beta)/2}) = \frac{1 - e^{i(\beta-\alpha)/2}}{1 - e^{i(\alpha-\beta)/2}} = e^{i(\beta-\alpha)/2}\frac{e^{i(\alpha-\beta)/2} - 1}{1 - e^{i(\alpha-\beta)/2}} = -e^{i(\beta-\alpha)/2},$$

T takes the arc $A = \{e^{i\theta} : \alpha < \theta < \beta\}$ to the ray through $-e^{i(\beta-\alpha)/2}$ and the complementary arc $B = \{e^{i\theta} : \beta < \theta < \alpha + 2\pi\}$ to the ray through $e^{i(\beta-\alpha)/2}$.

We observe that $\mathrm{Log}\, z$ is defined for all $z \in H$ and is one-to-one. The mapping properties of the logarithm (see Example 2.5.3) show that $\mathrm{Log}(H) = \Omega$, where

$$\Omega = \left\{ w \in \mathbb{C} : \frac{\beta - \alpha}{2} - \pi < \mathrm{Im}\, w < \frac{\beta - \alpha}{2} \right\}.$$

Thus the function given by

$$f(z) = \mathrm{Log}(T(z)) = \mathrm{Log}\left(\frac{1 - e^{-i\alpha}z}{1 - e^{-i\beta}z} \right)$$

is analytic and maps $\mathbb{D}$ one-to-one and onto the horizontal strip Ω with A mapped to the bottom boundary line and B mapped to the top.

The real-linear function $v \colon \mathbb{C} \to \mathbb{R}$ given by

$$v(w) = \frac{\beta - \alpha}{2\pi} - \frac{1}{\pi}\mathrm{Im}\, w$$

is harmonic and $v(w) = 0$ on the top line of $\partial\Omega$ and $v(w) = 1$ on the bottom. By Theorem 6.1.9, the function $u = v \circ f$ is harmonic on $\mathbb{D}$. We further see that it is continuous on $\overline{\mathbb{D}} \setminus \{e^{i\alpha}, e^{i\beta}\}$, $u(A) = \{1\}$, and $u(B) = \{0\}$. Hence

$$u(z) = \frac{\beta - \alpha}{2\pi} - \frac{1}{\pi}\mathrm{Arg}\left(\frac{1 - e^{-i\alpha}z}{1 - e^{-i\beta}z} \right)$$

is the solution to the Dirichlet problem on $\mathbb{D}$ with boundary function φ.

Using linearity, we can generalize this significantly. The result is a closed-form solution to the Dirichlet problem on $\mathbb{D}$, in the case of a piecewise constant boundary function, found in an easier manner than integrating the corresponding Poisson integral.

6.3.2 Theorem. *Let $n \in \mathbb{N}$, $n \geq 2$, $\omega \in \mathbb{R}$ and*

$$\omega = \alpha_0 < \alpha_1 < \cdots < \alpha_n = \omega + 2\pi$$

be a partition of $[\omega, \omega + 2\pi]$. If $c_k \in \mathbb{R}$ for $k = 1, \ldots, n$ and $\varphi \colon \partial\mathbb{D} \to \mathbb{R}$ is given by $\varphi(e^{i\theta}) = c_k$ for $\alpha_{k-1} \leq \theta < \alpha_k$, then the solution to the Dirichlet problem on $\mathbb{D}$ with boundary function φ is

$$u(z) = \sum_{k=1}^{n} c_k \left(\frac{\alpha_k - \alpha_{k-1}}{2\pi} - \frac{1}{\pi} \operatorname{Arg} \left(\frac{1 - e^{-i\alpha_{k-1}} z}{1 - e^{-i\alpha_k} z} \right) \right), \qquad z \in \mathbb{D}.$$

Proof. Write $\varphi_{\alpha,\beta}$ and $u_{\alpha,\beta}$ for the functions considered in Example 6.3.1. Then $\varphi = \sum_{k=1}^{n} c_k \varphi_{\alpha_{k-1}, \alpha_k}$ and $u = \sum_{k=1}^{n} c_k u_{\alpha_{k-1}, \alpha_k}$. By Theorem 6.1.9, u is harmonic on $\mathbb{D}$, and passing limits through finite sums, we see that u extends to equal φ at all points of continuity of φ. $\qquad\square$

We now use the mean value property (specifically Theorem 6.2.8) to show how it is possible to extend a harmonic function symmetrically across $\mathbb{R}$. We will therefore be considering an open set $\Omega \subseteq \mathbb{C}$ that is symmetric in the line $\mathbb{R}$, and we adopt the notation

$$\Omega^+ = \{z \in \Omega : \operatorname{Im} z > 0\}$$
$$\Omega^0 = \{z \in \Omega : \operatorname{Im} z = 0\}$$
$$\Omega^- = \{z \in \Omega : \operatorname{Im} z < 0\}$$

for the remainder of this section. See Figure 6.3.

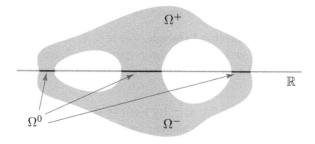

Figure 6.3 $\quad \Omega = \Omega^+ \cup \Omega^0 \cup \Omega^-$ is symmetric in $\mathbb{R}$

6.3.3 Harmonic Schwarz Reflection Principle. *Let $\Omega \subseteq \mathbb{C}$ be open and symmetric in $\mathbb{R}$. If $v \colon \Omega^+ \cup \Omega^0 \to \mathbb{R}$ is continuous, harmonic on Ω^+, and satisfies $v(z) = 0$ for all $z \in \Omega^0$, then the function $V \colon \Omega \to \mathbb{R}$ given by*

$$V(z) = \begin{cases} v(z) & \text{if } z \in \Omega^+ \cup \Omega^0, \\ -v(\bar{z}) & \text{if } z \in \Omega^- \end{cases} \tag{6.3.1}$$

is harmonic.

The function $V(z) = -v(\bar{z})$ defined for $z \in \Omega^-$ is called the *harmonic extension of v across* $\mathbb{R}$.

Proof. It is clear that V is harmonic on Ω^+. Writing V in the rectangular form

$$V(x, y) = -v(x, -y), \qquad (x, y) \in \Omega^-,$$

and directly taking second partial derivatives verifies Laplace's equation, and hence V is harmonic on Ω^-.

To show that V is harmonic on all of Ω, we use Theorem 6.2.8, so let $z_0 \in \Omega$. If $z_0 \in \Omega^+$ or $z_0 \in \Omega^-$, then there is $\varepsilon > 0$ such that $D(z_0; \varepsilon) \subseteq \Omega^+$ or $D(z_0; \varepsilon) \subseteq \Omega^-$, respectively. As V is harmonic on those sets, it satisfies (6.2.7) (in place of φ) for all $r \in (0, \varepsilon)$.

If $z_0 \in \Omega^0$, then let $\varepsilon > 0$ such that $D(z_0; \varepsilon) \subseteq \Omega$, and choose $r \in (0, \varepsilon)$. Noting that $z_0 \in \mathbb{R}$ and using the integral substitution $t = -\theta$ shows

$$\int_{-\pi}^{0} v\big(\overline{z_0 + re^{i\theta}}\big)\, d\theta = \int_{\pi}^{0} -v\big(z_0 + re^{it}\big)\, dt = \int_{0}^{\pi} v\big(z_0 + re^{it}\big)\, dt.$$

Hence

$$\int_{-\pi}^{\pi} V\big(z_0 + re^{i\theta}\big)\, d\theta = -\int_{-\pi}^{0} v\big(\overline{z_0 + re^{i\theta}}\big)\, d\theta + \int_{0}^{\pi} v\big(z_0 + re^{i\theta}\big)\, d\theta = 0.$$

Since $V(z_0) = 0$, the proof is complete. $\qquad\square$

This result on harmonic reflection is used to develop an analytic version.

6.3.4 Analytic Schwarz Reflection Principle. *Let $\Omega \subseteq \mathbb{C}$ be open and symmetric in $\mathbb{R}$, $f \in H(\Omega^+)$, and suppose that $v = \operatorname{Im} f$ extends continuously to equal 0 on Ω^0. Then there is $F \in H(\Omega)$ such that $F(z) = f(z)$ for $z \in \Omega^+$ and $F(z) = \overline{F(\bar{z})}$ for all $z \in \Omega$.*

Proof. Define $F(z) = f(z)$ for $z \in \Omega^+$. By Exercise 6 in Section 2.7, we see that defining $F(z) = \overline{f(\bar{z})}$ for $z \in \Omega^-$ is analytic on Ω^-. Hence it remains to address the definition of F on Ω^0.

We note that v satisfies the hypotheses of the harmonic Schwarz reflection principle, and let $V \colon \Omega \to \mathbb{R}$ be the resulting extension. For $a \in \Omega^0$, let $r_a > 0$ such that $D(a; r_a) \subseteq \Omega$. By Exercises 3 and 4 in Section 6.1, we see that $-\operatorname{Re} f$ is a harmonic conjugate of v on Ω^+ and the harmonic conjugates of V on $D(a; r_a)$ differ by an additive constant. Thus we may choose $-U_a$ to be the harmonic conjugate of V on $D(a; r_a)$ that agrees with $-\operatorname{Re} f$ on $D(a; r_a) \cap \Omega^+$. It follows that $F_a = U_a + iV \in H(D(a; r_a))$ and agrees with F on $D(a; r_a) \cap \Omega^+$. By hypothesis, F_a is real valued on $(a - r_a, a + r_a)$, and hence all of its derivatives at a are real. Accordingly,

$$\overline{F_a(\overline{z})} = \overline{\sum_{n=0}^{\infty} \frac{F_a^{(n)}(a)}{n!}(\overline{z} - a)^n} = \sum_{n=0}^{\infty} \frac{F_a^{(n)}(a)}{n!}(z - a)^n = F_a(z)$$

holds for all $z \in D(a; r_a)$, as z is the only nonreal expression in the series. Therefore F_a agrees with F on $D(a; r_a) \cap \Omega^-$. Define $F(z) = F_a(z)$ for all $z \in D(a; r_a)$.

To see that F is well defined, we must ensure it is consistently defined on any nonempty $D(a; r_a) \cap D(b; r_b)$ for $a, b \in \Omega^0$. In this case, we see that $F_a(z) = f(z) = F_b(z)$ for all $z \in D(a; r_a) \cap D(b; r_b) \cap \Omega^+$, a nonempty open set which therefore contains a limit point. Since $D(a; r_a) \cap D(b; r_b)$ is a domain, $F_a = F_b$ on this intersection by Corollary 3.4.7, thereby removing ambiguity from the definition of F. □

Through compositions, this concept may be generalized to reflections in sets other than $\mathbb{R}$. Linear fractional transformations provide one way to do so, and the reader should recall the meaning of *circles* in $\mathbb{C}_\infty$ and *symmetric points* with respect to these circles.

The key hypothesis in the analytic Schwarz reflection principle that $v = \operatorname{Im} f$ extend continuously to equal 0 on Ω^0, which enabled our use of the harmonic Schwarz reflection principle, can be rephrased to say that $f(z)$ "tends to $\mathbb{R}$" as z tends to any point on Ω^0. We formalize that notion here to withstand using curves other than $\mathbb{R}$ for reflection. Specifically, for a sequence $\{z_n\}_{n=1}^{\infty} \subseteq \mathbb{C}$ and set $E \subseteq \mathbb{C}$, we say that $z_n \to E$ as $n \to \infty$ if, for every $\varepsilon > 0$, there is $N \in \mathbb{N}$ such that $D(z_n; \varepsilon) \cap E \neq \varnothing$ for all $n \geq N$. (This is regular sequential convergence when E is a singleton set.)

If $C_1 \subseteq \mathbb{C}_\infty$ is a circle, an open set $\Omega \subseteq \mathbb{C}$ is *symmetric* in C_1 if $z^* \in \Omega$ for all $z \in \Omega$, where z^* denotes the point symmetric to z in C_1. Let U_1 and U_2 denote the connected components of $\mathbb{C} \setminus C_1$. (These sets are the inside and outside of C_1 if C_1 is a circle in $\mathbb{C}$ and are the two resulting half-planes when C_1 is a line. We purposefully do not note which components U_1 and U_2 are.) With this in place, we have $\Omega = \Omega_0 \cup \Omega_1 \cup \Omega_2$, where $\Omega_0 = \Omega \cap C_1$, $\Omega_1 = \Omega \cap U_1$, and $\Omega_2 = \Omega \cap U_2$. (See Figure 6.4.) We are now prepared to prove the following.

6.3.5 Schwarz Reflection in Circles. *Let C_1 and C_2 be circles in $\mathbb{C}_\infty$, $\Omega \subseteq \mathbb{C}$ be open and symmetric in C_1, and write $\Omega = \Omega_0 \cup \Omega_1 \cup \Omega_2$ as described above. If $f \in H(\Omega_1)$, $f(z_n) \to C_2$ for any sequence $\{z_n\}_{n=1}^{\infty} \subseteq \Omega_1$ converging to a point of Ω_0, and $f(\Omega_1)$ does not contain the center of C_2 in the case that C_2 is a circle in $\mathbb{C}$, then there exists $F \in H(\Omega)$ such that $F(z) = f(z)$ for all $z \in \Omega_1$ and $F(z)^* = F(z^*)$ for all $z \in \Omega$, where z^* is the point symmetric to z in C_1 and $F(z)^*$ is the point symmetric to $F(z)$ in C_2.*

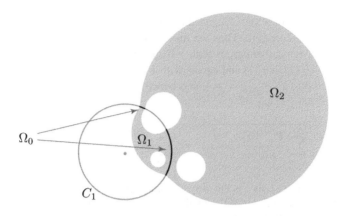

Figure 6.4 $\Omega = \Omega_0 \cup \Omega_1 \cup \Omega_2$ is symmetric in C_1

Proof. Let T_1 and T_2 be linear fractional transformations taking C_1 and C_2 respectively onto $\mathbb{R} \cup \{\infty\}$, where T_1 maps U_1 into the upper half-plane and $T_2(\infty) = \infty$ in the case that C_2 is a line in $\mathbb{C}$. (The possibility that $T_1^{-1}(\infty) \in \Omega_0$ cannot be avoided, so we will be careful about this.) Since T_1^{-1} is continuous, $V = T_1(\Omega) \cap \mathbb{C}$ is open. Preservation of symmetric points gives that V is symmetric in $\mathbb{R}$, and we see that

$$T_1(\Omega_0) \cap \mathbb{C} = V^0, \qquad T_1(\Omega_1) = V^+, \qquad T_1(\Omega_2) = V^-.$$

Our strategy is to use compositions involving T_1 and T_2 in order to reflect in $\mathbb{R}$. Figure 6.5 is helpful.

Define $g \in H(V^+)$ by $g = T_2 \circ f \circ T_1^{-1}$. By continuity, if $\{z_n\}_{n=1}^{\infty} \subseteq V^+$ converges to a value on V^0, then $T_1^{-1}(z_n)$ converges to a point of Ω_0. Accordingly, $f(T^{-1}(z_n)) \to C_2$ and $g(z_n) \to \mathbb{R}$. We conclude that $\operatorname{Im} g$ can be continuously extended so that $\operatorname{Im} g(z) = 0$ for all $z \in V^0$. Thus g meets the requirements for the analytic Schwarz reflection principle.

Let $G \in H(V)$ be the function that satisfies $G(z) = g(z)$ for all $z \in V^+$ and $G(z) = \overline{G(\bar{z})}$ for all $z \in V$. Note that $G(V)$ is symmetric in $\mathbb{R}$. The hypothesis that $f(\Omega_1)$ does not contain the center of C_2 if C_2 is a circle in $\mathbb{C}$ and the choice that $T_2(\infty) = \infty$ if C_2 is a line in $\mathbb{C}$ ensures that $T_2(\infty) \notin G(V)$. We may thus define $F_1 \in H(\Omega \setminus \{T_1^{-1}(\infty)\})$ by $F_1 = T_2^{-1} \circ G \circ T_1$. Then $F_1(z) = f(z)$ for all $z \in \Omega_1$ and, using preservation of symmetric points,

$$
\begin{aligned}
F_1(z^*) &= T_2^{-1} \circ G \circ T_1(z^*) \\
&= T_2^{-1} \circ G(\overline{T_1(z)}) \\
&= T_2^{-1}(\overline{G(T_1(z))}) \\
&= T_2^{-1}(G(T_1(z)))^* \\
&= F_1(z)^*
\end{aligned}
$$

for all $z \in \Omega \setminus \{T_1^{-1}(\infty)\}$.

If $T_1^{-1}(\infty) \notin \Omega_0$, then letting $F = F_1$ completes the proof. Otherwise, repeating the above steps, replacing T_1 with a linear fractional transformation S that takes C_1 into $\mathbb{R} \cup \{\infty\}$, U_1 into the upper half-plane, and satisfies $S^{-1}(\infty) \neq T_1^{-1}(\infty)$ gives a function $F_2 \in H(\Omega \setminus \{S^{-1}(\infty)\})$ such that $F_2(z) = f(z)$ for all $z \in \Omega_1$ and $F_2(z^*) = F_2(z)^*$ for all $z \in \Omega \setminus \{S^{-1}(\infty)\}$. Symmetry and continuity imply that $F_1(z) = F_2(z)$ for all $z \in \Omega \setminus \{T_1^{-1}(\infty), S^{-1}(\infty)\}$. Therefore, $F \colon \Omega \to \mathbb{C}$ given by

$$F(z) = \begin{cases} F_1(z) & \text{if } z \in \Omega \setminus \{T_1^{-1}(\infty)\}, \\ F_2(z) & \text{if } z \in \Omega \setminus \{S^{-1}(\infty)\} \end{cases}$$

is a well-defined analytic function giving the result. $\qquad\square$

We conclude this section with a result that shows that annuli are much more restrictive than disks when it comes to mapping one onto another with an analytic bijection. Unless the inner and outer radii of the two annuli share the same ratio, it is impossible to do so! Even when it is possible, there are pitifully few mappings that work. Here, we consider annuli centered at 0, noting simple translations can generalize the result to annuli centered elsewhere.

6.3.6 Theorem. *Let $0 < r_1 < R_1 < \infty$ and $0 < r_2 < R_2 < \infty$. If f is a one-to-one analytic function mapping $A(0; r_1, R_1)$ onto $A(0, r_2, R_2)$ then $r_1/R_1 = r_2/R_2$ and $f = g \circ h$, where $h(z) = R_2 c z / R_1$ for some $c \in \partial \mathbb{D}$ and $g(z) = z$ or $g(z) = r_2 R_2 / z$.*

Proof. We note that f^{-1} is well defined and analytic by Theorem 2.7.7. Let $\rho = \sqrt{r_2 R_2} \in (r_2, R_2)$. Since $K = f^{-1}(\partial D(0; \rho))$ is a compact subset of $A(0; r_1, R_1)$, we have that

$$s = \min\{|z| : z \in K\} > r_1.$$

(See Figure 6.6.) Now $f(A(0; r_1, s))$ is connected and $f(A(0; r_1, s)) \cap \partial D(0; \rho) = \varnothing$, and thus $f(A(0; r_1, s))$ is a subset of either $A(0; r_2, \rho)$ or $A(0; \rho, R_2)$. Assume the former.

Let $\{z_n\}_{n=1}^{\infty} \subseteq A(0; r_1, R_1)$ converge to a point of $\partial D(0; r_1)$, and let $\varepsilon > 0$. For some $m \in \mathbb{N}$, $\{z_n\}_{n=m}^{\infty} \subseteq A(0; r_1, s)$. If infinitely many terms of $\{|f(z_n)|\}_{n=m}^{\infty}$ are greater than or equal to $r_2 + \varepsilon$, then from these terms we may construct a subsequence $\{f(z_{n_k})\}_{k=1}^{\infty}$ converging to a point $a \in \overline{A}(0; r_2 + \varepsilon, \rho)$ by the Bolzano–Weierstrass theorem. But then $z_{n_k} \to f^{-1}(a)$, a contradiction. Hence there is $N \in \mathbb{N}$ such that $|f(z_n)| < r_2 + \varepsilon$ for all $n \geq N$, showing $f(z_n) \to \partial D(0; r_2)$. A similar argument shows that $f(z_n) \to \partial D(0; R_2)$ whenever $\{z_n\} \subseteq A(0; r_1, R_1)$ converges to a point on $\partial D(0; R_1)$.

We have met the hypotheses for the Schwarz reflection in circles. Therefore f extends across $\partial D(0; r_1)$ to an $F_1 \in H(A(0; r_1^2/R_1, R_1))$, which, by symmetry, maps one-to-one and onto $A(0; r_2^2/R_2, R_2)$. Since $\partial D(0; R_1)$ and $\partial D(0; R_2)$ reflect to $\partial D(0; r_1^2/R_1)$ and $\partial D(0; r_2^2/R_2)$, respectively, our above observation about convergence to outer circles allows for subsequent reflection in inner circles. This can be done ad infinitum, the mth reflection resulting in an analytic function F_m taking

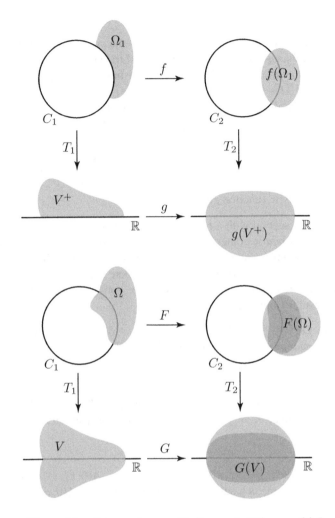

Figure 6.5   Constructions used in the proof of Theorem 6.3.5

$A(0; r_1^{m+1}/R_1^m, R_1)$ one-to-one and onto $A(0; r_2^{m+1}/R_2^m, R_2)$. The end result is a one-to-one analytic F mapping $A(0; 0, R_1)$ onto $A(0; 0, R_2)$.

Since F is bounded, it has a removable singularity at 0. Hence $\lim_{z \to 0} F(z)$ exists. But the correspondences of the circles in the reflection process give

$$\left| F\left(\frac{r_1^{m+1}}{R_1^m} \right) \right| = \frac{r_2^{m+1}}{R_2^m} \to 0$$

as $m \to \infty$. It follows that $\lim_{z \to 0} F(z) = 0$, and F can be extended by defining $F(0) = 0$.

We now have that $F \colon D(0; R_1) \to D(0; R_2)$ is one-to-one, onto, and analytic. If $G(z) = F(R_1 z)/R_2$, then $G \in \text{Aut } \mathbb{D}$ and $G(0) = 0$. By Theorem 3.5.7, $G(z) = cz$

for some $c \in \partial \mathbb{D}$. It follows that $F(z) = R_2 c z / R_1$. But F and f agree on the original domain, and hence

$$r_2 = |F(r_1)| = \frac{R_2 r_1}{R_1}.$$

We conclude that $r_1 / R_1 = r_2 / R_2$.

In the case that $f(A(0; r_1, s))$ is a subset of $A(0; \rho, R_2)$ (the second case above), let $g(z) = r_2 R_2 / z$. Then g exchanges $A(0; r_2, \rho)$ and $A(0; \rho, R_2)$, fixing $\partial D(0; \rho)$. Thus $g \circ f$ is of the case studied above, and hence $r_1 / R_1 = r_2 / R_2$ and $g \circ f(z) = R_2 c z / R_1$. The result follows because g is its own inverse. $\square$

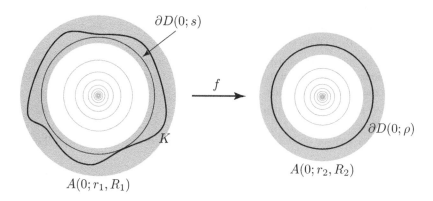

Figure 6.6   Constructions used in the proof of Theorem 6.3.6, together with boundaries of repeated reflections of the annuli in inner circles

Summary and Notes for Section 6.3.

The connection between harmonic and analytic functions can be exploited in the study of each. This was seen in Section 6.1, where our knowledge of analytic functions provided straightforward proofs of several results. We now see that careful use of geometric mapping properties of analytic functions can help to provide closed-form solutions to the Dirichlet problem, and the Schwarz reflection principle for analytic functions follows from an easier-to-prove reflection principle for harmonic functions.

The (analytic) Schwarz reflection principle provides us with an example of *analytic continuation*. That is, given an analytic function defined on an open set, the process of enlarging the domain and consistently defining the analytic function in the larger domain.

While we were impressed in Section 6.2 by the abundance of analytic functions provided by the solution to the Dirichlet problem, here we see that there exist no "interesting" analytic bijections from one annulus onto another.

Exercises for Section 6.3.

1. Let $\varphi\colon \partial\mathbb{D} \to \mathbb{R}$ be defined by

$$\varphi(e^{i\theta}) = \begin{cases} 2 & \text{if } 0 \le \theta \le \pi/2, \\ 3 & \text{if } \pi/2 < \theta < \pi, \\ -1 & \text{if } \pi \le \theta < 2\pi. \end{cases}$$

Find a formula for the solution u to the Dirichlet problem on $\mathbb{D}$ with boundary function φ. Calculate the following radial limits

$$\lim_{r\to 1^-} u(r), \qquad \lim_{r\to 1^-} u(ir), \qquad \lim_{r\to 1^-} u(-r).$$

What do you observe?

2. Let u be the harmonic function in the conclusion of Theorem 6.3.2. Find an $f \in H(\mathbb{D})$ such that $u = \text{Re } f$. What is a harmonic conjugate of u?

3. Remark 6.2.7 illustrates how the solution to the Dirichlet problem is generalized from $\mathbb{D}$ to an arbitrary disk. In that spirit, adapt Theorem 6.3.2 for a piecewise constant function on the boundary of a general disk.

4. Let $\Omega \subseteq \mathbb{C}$ be a domain such that $\text{Im } z > 0$ for all $z \in \Omega$ and there exists $a \in \partial\Omega \cap \mathbb{R}$ and $r > 0$ such that $D(a; r)^+ \subseteq \Omega$. Let $I = (a - r, a + r)$, and suppose $f\colon \Omega \cup I \to \mathbb{C}$ is continuous, analytic on Ω, and $f(x) = 0$ for all $x \in I$. Prove that $f(z) = 0$ for all $z \in \Omega$.

5. Let $f \in H(\mathbb{C})$ such that $f(z)$ is real when z is real and $f(z)$ is imaginary when z is imaginary. Prove that f is an odd function. (That is, verify $f(-z) = -f(z)$ for all $z \in \mathbb{C}$.)

6. Prove a meromorphic Schwarz reflection principle: Let $\Omega \subseteq \mathbb{C}$ be open and symmetric in $\mathbb{R}$, f be meromorphic on Ω^+ such that the set of singularities of f has no limit points in Ω^0, and suppose $v = \text{Im } f$ extends continuously to equal 0 on Ω^0. Then there is a meromorphic function F on Ω satisfying $F(z) = f(z)$ for each $z \in \Omega^+$ at which f has no singularity.

7. Let $f \in H(\mathbb{C})$, and suppose that $|f|$ is constant on some circle $\partial D(a; r)$. Prove that $f(z) = c(z - a)^n$ for some $c \in \mathbb{C}$ and $n \in \mathbb{N} \cup \{0\}$.

8. Let $S \subseteq \mathbb{C}$ be an open infinite strip and $f\colon \overline{S} \to \mathbb{C}$ be continuous, analytic on S, and such that $\text{Re } f(z) = 0$ for all $z \in \partial S$. Prove that f is unbounded or constant. (Consider the technique used in the proof of Theorem 6.3.6.) Give an example where f is unbounded.

6.4 Fourier Series

In this section, we are primarily concerned with analysis of functions defined on the unit circle $\partial\mathbb{D}$. (This is equivalent to studying 2π-periodic functions on $\mathbb{R}$; see Remark 6.4.4.)

Let φ be real valued and piecewise continuous on $\partial\mathbb{D}$ and $u\colon \mathbb{D} \to \mathbb{R}$ be the harmonic function that is the solution to the Dirichlet problem on $\mathbb{D}$ with boundary function φ. (See Theorem 6.2.6.) We now essentially reverse the work done at the start of Section 6.2, albeit in a much more general setting. As argued in the proof Theorem 6.2.6, φ is bounded. By decomposing as in (2.8.4) and applying the Weierstrass M-test (for fixed $r \in [0, 1)$) to integrate inside the series, we obtain

$$
\begin{aligned}
u(re^{i\theta}) &= \frac{1}{2\pi} \int_0^{2\pi} P_r(\theta - t)\varphi(e^{it})\, dt \\
&= \frac{1}{2\pi} \int_0^{2\pi} \left(\sum_{n=-\infty}^{\infty} r^{|n|} e^{in(\theta-t)} \right) \varphi(e^{it})\, dt \\
&= \sum_{n=-\infty}^{\infty} a_n r^{|n|} e^{in\theta}
\end{aligned}
\tag{6.4.1}
$$

for all $re^{i\theta} \in \mathbb{D}$, where

$$
a_n = \frac{1}{2\pi} \int_0^{2\pi} e^{-int} \varphi(e^{it})\, dt, \qquad n \in \mathbb{Z}.
\tag{6.4.2}
$$

When calculating these coefficients, it is often beneficial to remember that the integral can be taken over any interval of length 2π by Exercise 1 in Section 6.2.

We know that $\lim_{r \to 1^-} u(re^{i\theta}) = \varphi(e^{i\theta})$ for all $\theta \in \mathbb{R}$ such that φ is continuous at $e^{i\theta}$. We are, of course, tempted by the idea of taking the limit as $r \to 1^-$ inside the series (6.4.1) and concluding that the resulting series is equal to φ at its points of continuity. *We cannot do this!* This strategy (and exchanging limit processes, in general) always seems to have worked for us in the past, but not any more. (See Example 6.4.11.) We carefully phrase the following definition.

6.4.1 Definition. If φ is real valued and piecewise continuous on $\partial\mathbb{D}$, then the *Fourier series* of φ is

$$
\varphi(e^{i\theta}) \sim \sum_{n=-\infty}^{\infty} a_n e^{in\theta},
\tag{6.4.3}
$$

where $\{a_n\}_{n=-\infty}^{\infty}$ is the sequence of *Fourier coefficients* given in (6.4.2).

Again, we do not know for what $\theta \in \mathbb{R}$ the series converges or if it converges to φ when it does converge. Thus we use the symbol "$\sim$" to denote the relationship and understand this to be a "formal" series until convergence questions can be resolved.

The following result, addressing convergence under a certain circumstance, takes advantage of the above analysis.

6.4.2 Theorem. *If $\varphi \colon \partial\mathbb{D} \to \mathbb{R}$ is continuous and the series $\sum_{n=-\infty}^{\infty} a_n$ of Fourier coefficients of φ converges absolutely, then the Fourier series of φ converges to φ uniformly.*

Proof. Since $\sum_{n=-\infty}^{\infty} |a_n|$ converges, the Fourier series of φ converges uniformly on $\partial\mathbb{D}$ by the Weierstrass M-test. To conclude that the limit is actually φ, our above reasoning shows it is sufficient to verify

$$\lim_{r \to 1^-} \sum_{n=-\infty}^{\infty} a_n r^{|n|} e^{in\theta} = \sum_{n=-\infty}^{\infty} a_n e^{in\theta}$$

for all θ.

Let $\varepsilon > 0$. There is $N \in \mathbb{N}$ such that $\sum_{|n|>N} |a_n| < \varepsilon/2$. Since

$$\lim_{r \to 1^-} \sum_{n=-N}^{N} |a_n|(1 - r^{|n|}) = 0,$$

we may choose $\delta \in (0,1)$ such that for all $r \in (1 - \delta, 1)$, we have

$$\sum_{n=-N}^{N} |a_n|(1 - r^{|n|}) < \frac{\varepsilon}{2}.$$

It follows that for all $r \in (1 - \delta, 1)$ and $\theta \in \mathbb{R}$,

$$\left| \sum_{n=-\infty}^{\infty} a_n r^{|n|} e^{in\theta} - \sum_{n=-\infty}^{\infty} a_n e^{in\theta} \right| = \left| \sum_{n=-\infty}^{\infty} a_n (1 - r^{|n|}) e^{in\theta} \right|$$

$$\leq \sum_{n=-N}^{N} |a_n|(1 - r^{|n|}) + \sum_{|n|>N} |a_n|(1 - r^{|n|})$$

$$< \frac{\varepsilon}{2} + \sum_{|n|>N} |a_n|$$

$$< \varepsilon,$$

as desired.   □

We are now equipped to give an appealing example.

6.4.3 Example. Consider the continuous function $\varphi \colon \partial \mathbb{D} \to \mathbb{R}$ given by

$$\varphi(e^{i\theta}) = \theta^2, \qquad -\pi < \theta \leq \pi.$$

We directly calculate its Fourier coefficients. First, we have

$$a_0 = \frac{1}{2\pi} \int_{-\pi}^{\pi} t^2 \, dt = \frac{\pi^2}{3}.$$

For $n \neq 0$, we use two integrations by parts to calculate

$$
\begin{aligned}
a_n &= \frac{1}{2\pi} \int_{-\pi}^{\pi} t^2 e^{-int} \, dt \\
&= \frac{1}{2\pi} \left(\frac{-t^2 e^{-int}}{in} \bigg]_{t=-\pi}^{\pi} + \frac{2te^{-int}}{n^2} \bigg]_{t=-\pi}^{\pi} - \frac{2}{n^2} \int_{-\pi}^{\pi} e^{-int} \, dt \right) \\
&= \frac{(-1)^n \pi - (-1)^n (-\pi)}{\pi n^2} \\
&= \frac{2(-1)^n}{n^2},
\end{aligned}
$$

using symmetry and that $e^{in\pi} = (-1)^n$ for all $n \in \mathbb{Z}$. We know from calculus that the series $\sum_{n=1}^{\infty} 1/n^2$ converges, and hence the series of Fourier coefficients is absolutely convergent. For all θ, we now have

$$
\varphi(e^{i\theta}) = \frac{\pi^2}{3} + 2 \sum_{|n|>0} \frac{(-1)^n}{n^2} e^{in\theta} = \frac{\pi^2}{3} + 4 \sum_{n=1}^{\infty} \frac{(-1)^n}{n^2} \cos n\theta.
$$

Setting $\theta = \pi$ in the above gives

$$
\pi^2 = \frac{\pi^2}{3} + 4 \sum_{n=1}^{\infty} \frac{1}{n^2}.
$$

Solving for the series gives the famous summation identity

$$
\sum_{n=1}^{\infty} \frac{1}{n^2} = \frac{\pi^2}{6}. \tag{6.4.4}
$$

Other substitutions lead to other identities. For instance, letting $\theta = 0$ gives

$$
\sum_{n=1}^{\infty} \frac{(-1)^{n+1}}{n^2} = \frac{\pi^2}{12}.
$$

6.4.4 Remark. Before further consideration of the convergence of Fourier series, we must make note of the convention that convergence of the doubly infinite series is *taken to be symmetric* in this setting. That is, given a real-valued piecewise continuous φ on $\partial \mathbb{D}$ with Fourier series $\sum_{n=-\infty}^{\infty} a_n e^{in\theta}$, we consider the partial sums

$$
\varphi_N(e^{i\theta}) = \sum_{n=-N}^{N} a_n e^{in\theta} \tag{6.4.5}
$$

for $N \in \mathbb{N}$ and say the series converges to $s \in \mathbb{R}$ at a given $e^{i\theta}$ if $\varphi_N(e^{i\theta}) \to s$ as $N \to \infty$. Contrast this with Definition 5.1.1, where we emphasized that the series of negative- and nonnegative-indexed terms must separately converge. It can certainly be that a doubly infinite series $\sum_{n=-\infty}^{\infty} z_n$ diverges but $\lim_{N\to\infty} \sum_{n=-N}^{N} z_n$ exists. The latter limit is often called the *principal value* of the series.

As a partial explanation for this convention, we note that $a_{-n} = \bar{a}_n$ for all n by (6.4.2), and hence

$$\varphi_N(e^{i\theta}) = a_0 + \sum_{n=1}^{N} a_n e^{in\theta} + \sum_{n=1}^{N} \overline{a_n e^{in\theta}} = a_0 + 2\operatorname{Re} \sum_{n=1}^{N} a_n e^{in\theta}. \qquad (6.4.6)$$

Thus we may interpret the convergence of the Fourier series as the convergence of the singly infinite series formed by taking $N \to \infty$ on the right-hand side.

We may take this one step further and note

$$\varphi_N(e^{i\theta}) = a_0 + 2\sum_{n=1}^{N} [(\operatorname{Re} a_n) \cos n\theta - (\operatorname{Im} a_n) \sin n\theta].$$

It is therefore common, when considering Fourier series for a 2π-periodic function $f : \mathbb{R} \to \mathbb{R}$, to consider the complex-free trigonometric form

$$f(x) \sim \frac{\alpha_0}{2} + \sum_{n=1}^{\infty} (\alpha_n \cos nx + \beta_n \sin nx), \qquad (6.4.7)$$

where the partial sums match our convention and the Fourier coefficients are given by

$$\alpha_n = 2\operatorname{Re} a_n = \frac{1}{\pi} \int_0^{2\pi} f(t) \cos nt \, dt, \qquad (6.4.8)$$

$$\beta_n = -2\operatorname{Im} a_n = \frac{1}{\pi} \int_0^{2\pi} f(t) \sin nt \, dt. \qquad (6.4.9)$$

Now rewrite the partial sum (6.4.5) as

$$\varphi_N(e^{i\theta}) = \sum_{n=-N}^{N} \left(\frac{1}{2\pi} \int_0^{2\pi} e^{-int} \varphi(e^{it}) \, dt \right) e^{in\theta}$$

$$= \frac{1}{2\pi} \int_0^{2\pi} \left[\sum_{n=-N}^{N} e^{in(\theta-t)} \right] \varphi(e^{it}) \, dt.$$

This mimics our development of the Poisson kernel, except we cannot guarantee convergence of the full series, and hence only consider partial sums. The result is a new kernel that is the partial sum of the Poisson kernel with $r = 1$.

6.4.5 Definition. The family of 2π-periodic functions $D_N : \mathbb{R} \to \mathbb{R}$, $N = 0, 1, \ldots$, given by

$$D_N(t) = \sum_{n=-N}^{N} e^{int} \qquad (6.4.10)$$

is called the *Dirichlet kernel*.

With this in place, we see that our partial sum is

$$\varphi_N(e^{i\theta}) = \frac{1}{2\pi}\int_0^{2\pi} D_N(\theta - t)\varphi(e^{it})\,dt.$$

Using the second identity in Exercise 1 from Section 6.2, we have the alternative formulation

$$\varphi_N(e^{i\theta}) = \frac{1}{2\pi}\int_0^{2\pi} D_N(t)\varphi(e^{i(\theta-t)})\,dt.$$

6.4.6 Theorem. *The Dirichlet kernel satisfies the following properties.*

(a) *For all $N \in \mathbb{N}$ and $t \in \mathbb{R}$, $D_0(t) = 1$ and*

$$D_N(t) = 1 + 2\sum_{n=1}^{N}\cos nt.$$

(b) *For all $N \in \mathbb{N} \cup \{0\}$ and $t \in \mathbb{R}$ such that $t \neq 2m\pi$ for some $m \in \mathbb{Z}$,*

$$D_N(t) = \frac{\sin((2N+1)t/2)}{\sin(t/2)}.$$

For $m \in \mathbb{Z}$, $D_N(2m\pi) = 2N + 1$.

(c) *For all $N \in \mathbb{N} \cup \{0\}$ and $t \in \mathbb{R}$, $D_N(t) = D_N(-t)$.*

(d) *For all $N \in \mathbb{N} \cup \{0\}$,*

$$\frac{1}{2\pi}\int_{-\pi}^{0} D_N(t)\,dt = \frac{1}{2\pi}\int_0^{\pi} D_N(t)\,dt = \frac{1}{2}.$$

Proof. That $D_0(t) = 1$ is a direct calculation, and the rest of part (a) is an immediate consequence of

$$D_N(t) = 1 + \sum_{n=1}^{N} e^{int} + \sum_{n=1}^{N} e^{-int} = 1 + 2\,\mathrm{Re}\sum_{n=1}^{N} e^{int}.$$

From this, we have $D_N(2m\pi) = 2N + 1$ for $m \in \mathbb{Z}$.

The rest of part (b) follows by considering the telescoping sum

$$2i\sin\frac{t}{2} D_N(t) = (e^{it/2} - e^{-it/2})D_N(t)$$
$$= \sum_{n=-N}^{N} e^{i(n+1/2)t} - \sum_{n=-N}^{N} e^{i(n-1/2)t}$$
$$= e^{i(N+1/2)t} - e^{i(-N-1/2)t}$$
$$= 2i\sin\left(Nt + \frac{t}{2}\right).$$

(Figure 6.7 is useful for visualizing this form of $D_N(t)$.)

Part (c) follows from part (a). Moving the integral inside the sum yields

$$\frac{1}{2\pi} \int_{-\pi}^{\pi} D_N(t)\, dt = 1.$$

Part (d) results from using the symmetry in part (c).   □

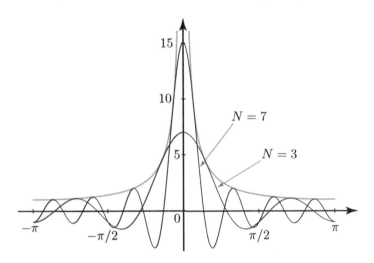

Figure 6.7   Graphs of $D_N(t)$ over $[-\pi, \pi]$ for $N = 3, 7$ together with (shaded) bounding function $|\csc(t/2)|$ (graph not to scale)

Before we can prove our main convergence theorem, we need one more result and its useful corollary.

6.4.7 Bessel's Inequality.   *Let φ be a real-valued piecewise continuous function on $\partial\mathbb{D}$ with Fourier coefficients $\{a_n\}_{n=-\infty}^{\infty}$. Then*

$$\sum_{n=-\infty}^{\infty} |a_n|^2 \leq \frac{1}{2\pi} \int_0^{2\pi} [\varphi(e^{i\theta})]^2 \, d\theta.$$

Proof. We know from (6.4.6) that each partial sum φ_N is real valued. Observe that for all θ such that $\varphi(e^{i\theta})$ exists,

$$0 \leq [\varphi(e^{i\theta}) - \varphi_N(e^{i\theta})]^2 = [\varphi(e^{i\theta})]^2 - 2\varphi(e^{i\theta})\varphi_N(e^{i\theta}) + [\varphi_N(e^{i\theta})]^2. \quad (6.4.11)$$

We pass the integral through the finite sums in the calculations

$$\frac{1}{2\pi} \int_0^{2\pi} \varphi(e^{i\theta}) \varphi_N(e^{i\theta}) \, d\theta = \sum_{n=-N}^{N} \frac{a_n}{2\pi} \int_0^{2\pi} e^{in\theta} \varphi(e^{i\theta}) \, d\theta$$

$$= \sum_{n=-N}^{N} a_n a_{-n}$$

$$= \sum_{n=-N}^{N} |a_n|^2$$

and

$$\frac{1}{2\pi} \int_0^{2\pi} [\varphi_N(e^{i\theta})]^2 \, d\theta = \frac{1}{2\pi} \int_0^{2\pi} \left(\sum_{n=-N}^{N} a_n e^{in\theta} \right) \left(\sum_{m=-N}^{N} a_m e^{im\theta} \right) d\theta$$

$$= \sum_{n=-N}^{N} \sum_{m=-N}^{N} \frac{a_n a_m}{2\pi} \int_0^{2\pi} e^{i(n+m)\theta} \, d\theta \qquad (6.4.12)$$

$$= \sum_{n=-N}^{N} a_n a_{-n}$$

$$= \sum_{n=-N}^{N} |a_n|^2,$$

where we observe the individual integrals in (6.4.12) are equal to 0 except when $m = -n$. Therefore integrating (6.4.11) gives

$$0 \le \frac{1}{2\pi} \int_0^{2\pi} [\varphi(e^{i\theta})]^2 \, d\theta - 2 \sum_{n=-N}^{N} |a_n|^2 + \sum_{n=-N}^{N} |a_n|^2,$$

and hence

$$\sum_{n=-N}^{N} |a_n|^2 \le \frac{1}{2\pi} \int_0^{2\pi} [\varphi(e^{i\theta})]^2 \, d\theta.$$

Since this holds for all N, the result follows. $\square$

6.4.8 Riemann–Lebesgue Lemma. *Let φ be real valued and piecewise continuous on $\partial \mathbb{D}$ with Fourier coefficients $\{a_n\}_{n=-\infty}^{\infty}$. Then $\lim_{|n| \to \infty} a_n = 0$.*

Proof. We know that $\sum_{n=-\infty}^{\infty} |a_n|^2$ converges, and hence $\lim_{|n| \to \infty} |a_n|^2 = 0$. $\square$

We are now equipped to prove the following strong pointwise convergence result, which gives that if φ is not just piecewise continuous, but is piecewise continuously differentiable, then the Fourier series converges to φ at all points where φ is continuous and furthermore, *the series converges to the midpoint of the jump at any discontinuity!*

6.4.9 Theorem. *If φ is real valued and piecewise continuously differentiable on $\partial\mathbb{D}$ with Fourier coefficients $\{a_n\}_{n=-\infty}^{\infty}$, then the Fourier series of φ converges on $\partial\mathbb{D}$ to*

$$\sum_{n=-\infty}^{\infty} a_n e^{in\theta} = \frac{1}{2}\left(\lim_{t\to\theta^-}\varphi(e^{it}) + \lim_{t\to\theta^+}\varphi(e^{it})\right).$$

In particular, the series converges to $\varphi(e^{i\theta})$ when φ is continuous at $e^{i\theta}$.

We note that φ is *piecewise continuously differentiable* on $\partial\mathbb{D}$ if $\varphi(e^{it})$ is piecewise continuously differentiable in the variable t on $[0, 2\pi]$. (See Definition 2.8.2.) Therefore the one-sided limits of $\varphi(e^{it})$ and $[d/dt]\varphi(e^{it})$ as t approaches any $\theta \in \mathbb{R}$ exist (and are equal when θ is not a multiple of 2π plus one of the finitely many points in $[0, 2\pi]$ of non–continuous differentiability), and we define the shorthand

$$\varphi^-(e^{i\theta}) = \lim_{t\to\theta^-}\varphi(e^{it}), \qquad \varphi^+(e^{i\theta}) = \lim_{t\to\theta^+}\varphi(e^{it})$$

for all θ. Now, on with the proof.

Proof. Fix $\theta \in \mathbb{R}$. For all $N \in \mathbb{N}$,

$$\varphi_N(e^{i\theta}) - \frac{\varphi^+(e^{i\theta}) + \varphi^-(e^{i\theta})}{2}$$

$$= \frac{1}{2\pi}\int_{-\pi}^{\pi} D_N(t)\varphi(e^{i(\theta-t)})\,dt - \frac{\varphi^+(e^{i\theta})}{2\pi}\int_{-\pi}^{0} D_N(t)\,dt$$

$$\qquad - \frac{\varphi^-(e^{i\theta})}{2\pi}\int_{0}^{\pi} D_N(t)\,dt$$

$$= \frac{1}{2\pi}\int_{-\pi}^{0} D_N(t)[\varphi(e^{i(\theta-t)}) - \varphi^+(e^{i\theta})]\,dt$$

$$\qquad + \frac{1}{2\pi}\int_{0}^{\pi} D_N(t)[\varphi(e^{i(\theta-t)}) - \varphi^-(e^{i\theta})]\,dt$$

$$= \frac{1}{2\pi}\int_{-\pi}^{\pi} g(e^{it})\sin\frac{(2N+1)t}{2}\,dt,$$

where g is given for $t \in [-\pi, \pi) \setminus \{0\}$ for which $\varphi(e^{i(\theta-t)})$ exists by

$$g(e^{it}) = \begin{cases} \dfrac{\varphi(e^{i(\theta-t)}) - \varphi^+(e^{i\theta})}{\sin(t/2)} & \text{if } -\pi \leq t < 0, \\[4mm] \dfrac{\varphi(e^{i(\theta-t)}) - \varphi^-(e^{i\theta})}{\sin(t/2)} & \text{if } 0 < t < \pi. \end{cases}$$

We will show that g is piecewise continuous. Since φ is piecewise continuous, we need only verify $\lim_{t\to 0^-} g(e^{it})$ and $\lim_{t\to 0^+} g(e^{it})$ each exist. Letting $u = \theta - t$, we find

$$\lim_{t\to 0^-}\frac{d}{dt}\varphi(e^{i(\theta-t)}) = -\lim_{u\to\theta^+}\frac{d}{du}\varphi(e^{iu}),$$

which we know exists. An application of l'Hôpital's rule gives

$$\lim_{t \to 0^-} g(e^{it}) = \lim_{t \to 0^-} \left(2 \sec \frac{t}{2} \right) \left(\frac{d}{dt} \varphi(e^{i(\theta - t)}) \right),$$

which therefore exists. The right-hand limit exists through similar reasoning.

The observation

$$\sin \frac{(2N+1)t}{2} = \sin \frac{t}{2} \cos Nt + \cos \frac{t}{2} \sin Nt = \sin \frac{t}{2} \operatorname{Re} e^{iNt} + \cos \frac{t}{2} \operatorname{Im} e^{iNt}$$

allows us to write

$$\varphi_N(e^{i\theta}) - \frac{\varphi^+(e^{i\theta}) + \varphi^-(e^{i\theta})}{2} = \operatorname{Re} \left(\frac{1}{2\pi} \int_0^{2\pi} \left[g(e^{it}) \sin \frac{t}{2} \right] e^{iNt} \, dt \right)$$

$$+ \operatorname{Im} \left(\frac{1}{2\pi} \int_0^{2\pi} \left[g(e^{it}) \cos \frac{t}{2} \right] e^{iNt} \, dt \right).$$

Each of the above integrals is equal to the $-N^{\text{th}}$ Fourier coefficient of the piecewise continuous function within the brackets in its integrand. Hence the integrals tend to 0 as $N \to \infty$ by the Riemann–Lebesgue lemma. This gives the result. □

6.4.10 Example. Consider the function $\varphi \colon \partial \mathbb{D} \to \mathbb{R}$ given by

$$\varphi(e^{i\theta}) = \theta, \qquad 0 \leq \theta < 2\pi.$$

We note that φ is piecewise continuously differentiable with its only discontinuity at 1. We calculate the Fourier coefficients of φ by first noting

$$a_0 = \frac{1}{2\pi} \int_0^{2\pi} t \, dt = \pi.$$

For $n \neq 0$, integration by parts yields

$$a_n = \frac{1}{2\pi} \int_0^{2\pi} t e^{-int} \, dt = \frac{1}{2\pi} \left(\frac{-t e^{-int}}{in} \bigg]_{t=0}^{2\pi} + \frac{1}{in} \int_0^{2\pi} e^{-int} \, dt \right) = -\frac{1}{in}.$$

It follows that

$$\varphi(e^{i\theta}) = \pi + \sum_{|n|>0} \frac{-1}{in} e^{in\theta}$$

$$= \pi + \sum_{n=1}^{\infty} \left(\frac{-1}{in} e^{in\theta} + \frac{1}{in} e^{-in\theta} \right)$$

$$= \pi - 2 \sum_{n=1}^{\infty} \frac{\sin n\theta}{n}.$$

at the points where φ is continuous. At the discontinuity (using $\theta = 0$), we see that the series converges to

$$\frac{1}{2}\left(\lim_{t \to 0^-} \varphi(e^{it}) + \lim_{t \to 0^+} \varphi(e^{it})\right) = \frac{2\pi + 0}{2} = \pi.$$

Note that the series converges to a discontinuous function, and hence the convergence cannot be uniform. It is also evident that the series of Fourier coefficients is not absolutely convergent, in concert with Theorem 6.4.2.

As in Example 6.4.3, we can use this representation to sum some numerical series. For instance, by letting $\theta = \pi/2$ and observing

$$\sin\frac{n\pi}{2} = \begin{cases} 0 & \text{if } n \text{ is even,} \\ (-1)^{k+1} & \text{if } n = 2k - 1 \text{ for } k \in \mathbb{N}, \end{cases}$$

we obtain

$$\frac{\pi}{2} = \pi - 2\sum_{k=1}^{\infty} \frac{(-1)^{k+1}}{2k - 1}.$$

Solving for the series gives

$$\sum_{k=1}^{\infty} \frac{(-1)^{k+1}}{2k - 1} = \frac{\pi}{4}. \tag{6.4.13}$$

We now come full circle back to our remarks at the start of the section with an example that shows that the Fourier series of a continuous function on $\partial \mathbb{D}$ may fail to converge. Such a function generates a harmonic extension inside $\mathbb{D}$ that is representable by a series. This harmonic function on $\mathbb{D}$ extends continuously to the boundary function, but *the series* (6.4.1) *that represents it does not!* This surprising illustration of the inability to move the limit as $r \to 1^-$ inside of that series is well worth its complexity.

6.4.11 Example. We begin by considering $\varphi \colon \partial \mathbb{D} \to \mathbb{R}$ given by $\varphi(e^{i\theta}) = \theta - \pi$ for all $0 \le \theta < 2\pi$. This is simply π subtracted from the function considered in Example 6.4.10, and hence the Fourier coefficients of φ are $a_n = i/n$ for all $n \in \mathbb{Z} \setminus \{0\}$ and $a_0 = 0$.

Our first step is to find a bound on the partial sums of φ. Let $u \colon \mathbb{D} \to \mathbb{R}$ be the harmonic function that is the solution to the Dirichlet problem on $\mathbb{D}$ with boundary values φ. Using properties (b) and (c) of Lemma 6.2.5, we have, for all $re^{i\theta} \in \mathbb{D}$,

$$|u(re^{i\theta})| \le \frac{1}{2\pi}\int_0^{2\pi} P_r(\theta - t)|\varphi(e^{it})|\, dt \le \frac{1}{2\pi}\int_0^{2\pi} \pi P_r(\theta - t)\, dt = \pi.$$

We write u as in (6.4.1) with Fourier coefficients $\{a_n\}$ so that, for a given $N \in \mathbb{N}$ and $\theta \in \mathbb{R}$,

$$
|\varphi_N(e^{i\theta}) - u(re^{i\theta})| = \left| \sum_{1 \leq |n| \leq N} \frac{i(1 - r^{|n|})e^{in\theta}}{n} - \sum_{|n| > N} \frac{ir^{|n|}e^{in\theta}}{n} \right|
$$

$$
\leq \sum_{1 \leq |n| \leq N} \frac{1 - r^{|n|}}{|n|} + \sum_{|n| > N} \frac{r^{|n|}}{|n|}
$$

$$
= 2 \sum_{n=1}^{N} \frac{1 - r^n}{n} + 2 \sum_{n=N+1}^{\infty} \frac{r^n}{n}
$$

$$
\leq 2 \sum_{n=1}^{N} (1 - r) + \frac{2}{N} \sum_{n=N+1}^{\infty} r^n
$$

$$
< 2N(1 - r) + \frac{2}{N} \frac{1}{1 - r},
$$

where we used the observation

$$
1 - r^n = (1 - r) \sum_{k=0}^{n-1} r^k \leq n(1 - r).
$$

Since the above calculation holds for all $r \in [0, 1)$, we use $r = 1 - 1/N$ and the triangle inequality to find

$$
|\varphi_N(e^{i\theta})| < \pi + 4
$$

for all N and θ. Let $M = \pi + 4$.

For all $N \in \mathbb{N}$, define $\Phi^N : \partial \mathbb{D} \to \mathbb{R}$ by

$$
\Phi^N(e^{i\theta}) = \text{Im}(e^{2iN\theta} \varphi_N(e^{i\theta}))
$$

$$
= \text{Im} \left(\sum_{n=N}^{3N} a_{n-2N} e^{in\theta} \right)
$$

$$
= \frac{1}{2i} \left(\sum_{n=-3N}^{-N} a_{|n|-2N} e^{in\theta} + \sum_{n=N}^{3N} a_{n-2N} e^{in\theta} \right) \tag{6.4.14}
$$

(We use the superscript to avoid confusion with the subscript used for partial sums.) Here, we used that $\bar{a}_n = -a_n$ for all n. Observe that (6.4.14) is the Fourier series of Φ^N. Take note that $|\Phi^N(e^{i\theta})| < M$ for all θ and

$$
\Phi_{2N}^N(1) = \frac{1}{i} \sum_{n=N}^{2N} a_{n-2N} = \sum_{n=-N}^{-1} \frac{1}{n}.
$$

Thus

$$
|\Phi_{2N}^N(1)| = \sum_{n=1}^{N} \frac{1}{n} > \sum_{n=1}^{N} \int_{n}^{n+1} \frac{dx}{x} > \int_{1}^{N} \frac{dx}{x} = \ln N.
$$

We can now construct our example. Set $N_k = 2^{k^3}$ for $k \in \mathbb{N}$. Then $N_{k+1} > 3N_k$ for all k, and therefore the indices in the expansion of $\Phi^{N_{k+1}}$ in (6.4.14) all have absolute value greater than those of Φ^{N_k}. Define $f : \partial\mathbb{D} \to \mathbb{R}$ by

$$f(e^{i\theta}) = \sum_{k=1}^{\infty} \frac{\Phi^{N_k}(e^{i\theta})}{k^2} = \sum_{k=1}^{\infty} \left(\sum_{n=-3N_k}^{-N_k} \frac{a_{|n|-2N_k}}{2ik^2} e^{in\theta} + \sum_{n=N_k}^{3N_k} \frac{a_{n-2N_k}}{2ik^2} e^{in\theta} \right).$$

Since the terms of the series are bounded by M/k^2, the series converges uniformly by the Weierstrass M-test, and hence f is continuous. Uniform convergence also implies that the integrals in the formula for the Fourier coefficients of f can be taken inside the sums, and thus f has Fourier series

$$f(e^{i\theta}) \sim \sum_{n=-\infty}^{\infty} b_n e^{in\theta},$$

where

$$b_n = \begin{cases} \dfrac{a_{|n|-2N_k}}{2ik^2} & \text{if } N_k \leq |n| \leq 3N_k \text{ for some } k \in \mathbb{N}, \\[2mm] 0 & \text{otherwise.} \end{cases}$$

Consider the particular Fourier partial sum

$$f_{2N_m}(1) = \sum_{k=1}^{m-1} \frac{\Phi^{N_k}(1)}{k^2} + \frac{\Phi^{N_m}_{2N_m}(1)}{m^2}$$

for a given m. We have

$$|f_{2N_m}(1)| \geq \frac{|\Phi^{N_m}_{2N_m}(1)|}{m^2} - \sum_{k=1}^{m-1} \frac{|\Phi^{N_k}(1)|}{k^2} > \frac{\ln N_m}{m^2} - M \sum_{k=1}^{m-1} \frac{1}{k^2}.$$

Using $N_m = 2^{m^3}$ and Example 6.4.3, we see that

$$|f_{2N_m}(1)| > m \ln 2 - \frac{M\pi^2}{6},$$

and hence the sequence $\{f_{2N_m}(1)\}_{m=1}^{\infty}$ diverges. Since the sequence of partial sums of $f(1)$ has a divergent subsequence, the Fourier series of f diverges at 1.

Summary and Notes for Section 6.4.

We arrive at a Fourier series for a piecewise continuous function on $\partial\mathbb{D}$ by calculating the solution to the corresponding Dirichlet problem, forming a series by expanding the Poisson kernel, and taking $r \to 1^-$ (whether or not this is allowed). This is seen to be equivalent to a trigonometric series in terms of cosines and sines of a 2π-periodic function on $\mathbb{R}$. It is remarkable how different this representation is to that of a power series. While the latter converges uniformly on compact sets to an analytic (and hence infinitely differentiable) function, the former, despite having infinitely differentiable terms, may converge to a discontinuous function!

The historical development of Fourier series was much different than ours. In the 18th century, Daniel Bernoulli proposed that solutions of the wave equation, the partial differential equation modeling the vibration of a string, could be expressed as infinite sums of products of trigonometric functions, and that the function describing the initial position of the string would then be expressible as a trigonometric series with appropriately chosen coefficients. This led to a long, unresolved debate between Bernoulli, d'Alembert, and Leonhard Euler about the validity of such a representation.

In the 19th century, Jean Baptiste Joseph Fourier arrived at similar series in his work on the heat equation but was able to go further by explicitly calculating the Fourier coefficients (6.4.8) and (6.4.9) for the series representation of a function. It was left to show that the Fourier series actually converges to the function it represents, and this confounded a number of mathematicians, including Poisson and Augustin-Louis Cauchy. Eventually Dirichlet came to the rescue, using the Dirichlet kernel to show that the series converges for certain piecewise continuous functions.

For decades, it remained the belief of prominent mathematicians that it must be the case that any integrable periodic function's Fourier series converges to the function at points of continuity, and so it was quite the surprise when Paul du Bois-Reymond provided an example, similar to ours given at the end of the section, of a *continuous* function for which this fails.

We recall from Section 2.2 that Cauchy had mistakenly written that the limit of a series of continuous functions is continuous. Niels Henrik Abel provided a counterexample in the form of the trigonometric series

$$\sin t - \frac{1}{2} \sin 2t + \frac{1}{3} \sin 3t - \cdots,$$

which we can now verify converges to $t/2$ on $(-\pi, \pi)$, and extends periodically. Predating the work of Dirichlet, Abel was able to verify this series' sum using his work on the binomial theorem.

Lastly, we remark on the series on the left-hand side of (6.4.4). Pietro Mengoli proposed finding a closed form of this series in 1644, a challenge promulgated by Jakob Bernoulli 45 years later that became known as the "Basel problem." (The Bernoulli family was from Basel, Switzerland.) Euler (also from Basel) gained great fame by being the first to solve the problem in 1735, although his proof relied on a fact from function theory that Karl Weierstrass would prove over a century later. Now myriad proofs exist (Euler, himself, found another) including two in this text. (See Exercise 11 in Section 5.3.)

Exercises for Section 6.4.

1. For the following functions $\varphi \colon \partial \mathbb{D} \to \mathbb{R}$, find the Fourier series of φ and use it to verify the given numerical series, as is done in Examples 6.4.3 and 6.4.10. In (b), consider incorporating (6.4.13) when addressing the numerical series, and in (d), $c \in \mathbb{R} \setminus \{0\}$ is a constant.

(a) $\varphi(e^{i\theta}) = |\theta|, \ -\pi < \theta \le \pi,$ $\displaystyle\sum_{n=1}^{\infty} \frac{1}{(2n-1)^2} = \frac{\pi^2}{8}$

(b) $\varphi(e^{i\theta}) = \theta^3, \ -\pi < \theta \le \pi,$ $\displaystyle\sum_{n=1}^{\infty} \frac{(-1)^{n+1}}{(2n-1)^3} = \frac{\pi^3}{32}$

(c) $\varphi(e^{i\theta}) = |\sin\theta|,$ $\displaystyle\sum_{n=1}^{\infty} \frac{(-1)^{n+1}}{4n^2 - 1} = \frac{\pi - 2}{4}$

(d) $\varphi(e^{i\theta}) = e^{c\theta}, \ -\pi < \theta \le \pi,$ $\displaystyle\sum_{n=1}^{\infty} \frac{1}{n^2 + c^2} = \frac{\pi}{2c} \coth \pi c - \frac{1}{2c^2}$

2. Let φ be a real-valued piecewise continuous function on $\partial \mathbb{D}$ with Fourier coefficients $\{a_n\}_{n=-\infty}^{\infty}$. For the following functions ψ, how do the Fourier coefficients $\{b_n\}_{n=-\infty}^{\infty}$ of ψ relate to $\{a_n\}$?

   (a) $\psi(e^{i\theta}) = \varphi(e^{i(\theta + \alpha)})$, where $\alpha \in \mathbb{R}$ is constant

   (b) $\psi(e^{i\theta}) = \varphi(e^{-i\theta})$

3. Examine the proof of Bessel's inequality and show that equality holds if the Fourier series of φ converges to φ uniformly. This is known as *Parseval's equality*. Apply Parseval's equality to the Fourier series in Example 6.4.3 to derive the summation identity

$$\sum_{n=1}^{\infty} \frac{1}{n^4} = \frac{\pi^4}{90}.$$

4. Let φ be a real-valued piecewise continuous function on $\partial \mathbb{D}$. If $\varphi(e^{i\theta}) = \varphi(e^{-i\theta})$ for all $\theta \in \mathbb{R}$, prove that φ has a Fourier series of the form

$$\varphi(e^{i\theta}) \sim \frac{\alpha_0}{2} + 2 \sum_{n=1}^{\infty} \alpha_n \cos n\theta,$$

where the Fourier coefficients are all real.

5. Let φ be real valued, continuous, and piecewise continuously differentiable on $\partial \mathbb{D}$ with Fourier coefficients $\{a_n\}_{n=-\infty}^{\infty}$.

   (a) Show that the Fourier coefficients $\{b_n\}_{n=-\infty}^{\infty}$ of $[d/d\theta]\varphi(e^{i\theta})$ are well defined and satisfy $b_n = ina_n$ for all $n \in \mathbb{Z}$.

   (b) State a theorem about term-by-term differentiation of Fourier series.

   (c) Apply the theorem to the Fourier series found in part (a) of Exercise 1. What function on $\partial \mathbb{D}$ has the resulting Fourier series?

6. Let φ be a real-valued piecewise continuous function on $\partial \mathbb{D}$ with Fourier coefficients $\{a_n\}_{n=-\infty}^{\infty}$, let $\alpha \in \mathbb{R}$, and define $\Phi \colon \partial \mathbb{D} \to \mathbb{R}$ by

$$\Phi(e^{i\theta}) = \int_{\alpha}^{\theta} (\varphi(e^{it}) - a_0) \, dt.$$

Show that Φ is well defined on $\partial \mathbb{D}$ (that is, show $\Phi(e^{i\theta})$ is unchanged when θ is replaced by $\theta + 2m\pi$ for $m \in \mathbb{Z}$), and use Exercise 5 to state and prove a theorem about term-by-term antidifferentiation of Fourier series. Apply this to the Fourier series found in Example 6.4.3 and use the result to obtain the numerical series given in part (b) of Exercise 1.

7. Let $m \in \mathbb{N}$ and $\varphi \colon \partial \mathbb{D} \to \mathbb{R}$ be such that $\varphi(e^{i\theta})$ is m-times differentiable with respect to θ and the mth derivative is piecewise continuous.

   (a) Show that the Fourier coefficients $\{a_n\}_{n=-\infty}^{\infty}$ of φ satisfy

   $$|a_n| \leq \frac{M}{|n|^m}, \qquad n \in \mathbb{Z} \setminus \{0\},$$

   where M is a constant. (*Hint*: Use Exercise 5.)

   (b) Verify that if $m \geq 2$, then the Fourier series of φ converges uniformly to φ.

8. Let $\varphi, \psi \colon \partial \mathbb{D} \to \mathbb{R}$ be continuous. The *convolution* of φ and ψ is the function $\varphi * \psi \colon \partial \mathbb{D} \to \mathbb{R}$ given by

   $$(\varphi * \psi)(e^{i\theta}) = \frac{1}{2\pi} \int_0^{2\pi} \varphi(e^{i(\theta-t)}) \psi(e^{it}) \, dt.$$

   Prove the following.

   (a) $\varphi * \psi$ is continuous.

   (b) $\varphi * \psi = \psi * \varphi$.

   (c) If $\{a_n\}_{n=-\infty}^{\infty}$ and $\{b_n\}_{n=-\infty}^{\infty}$ are the Fourier coefficients of φ and ψ, respectively, then $\{a_n b_n\}_{n=-\infty}^{\infty}$ are the Fourier coefficients of $\varphi * \psi$.

9. Prove that the Dirichlet kernel satisfies

   $$\lim_{N \to \infty} \int_0^{2\pi} |D_N(t)| \, dt = \infty.$$

10. It is surprisingly hard to prove that if $\sum_{n=-\infty}^{\infty} a_n e^{in\theta}$ and $\sum_{n=-\infty}^{\infty} b_n e^{in\theta}$ both converge to the same piecewise continuous function on $\partial \mathbb{D}$ at its points of continuity, then $a_n = b_n$ for all $n \in \mathbb{Z}$. Show that this is the case if $\sum_{n=-\infty}^{\infty} a_n$ and $\sum_{n=-\infty}^{\infty} b_n$ are assumed to converge absolutely.

11. In this exercise, we explore the Cesàro summability of Fourier series. Parts (e) and (f) together form *Fejér's theorem*.

    (a) Let $\{z_n\}_{n=0}^{\infty} \subseteq \mathbb{C}$. The sequence of *Cesàro means* of $\{z_n\}$ is $\{a_n\}_{n=0}^{\infty}$ defined by

    $$a_n = \frac{1}{n+1} \sum_{k=0}^{n} z_k.$$

    Prove that if $\{z_n\}$ is convergent, then $\{a_n\}$ converges to the same limit. Give an example illustrating the failure of the converse.

    (b) Let φ be real valued and piecewise continuous on $\partial \mathbb{D}$, and let φ_n denote the nth partial sum of the Fourier series of φ. Then the Nth Cesàro mean of φ is

    $$\sigma_N = \frac{1}{N+1} \sum_{n=0}^{N} \varphi_n.$$

Prove that

$$\sigma_N(e^{i\theta}) = \frac{1}{2\pi} \int_0^{2\pi} K_N(\theta - t)\varphi(e^{it})\, dt,$$

where K_N is the Cesàro mean of the Dirichlet kernel functions $D_0, \ldots, D_N$. The family of functions K_N is called the *Fejér kernel*.

(c) Prove the following properties of K_N.

   i. $\dfrac{1}{2\pi} \displaystyle\int_0^{2\pi} K_N(t)\, dt = 1.$

   ii. $K_N(t) = \dfrac{1}{N+1} \left(\dfrac{\sin((N+1)t/2)}{\sin(t/2)} \right)^2.$

   iii. For any $\delta \in (0, \pi)$,

$$\lim_{N\to\infty} \int_\delta^\pi K_N(t)\, dt = 0.$$

(d) Show that for any $\theta, L \in \mathbb{R}$,

$$\sigma_N(e^{i\theta}) - L = \frac{1}{\pi} \int_0^\pi K_N(t) \left(\frac{\varphi(e^{i(\theta-t)}) + \varphi(e^{i(\theta+t)})}{2} - L \right) dt.$$

(e) Prove that for all $\theta \in \mathbb{R}$,

$$\lim_{N\to\infty} \sigma_N(e^{i\theta}) = \frac{1}{2} \left(\lim_{t\to\theta^-} \varphi(e^{it}) + \lim_{t\to\theta^+} \varphi(e^{it}) \right).$$

Comment on how this result compares to Theorem 6.4.9.

(f) Suppose $[a, b] \subseteq \mathbb{R}$ is such that φ is continuous at the point $e^{i\theta}$ for each $\theta \in [a, b]$. Show that $\sigma_N \to \varphi$ uniformly on $\{e^{i\theta} : a \le \theta \le b\}$.

12. Use Fejér's theorem from Exercise 11 to prove the Weierstrass approximation theorem (see Exercise 8 in Section 5.5) by completing the following steps.

(a) Show that for any $n = 0, 1, \ldots$, $\cos n\theta$ can be expressed as a sum of the form

$$\cos n\theta = \sum_{k=0}^n \alpha_k \cos^k \theta,$$

for some $\alpha_0, \ldots, \alpha_n \in \mathbb{R}$.

(b) Prove that the Nth Cesàro mean of a function φ as in Exercise 4 has the form

$$\sigma_N(e^{i\theta}) = \sum_{n=0}^N b_n \cos n\theta$$

for some $b_0, \ldots, b_N \in \mathbb{R}$.

(c) Let $f: [0, 1] \to \mathbb{R}$ be continuous, and set $\varphi(e^{i\theta}) = f(\cos\theta)$ for $\theta \in \mathbb{R}$. Use parts (a) and (b) together with part (f) of Exercise 11 to construct polynomials $p_N: [0, 1] \to \mathbb{R}$ such that $p_N \to f$ uniformly.

(d) Upgrade the result in (c) to work for complex-valued continuous functions on a general interval $[a, b]$.

EPILOGUE

In this coda to our course on complex analysis, we combine a multitude of ideas encountered throughout the text to prove the following theorem, one of the most impressive results in function theory.

Ep.1 Riemann Mapping Theorem. *Let $\Omega \subseteq \mathbb{C}$ be a simply connected domain such that $\Omega \neq \mathbb{C}$, and let $a \in \Omega$. Then there exists a unique analytic bijection $f : \Omega \to \mathbb{D}$ such that $f(a) = 0$ and $f'(a) > 0$.*

No matter how complicated the simply connected domain Ω and how inconveniently located the point a is within Ω, it must be that there is a conformal one-to-one correspondence between Ω and $\mathbb{D}$ taking a to 0 with positive derivative! (See Figure Ep.1 for some illustrations of messy domains.) It is worth noting that $\Omega \neq \mathbb{C}$ is necessary due to Liouville's theorem.

The theorem, with certain constraints placed on $\partial\Omega$, was given by Bernhard Riemann in his doctoral dissertation of 1851. Subsequent analysis by several mathematicians found flaws in Riemann's argument. Proofs of the theorem for general simply connected domains were provided by William Fogg Osgood and Constantin Carathéodory in the early 1900s. The standard proof used today, due to Lipót Fejér and Frigyes Riesz, relies on advanced notions beyond the scope of this text, but an alternative approach, due to Paul Koebe (and outlined in an exercise in [26]), gives us access.

Complex Analysis: A Modern First Course in Function Theory, First Edition. Jerry R. Muir, Jr.
© 2015 John Wiley & Sons, Inc. Published 2015 by John Wiley & Sons, Inc.

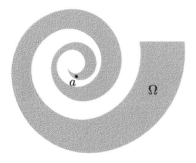
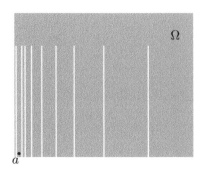

Figure Ep.1   Two domains Ω and points $a \in \Omega$ satisfying the hypotheses of the Riemann mapping theorem

In contrast to the regular chapters of this text, this epilogue is a short essay, written a bit more tersely and without exercises or examples. It should be thought of as a culmination of the preceding material, a tour de force of complex analysis that utilizes an abundance of topics from the previous chapters, including sequential compactness and connectedness from Chapter 1, uniform convergence, the complex exponential function and logarithm, and linear fractional transformations from Chapter 2, factorization using zeros, Schwarz's lemma, the open mapping theorem, and univalence from Chapter 3, Cauchy's theorem and integral formula from Chapter 4, the residue theorem and argument principle from Chapter 5, and harmonic functions and the Poisson integral formula from Chapter 6, to prove one of the most beautiful results in the field. Along the way, we will encounter a handful of other new and independently interesting developments.

Local Uniform Convergence

We begin with a notion of convergence that seems new, but will turn out to be familiar.

Ep.2 Definition. Let $E \subseteq \mathbb{C}$. A sequence of complex-valued functions $\{f_n\}_{n=1}^{\infty}$ on E converges to $f \colon E \to \mathbb{C}$ *locally uniformly* if, for every $a \in E$, there is an open set $U \subseteq \mathbb{C}$ such that $a \in U$ and $f_n \to f$ uniformly on $U \cap E$.

On a number of occasions, we have appreciated the power of uniform convergence on compact sets. Under such convergence, the limit function often inherits properties held by the functions in the sequence. We now prove the useful result that uniform convergence on compact sets is equivalent to local uniform convergence, which is sometimes an easier property to verify.

Ep.3 Theorem. *Let $\Omega \subseteq \mathbb{C}$ be open, $\{f_n\}_{n=1}^{\infty}$ be a sequence of functions on Ω, and $f \colon \Omega \to \mathbb{C}$. Then $\{f_n\}$ converges to f uniformly on compact subsets of Ω if and only if $\{f_n\}$ converges to f locally uniformly.*

Proof. First, suppose $f_n \to f$ uniformly on compact sets, and let $a \in \Omega$. There is $r > 0$ such that $\overline{D}(a; r) \subseteq \Omega$. Then $f_n \to f$ uniformly on $\overline{D}(a; r)$, and hence on $D(a; r)$.

Conversely, suppose $f_n \to f$ locally uniformly. Assume that $K \subseteq \Omega$ is compact, but that $\{f_n\}$ fails to converge to f uniformly on K. Then there exists $\varepsilon > 0$ such that for all $n \in \mathbb{N}$, $|f_k(a_n) - f(a_n)| \geq \varepsilon$ for some $k \geq n$ and $a_n \in K$. The sequence $\{a_n\}_{n=1}^{\infty} \subseteq K$ has a subsequence $\{a_{n_j}\}_{j=1}^{\infty}$ converging to some $a \in K$ by the Bolzano–Weierstrass theorem. There exists $r > 0$ such that $D(a; r) \subseteq \Omega$ and $f_n \to f$ uniformly on $D(a; r)$. Thus there is $N \in \mathbb{N}$ such that $|f_k(z) - f(z)| < \varepsilon$ for all $k \geq N$ and $z \in D(a; r)$. For some $j \in \mathbb{N}$, $n_j \geq N$ and $a_{n_j} \in D(a; r)$, giving a contradiction. Therefore $f_n \to f$ uniformly on K. $\qquad\square$

Here is a convenient consequence, another instance of exchanging limit processes.

Ep.4 Theorem. *Suppose $\Omega \subseteq \mathbb{C}$ is open and $\{f_n\}_{n=1}^{\infty} \subseteq H(\Omega)$ converges uniformly on compact subsets of Ω to $f \in H(\Omega)$. Then $f_n' \to f'$ uniformly on compact subsets of Ω.*

Proof. Let $a \in \Omega$ and $r > 0$ such that $\overline{D}(a; r) \subseteq \Omega$. For $\varepsilon > 0$, choose $N \in \mathbb{N}$ such that $|f_n(\zeta) - f(\zeta)| < r\varepsilon/4$ for all $\zeta \in \partial D(a; r)$ and $n \geq N$. Then for all $z \in D(a; r/2)$ and $n \geq N$, we use the Cauchy integral formula to calculate

$$
\begin{aligned}
|f_n'(z) - f'(z)| &= \left| \frac{1}{2\pi i} \int_{\partial D(a;r)} \frac{f_n(\zeta) - f(\zeta)}{(\zeta - z)^2} \, d\zeta \right| \\
&\leq \frac{1}{2\pi} \int_{\partial D(a;r)} \frac{|f_n(\zeta) - f(\zeta)|}{|\zeta - z|^2} \, |d\zeta| \\
&< \frac{1}{2\pi} \frac{r\varepsilon/4}{(r/2)^2} \int_{\partial D(a;r)} |d\zeta| \\
&= \frac{\varepsilon}{2\pi r} 2\pi r \\
&= \varepsilon.
\end{aligned}
$$

We see that $f_n' \to f'$ uniformly on $D(a; r/2)$. Since a was chosen arbitrarily, we have $f_n' \to f'$ locally uniformly, and the result follows from Theorem Ep.3. $\qquad\square$

After using Cauchy's integral formula in the previous result, we turn to the Poisson integral formula in the next.

Ep.5 Theorem. *Let $\Omega \subseteq \mathbb{C}$ be a domain and $\{f_n\}_{n=1}^{\infty} \subseteq H(\Omega)$. If $\{\operatorname{Re} f_n\}_{n=1}^{\infty}$ converges uniformly on compact subsets of Ω and $\{f_n(z_0)\}_{n=1}^{\infty}$ converges for some $z_0 \in \Omega$, then $\{f_n\}$ converges uniformly on compact subsets of Ω to a member of $H(\Omega)$.*

Proof. Let $u_n = \operatorname{Re} f_n$ for all n. Then $\{u_n\}_{n=1}^{\infty}$ is a sequence of harmonic functions and converges uniformly on compact subsets of Ω to a harmonic $u : \Omega \to \mathbb{R}$ by Exercise 5 in Section 6.2. Let $a \in \Omega$ and $r > 0$ such that $\overline{D}(a; r) \subseteq \Omega$. Performing a change of variables in Theorem 6.2.6 in a manner similar to what is done in Remark 6.2.7, we see that, in $D(a; r)$, each u_n is the real part of the analytic function $g_n \in H(D(a; r))$ given by

$$g_n(z) = \frac{1}{2\pi} \int_0^{2\pi} \frac{re^{it} + z - a}{re^{it} - z + a} u_n(a + re^{it}) \, dt.$$

Define $g \colon D(a; r) \to \mathbb{C}$ by

$$g(z) = \frac{1}{2\pi} \int_0^{2\pi} \frac{re^{it} + z - a}{re^{it} - z + a} u(a + re^{it}) \, dt.$$

For $z \in D(a; r/2)$, we see that

$$\left| \frac{re^{it} + z - a}{re^{it} - z + a} \right| \leq \frac{r + |z - a|}{r - |z - a|} < \frac{r + r/2}{r - r/2} = 3.$$

Therefore, given $\varepsilon > 0$, we may choose $N \in \mathbb{N}$ such that for all $n \geq N$, $|u_n(\zeta) - u(\zeta)| < \varepsilon/3$ for all $\zeta \in \partial D(a; r)$, and observe that

$$|g_n(z) - g(z)| \leq \frac{1}{2\pi} \int_0^{2\pi} \left| \frac{re^{it} + z - a}{re^{it} - z + a} \right| |u_n(a + re^{it}) - u(a + re^{it})| \, dt$$

$$< \frac{1}{2\pi} \int_0^{2\pi} 3 \frac{\varepsilon}{3} \, dt$$

$$= \varepsilon$$

for all $z \in D(a; r/2)$. Thus $g_n \to g$ uniformly on $D(a; r/2)$.

For each n, $\mathrm{Re}(f_n - g_n) = 0$, and hence $f_n - g_n$ maps $D(a; r)$ onto a subset of the imaginary axis. By the open mapping theorem, $f_n - g_n$ is constant. By the mean value property,

$$g_n(a) = \frac{1}{2\pi} \int_0^{2\pi} u_n(a + re^{it}) \, dt = u_n(a) = \mathrm{Re}\, f_n(a).$$

From this, we conclude

$$f_n(z) = g_n(z) + i \operatorname{Im} f_n(a), \qquad z \in D(a; r).$$

Hence if $\{f_n(a)\}$ converges, then $\{f_n\}$ converges uniformly on $D(a; r/2)$, and if $\{f_n(a)\}$ diverges, then $\{f_n(z)\}$ diverges for all $z \in D(a; r)$. If E is the set of all $z \in \Omega$ for which $\{f_n(z)\}$ converges, then this observation shows that both E and $\Omega \setminus E$ are open. Because Ω is a domain and $E \neq \varnothing$ by hypothesis, we see that $E = \Omega$ by Theorem 1.4.16. It follows that $\{f_n\}$ converges locally uniformly on Ω, and hence converges uniformly on compact sets by Theorem Ep.3. The limit function is analytic by Exercise 5 in Section 4.3.   □

We now consider the following consequence of the argument principle concerning the zeros of a sequence of analytic functions. (See Exercise 10 in Section 5.3.)

Ep.6 Hurwitz's Theorem. *Let* $\Omega \subseteq \mathbb{C}$ *be open and* $\{f_n\}_{n=1}^{\infty} \subseteq H(\Omega)$ *be a sequence converging to* $f \in H(\Omega)$ *uniformly on compact subsets of* Ω. *If* $C \subseteq \Omega$ *is a simple closed contour such that the inside of* C *lies in* Ω *and* $f(z) \neq 0$ *for all* $z \in C$, *then there is* $N \in \mathbb{N}$ *such that* f_n *and* f *have the same number of zeros inside of* C, *counting order, for all* $n \geq N$.

By "counting order," we mean that the sums of the orders of the zeros of f_n and f inside C are equal.

Proof. Since C is compact and f is nonzero on C, we can define

$$r = \min\{|f(z)| : z \in C\} > 0.$$

There is $N_1 \in \mathbb{N}$ such that $|f_n(z) - f(z)| < r$ for all $n \geq N_1$ and $z \in C$. Hence for $n \geq N_1$, $f_n(z) \neq 0$ for all $z \in C$. Theorem Ep.4 gives $f_n' \to f'$ uniformly on C, and it follows from Exercise 6 in Section 2.2 that $f_n'/f_n \to f'/f$ uniformly on C. If M_n and M denote the sums of the orders of any zeros of f_n and f, respectively, inside of C, then the argument principle gives

$$\lim_{n \to \infty} M_n = \lim_{n \to \infty} \frac{1}{2\pi i} \int_C \frac{f_n'(z)}{f_n(z)} \, dz = \frac{1}{2\pi i} \int_C \frac{f'(z)}{f(z)} \, dz = M.$$

Since $\{M_n\} \subseteq \mathbb{Z}$ and $M \in \mathbb{Z}$ there must be $N \in \mathbb{N}$ $(N \geq N_1)$ such that $M_n = M$ for all $n \geq N$. $\qquad\square$

Harnack's Theorem

We consider the following pair of results for harmonic functions.

Ep.7 Harnack's Inequalities. *Let* $a \in \mathbb{C}$ *and* $R > 0$. *For any harmonic function* $u \colon D(a; R) \to [0, \infty)$, *we have*

$$\frac{R - r}{R + r} u(a) \leq u(a + re^{i\theta}) \leq \frac{R + r}{R - r} u(a) \tag{Ep.1}$$

for all $\theta \in \mathbb{R}$ *and* $r \in [0, R)$.

Proof. We consider the form of the Poisson integral formula given in Remark 6.2.7. Let $r \in [0, R)$, $\theta \in \mathbb{R}$, and $s \in (r, R)$. We note that for all $t \in [0, 2\pi]$,

$$\frac{s - r}{s + r} = \frac{s^2 - r^2}{(s + r)^2} \leq \frac{s^2 - r^2}{s^2 - 2rs\cos(\theta - t) + r^2} \leq \frac{s^2 - r^2}{(s - r)^2} = \frac{s + r}{s - r}.$$

Since u is harmonic on $D(a; s)$ and continuous on $\overline{D}(a; s)$, we can then multiply the above inequality by $u(a + se^{it})/2\pi$, integrate from 0 to 2π, and use the Poisson integral formula to obtain

$$\frac{1}{2\pi} \int_0^{2\pi} \frac{s - r}{s + r} u(a + se^{it}) \, dt \leq u(a + re^{i\theta}) \leq \frac{1}{2\pi} \int_0^{2\pi} \frac{s + r}{s - r} u(a + se^{it}) \, dt.$$

By the mean value property, we know that

$$\frac{1}{2\pi} \int_0^{2\pi} u(a + se^{it})\, dt = u(a),$$

and therefore we remove constants from the outer integrals to see that

$$\frac{s - r}{s + r} u(a) \leq u(a + re^{i\theta}) \leq \frac{s + r}{s - r} u(a).$$

Since this holds true for all $s \in (r, R)$, we lake the limit of the above as $s \to R^-$ to complete the proof. $\square$

Ep.8 Harnack's Theorem. *Let $\Omega \subseteq \mathbb{C}$ be a domain and $\{u_n\}_{n=1}^{\infty}$ be an increasing sequence of harmonic functions on Ω. That is, $u_n(z) \leq u_{n+1}(z)$ for all $z \in \Omega$ and $n \in \mathbb{N}$. Then either $\{u_n\}$ converges uniformly on compact subsets of Ω to a harmonic function $u \colon \Omega \to \mathbb{R}$ or $\{u_n(z)\}$ diverges for every $z \in \Omega$.*

Proof. For each $n \in \mathbb{N}$, let $v_n = u_{n+1} - u_n$. Then each v_n is a nonnegative harmonic function. It follows that $\{u_n\}$ converges pointwise (resp. uniformly) on a set $A \subseteq \Omega$ if and only if $\sum_{n=1}^{\infty} v_n$ converges pointwise (resp. uniformly) on A.

Set

$$E = \left\{ z \in \Omega : \sum_{n=1}^{\infty} v_n(z) \text{ converges} \right\}.$$

Let $a \in E$, and choose $R_a > 0$ such that $D(a; R_a) \subseteq \Omega$. By Harnack's inequalities, we have

$$v_n(z) \leq \frac{R_a + |z - a|}{R_a - |z - a|} v_n(a) \leq \frac{R_a + R_a/2}{R_a - R_a/2} v_n(a) = 3v_n(a)$$

for all $z \in D(a; R_a/2)$ and $n \in \mathbb{N}$. It follows that $\sum_{n=1}^{\infty} v_n$ converges uniformly on $D(a; R_a/2)$ by the Weierstrass M-test, and hence $D(a; R_a/2) \subseteq E$. We conclude that E is open.

Similarly, if $b \in \Omega \setminus E$, then

$$v_n(z) \geq \frac{R_b - |z - b|}{R_b + |z - b|} v_n(b) \geq \frac{R_b - R_b/2}{R_b + R_b/2} v_n(b) = \frac{v_n(b)}{3}$$

for all $z \in D(b; R_b/2)$ and $n \in \mathbb{N}$, where $D(b; R_b) \subseteq \Omega$. Then $\sum_{n=1}^{\infty} v_n(z)$ diverges for $z \in D(b; R_b/2)$, showing $D(b; R_b/2) \subseteq \Omega \setminus E$. It follows that $\Omega \setminus E$ is open. By Theorem 1.4.16, one of E or $\Omega \setminus E$ is empty, and so $\sum_{n=1}^{\infty} v_n(z)$ either converges for all $z \in \Omega$ or diverges for all $z \in \Omega$. In the case of convergence, we see that the convergence is locally uniform, and hence is uniform on compact sets by Theorem Ep.3. That the limit is harmonic comes from Exercise 5 in Section 6.2. $\square$

Results for Simply Connected Domains

Recall the definition of *simply connected domain* given in Exercise 3 of Section 4.3. Note that the domains Ω in Figure Ep.1 are simply connected, while those in Figures 4.2 and 4.4 are not.

We now consider a sequence of lemmas.

Ep.9 Lemma. *Let $\Omega \subseteq \mathbb{C}$ be a simply connected domain. If $f \in H(\Omega)$ is univalent, then $f(\Omega)$ is simply connected. Furthermore, for any $a \in \Omega$ and closed contour γ in $\Omega \setminus \{a\}$, $\mathrm{Ind}_{f \circ \gamma} f(a) = \mathrm{Ind}_\gamma a$.*

Proof. Note that $f(\Omega)$ is open (open mapping theorem) and connected, and hence is a domain. Let $b \in \mathbb{C} \setminus f(\Omega)$ and Γ be a closed contour in $f(\Omega)$. We must show that $\mathrm{Ind}_\Gamma b = 0$. We do so through a change of variables. We know that f^{-1} exists and is analytic (and hence continuous), and therefore $f^{-1} \circ \Gamma$ is a closed contour in Ω. By hypothesis, $\mathrm{Ind}_{f^{-1} \circ \Gamma} z = 0$ for all $z \in \mathbb{C} \setminus \Omega$. Therefore, by Cauchy's theorem, we have

$$\mathrm{Ind}_\Gamma b = \frac{1}{2\pi i} \int_\Gamma \frac{dw}{w - b} = \frac{1}{2\pi i} \int_{f^{-1} \circ \Gamma} \frac{f'(z)}{f(z) - b} \, dz = 0,$$

where we used the substitution $w = f(z)$.

Now let a and γ be as in the statement of the lemma. Then, similar to above, we have

$$\mathrm{Ind}_{f \circ \gamma} f(a) = \frac{1}{2\pi i} \int_{f \circ \gamma} \frac{dw}{w - f(a)} = \frac{1}{2\pi i} \int_\gamma \frac{f'(z)}{f(z) - f(a)} \, dz.$$

Since f is univalent, the integrand has a as its only singularity in Ω. Furthermore, univalence gives that $f'(a) \neq 0$, and hence the singularity is a simple pole. Thus the above expression is seen to equal

$$\mathrm{Res}_{z=a} \frac{f'(z)}{f(z) - f(a)} \, \mathrm{Ind}_\gamma a = \lim_{z \to a} \left(\frac{z - a}{f(z) - f(a)} f'(z) \right) \mathrm{Ind}_\gamma a = \mathrm{Ind}_\gamma a,$$

completing the proof. $\qquad\square$

In the following, we show that an analytic function on a simply connected domain with no zeros has a logarithm.

Ep.10 Lemma. *If $\Omega \subseteq \mathbb{C}$ is a simply connected domain and $f \in H(\Omega)$ has no zeros in Ω, then there exists $F \in H(\Omega)$ such that $e^{F(z)} = f(z)$ for all $z \in \Omega$.*

Proof. The function f'/f is well defined and analytic on Ω. By Exercise 3 of Section 4.3, f'/f has an antiderivative on Ω. Fix $a \in \Omega$. By adding a constant, we see that there is $F \in H(\Omega)$ such that $F' = f'/f$ and $F(a) = \log f(a)$ for some branch of the logarithm. The calculation

$$\frac{d}{dz} f(z) e^{-F(z)} = f'(z) e^{-F(z)} - F'(z) f(z) e^{-F(z)} = 0$$

shows that $f(z)e^{-F(z)}$ is constant for $z \in \Omega$, and the observation $f(a)e^{-F(a)} = 1$ implies this constant is 1. It follows that $e^{F(z)} = f(z)$ for all $z \in \Omega$. □

Ep.11 Corollary. *If $\Omega \subseteq \mathbb{C}$ is a simply connected domain and $f \in H(\Omega)$ has no zeros in Ω, then there exists $g \in H(\Omega)$ such that $g^2 = f$.*

In this context, the function g is called an *analytic square root* of f.

Proof. Defining $g \in H(\Omega)$ by $g(z) = e^{F(z)/2}$, where F is as in Lemma Ep.10, gives $[g(z)]^2 = f(z)$ for all $z \in \Omega$. □

The following lemma is essentially the Riemann mapping theorem with an injection instead of a bijection. As we will see, there is still much work to be done.

Ep.12 Lemma. *Let $\Omega \subseteq \mathbb{C}$ be a simply connected domain such that $\Omega \neq \mathbb{C}$, and let $a \in \Omega$. Then there exists a univalent $f_0 \colon \Omega \to \mathbb{D}$ such that $f_0(a) = 0$ and $f_0'(a) > 0$.*

Proof. Let $\alpha \in \mathbb{C} \setminus \Omega$. By Corollary Ep.11, there is $g \in H(\Omega)$ such that $[g(z)]^2 = z - \alpha$ for all $z \in \Omega$. It is immediate that if $z_1, z_2 \in \Omega$ satisfy $g(z_1) = \pm g(z_2)$, then $z_1 = z_2$. Thus g is one-to-one and hence nonconstant. By the open mapping theorem, $g(\Omega)$ is open. Note that $0 \notin g(\Omega)$.

Let $c \in g(\Omega)$. We may then choose $r > 0$ such that $\overline{D}(c; r) \subseteq g(\Omega)$. If $w \in \overline{D}(-c; r)$, then $-w \in \overline{D}(c; r)$, and hence $-w = g(z_1)$ for some $z_1 \in \Omega$. If it were the case that $w = g(z_2)$ for some $z_2 \in \Omega$, then $z_1 = z_2$ from the above, which would lead to the contradiction $-w = w$. We conclude that $\overline{D}(-c; r) \cap g(\Omega) = \varnothing$. There exists a linear fractional transformation T taking $\partial D(-c; r)$ onto $\partial \mathbb{D}$ and $g(a)$ to 0. It follows that T maps $g(\Omega)$ into $\mathbb{D}$. Let $f_0 = e^{-i\theta} T \circ g$, where $\theta = \arg[(T \circ g)'(a)]$. Then f_0 is one-to-one, analytic, and satisfies $f_0(\Omega) \subseteq \mathbb{D}$, $f_0(a) = 0$, and $f_0'(a) > 0$. □

The Riemann Mapping Theorem

We are now ready to address the proof of the Riemann mapping theorem. As a preparatory step, we construct a particular type of univalent function on simply connected subdomains of $\mathbb{D}$ containing 0 that will serve as building blocks in the upcoming proof. We refer to these functions as *Carathéodory–Koebe expansions*.

Ep.13 Lemma. *Let $U \subseteq \mathbb{D}$ be a simply connected domain such that $0 \in U$. There exists a univalent $g \colon U \to \mathbb{D}$ satisfying $g(0) = 0$, $|g(z)| \geq |z|$ for all $z \in U$, and*

$$g'(0) = \frac{1+r}{2\sqrt{r}},$$

where

$$r = \min\{|z| : z \in \partial U\}.$$

It is tempting to suspect that $U \subseteq g(U)$ holds for these expansions, but it need not.

Proof. The proof is completely constructive. We note that r exists because ∂U is compact, and there exists $\alpha \in \partial U$ such that $|\alpha| = r$. If $U = \mathbb{D}$, then we may take g to be the identity. Otherwise, $\alpha \in \mathbb{D}$. Since U contains a disk centered at 0, $r > 0$.

To construct g, let $\varphi_\alpha \in \operatorname{Aut} \mathbb{D}$ be given as in (3.5.1). Since $\alpha \notin U$ and $\varphi_\alpha(\alpha) = 0$, we see that there exists $\psi \in H(U)$ such that $\psi^2 = \varphi_\alpha$ on U, by invoking Corollary Ep.11. For any $z \in U$, $[\psi(z)]^2 \in \mathbb{D}$ implies $\psi(z) \in \mathbb{D}$, and an argument as in the proof of Lemma Ep.12 shows ψ is one-to-one. Now let $\beta = \psi(0) \in \mathbb{D}$, and define $g = e^{-i\theta} \varphi_\beta \circ \psi$, where $\theta = \arg[(\varphi_\beta \circ \psi)'(0)]$. Then $g \colon U \to \mathbb{D}$ is univalent, $g(0) = 0$, and $g'(0) > 0$.

Note that for any $\gamma \in \mathbb{D}$,

$$\varphi'_\gamma(z) = \frac{1 - |\gamma|^2}{(1 - \overline{\gamma}z)^2}, \qquad z \in \mathbb{D}.$$

We differentiate both sides of $\psi^2 = \varphi_\alpha$ at the origin to obtain

$$2\beta\psi'(0) = \varphi'_\alpha(0) = 1 - |\alpha|^2 = 1 - r^2.$$

From the observation $\beta^2 = \varphi_\alpha(0) = -\alpha$, it follows that

$$|\psi'(0)| = \frac{1 - r^2}{2\sqrt{r}}.$$

Then we have

$$g'(0) = |g'(0)| = |\varphi'_\beta(\beta)||\psi'(0)| = \frac{1}{1 - |\beta|^2} \frac{1 - r^2}{2\sqrt{r}} = \frac{1 + r}{2\sqrt{r}}.$$

Finally, define $\omega \colon \mathbb{D} \to \mathbb{D}$ by $\omega(z) = \varphi_\alpha^{-1}([\varphi_\beta^{-1}(z)]^2)$. Then

$$\omega(0) = \varphi_\alpha^{-1}(\beta^2) = \varphi_\alpha^{-1}(-\alpha) = 0,$$

and hence ω satisfies the hypotheses of Schwarz's lemma. Thus $|\omega(z)| \leq |z|$ for $z \in \mathbb{D}$. It follows that

$$|z| = |\omega(e^{i\theta}g(z))| \leq |e^{i\theta}g(z)| = |g(z)|$$

for all $z \in U$. $\qquad\qquad\square$

We are now equipped to finally prove the Riemann mapping theorem.

Proof of the Riemann mapping theorem. Let f_0 be as given in Lemma Ep.12, and set $U_0 = f_0(\Omega)$. We note that U_0 is a simply connected domain by Lemma Ep.9. We will construct an analytic bijection $h \colon U_0 \to \mathbb{D}$ such that $h(0) = 0$ and $h'(0) > 0$. Then $f = h \circ f_0$ is an analytic bijection from Ω onto $\mathbb{D}$ with $f(a) = 0$ and $f'(a) = h'(0)f_0'(a) > 0$. If $F \colon \Omega \to \mathbb{D}$ is any analytic bijection with $F(a) = 0$ and $F'(a) > 0$, then $\varphi = f \circ F^{-1} \in \operatorname{Aut} \mathbb{D}$ satisfies $\varphi(0) = 0$ and

$$\varphi'(0) = \frac{f'(a)}{F'(a)} > 0.$$

By Theorem 3.5.7, the only such automorphism is the identity, giving that $F = f$. This establishes uniqueness, and hence the proof is reduced to the construction of h.

We inductively apply Lemma Ep.13 to obtain a sequence $U_0, U_1, \ldots$ of simply connected domains in $\mathbb{D}$, each containing 0, and Carathéodory–Koebe expansions $g_n \colon U_{n-1} \to \mathbb{D}$ such that $U_n = g_n(U_{n-1})$ for all $n \in \mathbb{N}$. Let

$$r_n = \min\{|z| : z \in \partial U_n\} > 0, \qquad n = 0, 1, \ldots.$$

For $n \in \mathbb{N}$, set $h_n = g_n \circ \cdots \circ g_1$. Then h_n is a univalent mapping of U_0 into $\mathbb{D}$ such that $h_n(0) = 0$. Since $D(0; r_0) \subseteq U_0$, we see that $z \mapsto h_n(r_0 z)$ satisfies the hypotheses of Schwarz's lemma, and we conclude that $r_0 h'_n(0) \leq 1$. We then have

$$h'_n(0) = \prod_{k=1}^{n} g'_k(0) = \prod_{k=0}^{n-1} \frac{1 + r_k}{2\sqrt{r_k}} \leq \frac{1}{r_0}.$$

Applying the natural logarithm yields

$$\ln h'_n(0) = \sum_{k=0}^{n-1} \ln \frac{1 + r_k}{2\sqrt{r_k}} \leq \ln \frac{1}{r_0}.$$

Differentiation reveals the function $t \mapsto (1 + t)/(2\sqrt{t})$ to be strictly decreasing on $(0, 1]$, and it is equal to 1 at $t = 1$. Therefore the logarithms in the above sum are all nonnegative, meaning the sum increases as n increases. The bound implies the resulting series is convergent, meaning its terms go to 0. It follows that

$$\lim_{n \to \infty} \frac{1 + r_n}{2\sqrt{r_n}} = 1.$$

The previous monotonicity argument then forces $r_n \to 1$.

We know that for each n, $h_n(0) = 0$ and $h'_n(0) \neq 0$. Therefore h_n has a simple zero at the origin, and we can write $h_n(z) = z H_n(z)$ for $H_n \in H(U_0)$ such that $H_n(0) \neq 0$. Since h_n is one-to-one, it has no other zeros, and hence H_n has no zeros. Thus by Lemma Ep.10, there is $G_n \in H(U_0)$ such that $e^{G_n(z)} = H_n(z)$ for all $z \in U_0$. Then $e^{G_n(0)} = H_n(0) = h'_n(0) > 0$, and therefore we may assume $G_n(0) = \ln h'_n(0)$ by adding the appropriate multiple of $2\pi i$ to G_n.

For each $n \in \mathbb{N}$, let $u_n = \operatorname{Re} G_n$ on U_0. Then u_n is harmonic, and $u_n(z) = \ln |H_n(z)|$ for all $z \in U_0$ because $e^{u_n(z)} = |e^{G_n(z)}|$. Since $|z| \leq |g_n(z)|$ for all n, we have $|h_n(z)| \leq |h_{n+1}(z)|$, and hence $|H_n(z)| \leq |H_{n+1}(z)|$ for all n. Therefore the sequence $\{u_n\}_{n=1}^{\infty}$ is increasing. Our work above shows $\{G_n(0)\} = \{u_n(0)\}$ converges. Therefore $\{u_n\}$ converges uniformly on compact subsets of U_0 by Harnack's theorem. But then $\{G_n\}$ converges uniformly on compact subsets of U_0 to some $G \in H(U_0)$ by Theorem Ep.5. By continuity, $h_n \to h$ uniformly on compact sets, where $h \in H(U_0)$ is given by $h(z) = z e^{G(z)}$. Note that h has an isolated zero at 0 and hence is nonconstant.

We complete the proof by showing h is as desired. We first show h is univalent. If $z_1, z_2 \in U_0$ are distinct such that $h(z_1) = h(z_2) = b$ for $b \in \mathbb{D}$, then z_1

and z_2 are isolated zeros of $h - b$. There is then $0 < r < |z_1 - z_2|/2$ such that $\overline{D}(z_1; r), \overline{D}(z_2; r) \subseteq U_0$ and $h - b$ is nonzero on $\partial D(z_1; r)$ and $\partial D(z_2; r)$. Since $(h_n - b) \to (h - b)$ uniformly on compact sets, there is $n \in \mathbb{N}$ such that $h_n - b$ has the same number of zeros (up to order) as $h - b$ in each of $D(z_1; r)$ and $D(z_2; r)$ by Hurwitz's theorem. Thus, since $h - b$ has zeros in both $D(z_1; r)$ and $D(z_2; r)$, so too does $h_n - b$, a contradiction to the univalence of h_n. Hence h is univalent.

Theorem Ep.4 gives $h'(0) \geq 0$ because $h_n'(0) > 0$ for all n. Since h is univalent, we conclude $h'(0) > 0$. From convergence, we also get $h(U_0) \subseteq \overline{\mathbb{D}}$. Since h is nonconstant, the open mapping theorem then gives $h(U_0) \subseteq \mathbb{D}$. To see that h is onto, let $b \in \mathbb{D}$, $r \in (|b|, 1)$ and $n \in \mathbb{N}$ such that $r_n > r$. (Recall that $r_n \to 1$.) Then $\overline{D}(0; r) \subseteq U_n = h_n(U_0)$. By Lemma Ep.9, $C = h_n^{-1}(\partial D(0; r))$ is a simple closed contour in U_0 surrounding 0. But then $h_k(C)$ is a simple closed contour in $\mathbb{D}$ surrounding 0 for $k \geq n$. Furthermore $|h_k(z)| \geq |h_n(z)| \geq r$ for all $z \in C$, means b and 0 are in the same connected component of $\mathbb{C} \setminus h_k(C)$, and hence b is inside $h_k(C)$, and $h_k^{-1}(b)$ is inside of C, for all $k \geq n$. Since $h_k - b$ has a zero inside of C for all $k \geq n$, so too does $h - b$ by another application of Hurwitz's theorem. Therefore $b \in h(U_0)$. $\qquad \square$

APPENDIX A

SETS AND FUNCTIONS

All areas of modern mathematics follow axiomatically from the principles of set theory. The purpose of this appendix is to introduce the basic language and notation of sets and functions to the extent that they are used in this text. While far from a thorough treatment of the topic, it should help those who find the notation and terminology at the start of first chapter unfamiliar.

Sets and Elements

A *set* is the basic building block of mathematics, and as such, is difficult to define. We may think of a set as a collection of objects, which are called the *elements* of the set. Of course, using the word "collection" to define the word "set" is vague, since we would then have to define the word "collection." The true definition of a set requires a list of axioms, resolving this conundrum. Reviewing these axioms is beyond what we shall do here. Instead, we shall accept the words "set" and "element" as understood and work from there.

Complex Analysis: A Modern First Course in Function Theory, First Edition. Jerry R. Muir, Jr.
© 2015 John Wiley & Sons, Inc. Published 2015 by John Wiley & Sons, Inc.

A.1 Definition. If A is a set and x is an element of A, then we denote this relationship symbolically by $x \in A$. The negation of this statement is $x \notin A$. There is one set that contains no elements, namely the *empty set*, denoted $\varnothing$.

A.2 Example. As discussed in Section 1.1, the familiar sets of natural numbers, integers, rational numbers, and real numbers are denoted by the special symbols, $\mathbb{N}$, $\mathbb{Z}$, $\mathbb{Q}$, and $\mathbb{R}$, respectively. The statements $5 \in \mathbb{N}$, $-3 \in \mathbb{Z}$, $-3 \notin \mathbb{N}$, $1/2 \in \mathbb{Q}$, $1/2 \notin \mathbb{Z}$, $\pi \in \mathbb{R}$, and $\pi \notin \mathbb{Q}$ show how the notation concerning elements and sets works.

A.3 Example. Some sets can easily be denoted by listing their elements within braces. For more complicated sets, it is often convenient to use *set-builder notation*. If A is a set and $P(x)$ is a statement that is true or false for each element $x \in A$, then we write $\{x \in A : P(x)\}$ for the set of all $x \in A$ for which $P(x)$ is true. This is shortened to $\{x : P(x)\}$ when A is understood, and other simplifications are commonly made. For instance, $\mathbb{Q}$, written formally as

$$\mathbb{Q} = \left\{ x \in \mathbb{R} : x = \frac{m}{n} \text{ for some } m \in \mathbb{Z} \text{ and } n \in \mathbb{N} \right\},$$

can be expressed in the simpler form

$$\mathbb{Q} = \left\{ \frac{m}{n} : m \in \mathbb{Z} \text{ and } n \in \mathbb{N} \right\}.$$

The open, closed, and half-open intervals encountered in calculus are familiar examples of sets. For instance, if $a, b \in \mathbb{R}$ and $a < b$, we write

$$[a, b) = \{x \in \mathbb{R} : a \leq x < b\}.$$

New sets can be created from sets already defined.

A.4 Definition. Let A and B be sets.

(a) The *union* of A and B is the set

$$A \cup B = \{x : x \in A \text{ or } x \in B\}. \tag{A.1}$$

(b) The *intersection* of A and B is the set

$$A \cap B = \{x : x \in A \text{ and } x \in B\}. \tag{A.2}$$

(c) The *complement* of B in A is the set

$$A \setminus B = \{x \in A : x \notin B\}. \tag{A.3}$$

A.5 Example. We turn to intervals to observe that $[0, 4] \cup (2, 5) = [0, 5)$, $(-3, 7) \cap [-10, 2] = (-3, 2]$, and $(5, 7] \setminus [6, 12] = (5, 6)$.

We have the following relationships between sets.

A.6 Definition. Let A and B be sets.

(a) We say that A is a *subset* of B provided that every $x \in A$ is such that $x \in B$. Denote this relationship by $A \subseteq B$.

(b) We say that A and B are *disjoint* if $A \cap B = \varnothing$.

A.7 Example. By their definitions, $\mathbb{N} \subseteq \mathbb{Z} \subseteq \mathbb{Q} \subseteq \mathbb{R}$. Notice that the sets $[1, 3]$ and $(3, 7)$ are disjoint.

Two sets are equal if they contain the same elements. It is often best, when attempting to prove two sets equal, to show that each set is a subset of the other. To establish that one set is a subset of another, it is often the case that "element chasing" is the best method. Consider the following.

A.8 Example. Let A, B, and C be sets. We will show that

$$(A \cup B) \cap C = (A \cap C) \cup (B \cap C).$$

To do so, we shall show that each set is a subset of the other.

Let $x \in (A \cup B) \cap C$. Then $x \in A \cup B$ and $x \in C$. It follows that $x \in A$ or $x \in B$. If $x \in A$, then since $x \in C$, $x \in A \cap C$. If $x \in B$, then since $x \in C$, $x \in B \cap C$. Thus $x \in A \cap C$ or $x \in B \cap C$, showing that $x \in (A \cap C) \cup (B \cap C)$. We have proved that $(A \cup B) \cap C \subseteq (A \cap C) \cup (B \cap C)$.

Now let $x \in (A \cap C) \cup (B \cap C)$. Then $x \in A \cap C$ or $x \in B \cap C$. If $x \in A \cap C$, then $x \in A$ and $x \in C$. Since $x \in A$, $x \in A \cup B$. Therefore $x \in (A \cup B) \cap C$. If $x \in B \cap C$, then $x \in B$ and $x \in C$. As $x \in B$, $x \in A \cup B$. It follows that $x \in (A \cup B) \cap C$. So in either case $x \in (A \cup B) \cap C$. This proves that $(A \cap C) \cup (B \cap C) \subseteq (A \cup B) \cap C$. We conclude that the two sets are equal.

We finish this section by discussing *sets of sets*.

A.9 Definition. Let I be a set and A_α be a set for each $\alpha \in I$. Then I is said to be an *index set* and $\mathcal{A} = \{A_\alpha : \alpha \in I\}$ is a *set of sets*.

(a) The *union* of the members of $\mathcal{A}$ is the set

$$\bigcup_{\alpha \in I} A_\alpha = \{x : x \in A_\alpha \text{ for some } \alpha \in I\}. \tag{A.4}$$

(b) The *intersection* of the members of $\mathcal{A}$ is the set

$$\bigcap_{\alpha \in I} A_\alpha = \{x : x \in A_\alpha \text{ for every } \alpha \in I\}. \tag{A.5}$$

The notation may change for familiar index sets. For instance, if $I = \mathbb{N}$, then we prefer to write

$$\bigcup_{n=1}^{\infty} A_n = \bigcup_{n \in \mathbb{N}} A_n$$

for clarity. In the finite case $I = \{1, \ldots, n\}$, we write

$$\bigcup_{k=1}^{n} A_k = \bigcup_{k \in \{1, \ldots, n\}} A_k.$$

Analogous notation holds for intersections.

A.10 Example. Observe the following equalities using the above definitions:

$$\bigcup_{n=1}^{\infty} [-n, n] = \mathbb{R}, \qquad \bigcap_{n=1}^{\infty} \left(-\frac{1}{n}, \frac{1}{n} \right) = \{0\}.$$

Functions

As sets form the foundation of all areas of modern mathematics, we need some way of operating between them. The concept of a function answers this call.

A.11 Definition. Let A and B be sets. A *function* from A to B is a rule by which each element of A is assigned a unique element of B. If f is such a function, then we write $f: A \to B$. If $x \in A$, then the unique element of B assigned to x by f is denoted $f(x)$. The sets A and B are called the *domain* and *codomain* of f, respectively. On occasion, the notation $x \mapsto f(x)$ is a nice alternative to describe a function f.

Since the study of functions defined on subsets of $\mathbb{R}$ lies at the heart of calculus, this idea should be familiar. Before considering examples, let us add a few more terms to the discussion.

A.12 Definition. Let A and B be sets, and let $f: A \to B$.

(a) The *image* of the set $C \subseteq A$ under f is the set

$$f(C) = \{f(x) : x \in C\}, \tag{A.6}$$

and the *range* of f is the image of A.

(b) The *inverse image* of the set $D \subseteq B$ under f is the set

$$f^{-1}(D) = \{x \in A : f(x) \in D\}. \tag{A.7}$$

(c) The function f is *one-to-one* (or an *injection*) if $f(x) \neq f(y)$ whenever x and y are distinct elements of A.

(d) The function f is *onto* (or a *surjection*) if $f(A) = B$.

(e) The function f is a *bijection* if it is both one-to-one and onto.

A.13 Example. Let $f\colon \mathbb{R} \to \mathbb{R}$ be defined by $f(x) = x^2$. Then we see that

$$f([1,3]) = [1,9], \qquad f^{-1}((4,25)) = (-5,-2) \cup (2,5).$$

Since $f(1) = f(-1)$, f is not one-to-one. Since $f(\mathbb{R}) = [0,\infty)$, f is not onto.

Now consider $g\colon [0,\infty) \to [0,\infty)$ defined by $g(x) = x^2$. Then $g(x) = g(y) = a$ for some x, y, and a means $x^2 = a = y^2$. Since a has only one nonnegative square root, $x = y$, showing g is one-to-one. Furthermore, one may verify that $g([0,\infty)) = [0,\infty)$, and therefore g is onto.

This example illustrates how the domain and codomain are both important components of a function.

A.14 Definition. Let A, B, and C be sets. If $f\colon A \to B$ and $g\colon B \to C$, then the *composition* of f and g is the function $g \circ f\colon A \to C$ defined by

$$(g \circ f)(x) = g(f(x)). \tag{A.8}$$

Suppose that $f\colon A \to B$ is a bijection. We may then define a function $g\colon B \to A$ as follows: For $y \in B$, f onto implies that there is some element $x \in A$ such that $f(x) = y$, and f one-to-one implies that this element is unique in that regard. Therefore define $g(y) = x$. This is the basis of the following definition.

A.15 Definition. Let A and B be sets, and let $f\colon A \to B$ be a bijection. Then the function g described in the preceding paragraph is called the *inverse* of f and is denoted $g = f^{-1}$.

Note that, given the circumstances of Definition A.15, $(f^{-1} \circ f)(x) = x$ for all $x \in A$ and $(f \circ f^{-1})(y) = y$ for all $y \in B$. By this reasoning, $(f^{-1})^{-1} = f$.

A.16 Example. We consider two examples from calculus. Let $f\colon \mathbb{R} \to \mathbb{R}$ be given by $f(x) = e^x$. Then f is one-to-one but not onto. However, if we restrict the codomain and define $g\colon \mathbb{R} \to (0,\infty)$ by $g(x) = e^x$, then g is one-to-one and onto. Accordingly, g^{-1} exists, and it is given by $g^{-1}(x) = \ln x$.

Now consider $\varphi\colon \mathbb{R} \to [-1,1]$ given by $\varphi(x) = \sin x$. Then φ is onto but not one-to-one. However, if we restrict the domain and define $\psi\colon [-\pi/2, \pi/2] \to [-1,1]$ by $\psi(x) = \sin x$, then ψ is one-to-one and onto, and ψ^{-1} is given by $\psi^{-1}(x) = \sin^{-1} x = \arcsin x$.

Lastly, we consider an example that demonstrates how "element chasing" can be used in the context of functions.

A.17 Example. Let A and B be sets, and let $f\colon A \to B$. For $C, D \subseteq A$ we will prove that

$$f(C \cap D) \subseteq f(C) \cap f(D).$$

Let $y \in f(C \cap D)$. Then $y = f(x)$ for some $x \in C \cap D$. Since $x \in C$, $f(x) \in f(C)$. Since $x \in D$, $f(x) \in f(D)$. This shows that $y \in f(C)$ and $y \in f(D)$. Hence $y \in f(C) \cap f(D)$, proving the result.

It is worth noting that if $f\colon \mathbb{R} \to \mathbb{R}$ is given by $f(x) = x^2$, then letting $C = \{-1\}$ and $D = \{1\}$ illustrates that containment in the opposite direction need not hold, in general.

Exercises for Appendix A.

1. Simplify the following sets by expressing them as intervals.

   (a) $[-4, 6) \cap (3, 9] \setminus [1, 5)$

   (b) $\{x \in \mathbb{R} : 4x - 2x^2 > 0\}$

   (c) $\displaystyle\bigcup_{n=1}^{\infty} \left[\frac{1}{n}, n\right]$

   (d) $\displaystyle\bigcap_{n=1}^{\infty} \left(\frac{n-1}{n^2}, \frac{n+1}{n}\right)$

2. Let A, B, and C be sets. Prove the following equalities.

   (a) $(A \cap B) \cup C = (A \cup C) \cap (B \cup C)$

   (b) $A \setminus (B \cup C) = (A \setminus B) \cap (A \setminus C)$

   (c) $A \setminus (B \cap C) = (A \setminus B) \cup (A \setminus C)$

   (d) $(A \cup B) \setminus (A \cap B) = (A \setminus B) \cup (B \setminus A)$

3. Let A and B be sets. The *Cartesian product* of A and B is the set of *ordered pairs*

$$A \times B = \{(x, y) : x \in A \text{ and } y \in B\}.$$

   (a) Let $A = \{1, 2, 3\}$ and $B = \{1, 2\}$. List all of the elements of $A \times B$.

   (b) Explain why the plane $\mathbb{R}^2$ is equal to $\mathbb{R} \times \mathbb{R}$. Describe the sets $[1, 2] \times [-2, 5]$ and $\{1, 4\} \times [0, \infty)$ in $\mathbb{R}^2$.

   (c) Determine whether each of the following is necessarily true for sets A, B, C, and D. Give either a proof or a counterexample to support your answer.

      i. $A \times (B \cap C) = (A \times B) \cap (A \times C)$
      ii. $A \times (B \cup C) = (A \times B) \cup (A \times C)$
      iii. $(A \times B) \cap (C \times D) = (A \cap C) \times (B \cap D)$
      iv. $(A \times B) \cup (C \times D) = (A \cup C) \times (B \cup D)$

4. Let A be a set, and let $\mathcal{B} = \{B_\alpha : \alpha \in I\}$ be a set of sets. Verify the following

   (a) $\displaystyle A \cap \left(\bigcup_{\alpha \in I} B_\alpha\right) = \bigcup_{\alpha \in I} (A \cap B_\alpha)$

   (b) $\displaystyle A \cup \left(\bigcap_{\alpha \in I} B_\alpha\right) = \bigcap_{\alpha \in I} (A \cup B_\alpha)$

   (c) $\displaystyle A \setminus \left(\bigcup_{\alpha \in I} B_\alpha\right) = \bigcap_{\alpha \in I} (A \setminus B_\alpha)$

   (d) $\displaystyle A \setminus \left(\bigcap_{\alpha \in I} B_\alpha\right) = \bigcup_{\alpha \in I} (A \setminus B_\alpha)$

5. A subset $E \subseteq \mathbb{R}$ is called *cofinite* if $\mathbb{R} \setminus E$ is finite. Let $\mathcal{A} = \{A_\alpha : \alpha \in I\}$ be a collection of cofinite subsets of $\mathbb{R}$. Prove that $\bigcup_{\alpha \in I} A_\alpha$ is cofinite.

6. Let A be a set. The *power set* of A is the collection $\mathcal{P}(A)$ of all subsets of A.

   (a) Let $A = \{1, 2, 3\}$. List all of the elements of $\mathcal{P}(A)$.

   (b) Let A and B be sets that satisfy $A \subseteq B$. Show that $\mathcal{P}(A) \subseteq \mathcal{P}(B)$.

   (c) Let A and B be sets. For each of the following, prove the equality or provide a counterexample showing it need not hold.

        i. $\mathcal{P}(A) \cap \mathcal{P}(B) = \mathcal{P}(A \cap B)$

        ii. $\mathcal{P}(A) \cup \mathcal{P}(B) = \mathcal{P}(A \cup B)$

7. Give an example of sets $A_k \subseteq \mathbb{N}$ for each $k \in \mathbb{N}$ such that $\bigcap_{k=1}^{n} A_k \neq \varnothing$ for all $n \in \mathbb{N}$, yet $\bigcap_{k=1}^{\infty} A_k = \varnothing$.

8. Let A_{mn} be a set for all $n, m \in \mathbb{N}$. Define the sets B and C by

$$B = \bigcap_{n=1}^{\infty} \bigcup_{m=1}^{\infty} A_{nm}, \qquad C = \bigcup_{m=1}^{\infty} \bigcap_{n=1}^{\infty} A_{nm}.$$

   For each of the following, prove the containment holds or give a counterexample showing it need not.

   (a) $B \subseteq C$

   (b) $C \subseteq B$

9. Determine the specified sets for the given function $f \colon \mathbb{R} \to \mathbb{R}$.

   (a) $f(x) = \cos x, \qquad f((0, \infty)), \; f^{-1}\left(\left[\frac{1}{2}, 2\right]\right)$

   (b) $f(x) = e^{x^2}, \qquad f([0, 6]), \; f^{-1}(f((2, 5)))$

10. In the following, $A \subseteq \mathbb{R}$ and $f \colon A \to \mathbb{R}$ are given. Determine $f(A)$, and find a set $B \subseteq A$ such that if $g \colon B \to \mathbb{R}$ is defined by $g(x) = f(x)$, then g is one-to-one and $g(B) = f(A)$.

   (a) $A = \mathbb{R}, \qquad f(x) = x^2 + 4x + 6$

   (b) $A = \mathbb{R}, \qquad f(x) = 3x^2 - 2x^3$

   (c) $A = \mathbb{R} \setminus \{(2n + 1)\pi/2 : n \in \mathbb{Z}\}, \qquad f(x) = \tan^2 x$

   (d) $A = (0, \infty), \qquad f(x) = (\ln x)^2 - 6\ln x + 2$

11. Let A and B be sets, and let $f \colon A \to B$. For $C, D \subseteq A$ and $E, F \subseteq B$, prove the following.

   (a) $C \subseteq f^{-1}(f(C))$

   (b) $f(f^{-1}(E)) \subseteq E$

   (c) $f(C \cup D) = f(C) \cup f(D)$

   (d) $f^{-1}(E \cap F) = f^{-1}(E) \cap f^{-1}(F)$

   (e) $f^{-1}(E \cup F) = f^{-1}(E) \cup f^{-1}(F)$

   (f) $f^{-1}(B \setminus E) = A \setminus f^{-1}(E)$

12. Restate the rules in Example A.17 and in Exercise 11, parts (c), (d), and (e), for a family $\mathcal{C}$ of subsets of A and a family $\mathcal{D}$ of subsets of B and prove them.

13. Does adding either the hypothesis that f is one-to-one or the hypothesis that f is onto give equality in either Example A.17 or in parts (a) and (b) of Exercise 11? Support your answer with proofs or counterexamples.

14. Let A, B, and C be sets, $f\colon A \to B$, and $g\colon B \to C$. Show that the following hold.

    (a) If f and g are one-to-one, then $g \circ f$ is one-to-one.

    (b) If f and g are onto, then $g \circ f$ is onto.

    (c) If f and g are both bijections, then $g \circ f$ has an inverse. What is it?

15. Here, we consider what help the derivative can be in analyzing whether a function is one-to-one.

    (a) Suppose $f\colon [a, b] \to \mathbb{R}$ is continuous and is differentiable on (a, b). Use the mean value theorem to show that if $f'(x) > 0$ for all $x \in (a, b)$, then f is one-to-one.

    (b) Let $f\colon \mathbb{R} \to \mathbb{R}$ be given by $f(x) = x^3 - 3x^2 + 3x - 4$. Use reasoning similar to that in part (a) to show that f is one-to-one.

16. For each pair of sets A and B, find a bijection $f\colon A \to B$.

    (a) $A = [0, 1]$, $B = [-4, 9]$

    (b) $A = (0, 1)$, $B = \mathbb{R}$

    (c) $A = [0, 1]$, $B = (0, 1)$ (*Hint*: The function f will not be continuous.)

17. This exercise is more advanced than the preceding. It introduces a concept that, while used in this text only for solving a couple of exercises in Section 3.4, has applications to many areas of mathematics. A nonempty set A is called *countable* if there exists a function $f\colon A \to \mathbb{N}$ that is one-to-one.

    (a) Show that A is countable if and only if there is a function $g\colon \mathbb{N} \to A$ that is onto.

    (b) If A is infinite, prove that A is countable if and only if there is a bijection $h\colon A \to \mathbb{N}$. In this case, A is said to be *denumerable*.

    (c) Show that $\mathbb{Z}$ is denumerable.

    (d) Prove that if A and B are countable, then $A \times B$ is countable.

    (e) Show that $\mathbb{Q}$ is denumerable.

    (f) Show that if $\{A_n : n \in \mathbb{N}\}$ is a collection of countable sets, then $\bigcup_{n=1}^{\infty} A_n$ is countable.

    (g) Verify that $\mathbb{R}$ is not countable. To do this, use contradiction. Assume that there is a function $g\colon \mathbb{N} \to \mathbb{R}$ that is onto. Begin by letting $I_0 = [0, 1]$. Choose a closed subinterval I_1 of I_0 with positive length that does not contain $g(1)$. Carry on in this manner, generating a sequence of intervals, and then invoke Theorem 1.6.15.

APPENDIX B

TOPICS FROM ADVANCED CALCULUS

All areas of analysis are connected, and there are items from real analysis (or advanced calculus) that we use to develop complex function theory. Our approach in this text is to assume those items covered in the standard calculus courses, leaving the task of their proofs to courses on real analysis. For completeness, we address a few useful items here that are not part of the standard calculus sequence. Some of these topics may be familiar to the reader and others are not used extensively, hence their relegation to an appendix.

The Supremum and Infimum

We begin by discussing notation and properties related to least upper bounds and greatest lower bounds.

B.1 Definition. Let $A \subseteq \mathbb{R}$ be nonempty. If $M \in \mathbb{R}$ and $x \leq M$ for all $x \in A$, then M is an *upper bound* of A. If an upper bound of A exists, then A is said to be *bounded above*. If $\alpha \in \mathbb{R}$ is an upper bound of A and $M \geq \alpha$ for any upper bound M of A, then α is the *least upper bound* or *supremum* of A, denoted $\alpha = \sup A$.

Complex Analysis: A Modern First Course in Function Theory, First Edition. Jerry R. Muir, Jr.
© 2015 John Wiley & Sons, Inc. Published 2015 by John Wiley & Sons, Inc.

The construction of the set $\mathbb{R}$ from the set $\mathbb{Q}$ is a very delicate process that we will not address. At the conclusion of the process, $\mathbb{R}$ is seen to be *the* ordered field containing $\mathbb{Q}$ such that the following axiom holds.

B.2 Axiom of Completeness. *Every nonempty $A \subseteq \mathbb{R}$ that is bounded above has a supremum.*

B.3 Example. There does not exist a number $a \in \mathbb{Q}$ such that $a^2 = 2$, yet the set $A = \{x \in \mathbb{Q} : x^2 < 2\}$ is easily seen to be bounded above. Some careful calculations show that if $\alpha = \sup A$ exists, then $\alpha^2 = 2$. It follows that A does not have a supremum in $\mathbb{Q}$. But α is guaranteed to exist in $\mathbb{R}$, showing $\sqrt{2} \in \mathbb{R} \setminus \mathbb{Q}$ is "added in" by the completion of $\mathbb{Q}$ to $\mathbb{R}$.

The following result is often useful.

B.4 Theorem. *Let $A \subseteq \mathbb{R}$ be nonempty, and let $\alpha \in \mathbb{R}$ be an upper bound of A. Then $\alpha = \sup A$ if and only if for every $\varepsilon > 0$, there exists $x \in A$ such that $x > \alpha - \varepsilon$.*

Proof. If $\alpha = \sup A$ and $\varepsilon > 0$, then $\alpha - \varepsilon$ is not an upper bound of A. Thus there is $x \in A$ such that $x > \alpha - \varepsilon$.

Inversely, suppose $\alpha \neq \sup A$. Now $\beta = \sup A$ exists, $\varepsilon = \alpha - \beta > 0$, and there does not exist $x \in A$ such that $x > \beta = \alpha - \varepsilon$. $\qquad\qquad\square$

B.5 Example. Consider the set

$$A = \left\{ \frac{m}{n} : m, n \in \mathbb{N} \text{ and } m < n \right\}.$$

We will verify that $\sup A = 1$. First $m < n$ implies $m/n < 1$, and hence 1 is an upper bound of A. Let $\varepsilon > 0$. For $n \in \mathbb{N}$ such that $n > \max\{1, 1/\varepsilon\}$, we see that $(n-1)/n \in A$ and

$$\frac{n-1}{n} = 1 - \frac{1}{n} > 1 - \varepsilon.$$

The result follows by Theorem B.4.

We will also make use of the following.

B.6 Theorem. *Let A be a nonempty set and $f, g \colon A \to \mathbb{R}$. If $f(x) \leq g(x)$ for all $x \in A$ and $g(A)$ is bounded above, then $\sup f(A) \leq \sup g(A)$.*

Proof. Since $g(A)$ is bounded above, $\sup g(A)$ exists. For all $x \in A$, $f(x) \leq g(x) \leq \sup g(A)$. This shows that $\sup g(A)$ is an upper bound of $f(A)$, and therefore $\sup f(A)$ exists and $\sup f(A) \leq \sup g(A)$. $\qquad\qquad\square$

The above definitions and theorems have analogs for the lower end of a set; the proofs are left as exercises, and the definition is as follows.

B.7 Definition. The terms *lower bound, bounded below,* and *greatest lower bound* for a nonempty set $A \subseteq \mathbb{R}$ are defined in a manner similar to Definition B.1. The greatest lower bound of A is called the *infimum* of A and is denoted $\inf A$.

In some circumstances, it is convenient to be more flexible with our notation for suprema and infima or for divergent sequences. This prompts the following definition, in which we adjoin the symbols $-\infty$ and ∞ to $\mathbb{R}$.

B.8 Definition. We call $[-\infty, \infty]$ the set of *extended real numbers*. It is endowed with the natural ordering.

B.9 Remark. The clear context will distinguish the symbol ∞ here as different from the complex infinity defined in Section 1.5.

The naturally inherited order properties of the extended real numbers are what we will find especially useful. For instance, every nonempty set $A \subseteq \mathbb{R}$ has a supremum in $[-\infty, \infty]$, where $\sup A = \infty$ when A is not bounded above. The analogous observation holds for the infimum.

We conclude with a concept that is useful for considering the radius of convergence of a power series in Section 2.3.

B.10 Definition. Let $\{x_n\}_{n=1}^{\infty}$ be a sequence of real numbers. For each $n \in \mathbb{N}$, define the extended real numbers

$$\alpha_n = \inf\{x_k : k \geq n\}, \qquad \beta_n = \sup\{x_k : k \geq n\}.$$

We define the *limit inferior* and *limit superior* of $\{x_n\}$ by

$$\liminf_{n\to\infty} x_n = \lim_{n\to\infty} \alpha_n, \qquad \limsup_{n\to\infty} x_n = \lim_{n\to\infty} \beta_n. \tag{B.1}$$

B.11 Remark. Either $\{x_n\}$ is not bounded above, in which case $\beta_n = \infty$ for all $n \in \mathbb{N}$, or it is, in which case β_n is finite for all $n \in \mathbb{N}$. In the former case, $\limsup_{n\to\infty} x_n = \infty$. In the latter case, $\{\beta_n\}_{n=1}^{\infty}$ is a decreasing sequence of real numbers. If it is bounded below, it converges to a finite value $\beta \in \mathbb{R}$ (see Exercise 6 in Section 1.6), and $\limsup_{n\to\infty} x_n = \beta$. Otherwise, $\limsup_{n\to\infty} x_n = -\infty$. Similar logic holds for $\liminf_{n\to\infty} x_n$. Further properties are given in the exercises.

Uniform Continuity

We now consider a stronger version of continuity. In the definition of continuity (see Definition 2.1.1), δ is dependent upon both ε and the point a chosen. That is, given a fixed $\varepsilon > 0$, the choice of $\delta > 0$ made for some a in the domain may not be satisfactory if a is replaced by a different point b. That is not the case in the following.

B.12 Definition. Let $A \subseteq \mathbb{C}$ and $f : A \to \mathbb{C}$. We say that f is *uniformly continuous* on A if for every $\varepsilon > 0$, there exists some $\delta > 0$ such that $|f(z) - f(w)| < \varepsilon$ whenever $z, w \in A$ and $|z - w| < \delta$.

This is similar to the improvement uniform convergence is over pointwise convergence, as seen in Section 2.2. It should be apparent that every uniformly continuous function is continuous. See the exercises for a counterexample to the converse. Here is special case when the converse is actually true.

B.13 Theorem. *Let $K \subseteq \mathbb{C}$ be compact. If $f : K \to \mathbb{C}$ is continuous, then f is uniformly continuous.*

Proof. Suppose that f is not uniformly continuous. Then there exists some $\varepsilon > 0$ such that for any $\delta > 0$, there are $z, w \in K$ such that $|z - w| < \delta$ and $|f(z) - f(w)| \geq \varepsilon$. Choose, for each $n \in \mathbb{N}$, $z_n, w_n \in K$ such that $|z_n - w_n| < 1/n$ and $|f(z_n) - f(w_n)| \geq \varepsilon$. A subsequence $\{z_{n_k}\}_{k=1}^{\infty}$ of $\{z_n\}_{n=1}^{\infty}$ then converges to a point $a \in K$ by the Bolzano–Weierstrass theorem. Theorems 1.6.8 and 1.6.9 give

$$\lim_{k \to \infty} w_{n_k} = \lim_{k \to \infty} z_{n_k} + \lim_{k \to \infty} (w_{n_k} - z_{n_k}) = a + 0 = a,$$

where we note $|(w_{n_k} - z_{n_k}) - 0| < 1/n_k$ and $1/n_k \to 0$. By continuity, $(f(z_{n_k}) - f(w_{n_k})) \to (f(a) - f(a)) = 0$, a contradiction. □

Uniform continuity in two variables is addressed in Exercise 16 in Section 2.1 and is necessary for certain exercises throughout the text.

The Cauchy Product

The basic linearity rules for complex series are given in Theorem 1.7.4. While not addressed in Section 1.7, it is useful to be able to handle taking the product of series in certain circumstances. We begin with the following definition, noting that, since these products are so often applied to power series, we use indices starting at $n = 0$.

B.14 Definition. Let $\{a_n\}_{n=0}^{\infty}$ and $\{b_n\}_{n=0}^{\infty}$ be sequences of complex numbers. The *Cauchy product* of $\{a_n\}$ and $\{b_n\}$ is the sequence $\{c_n\}_{n=0}^{\infty}$ given by

$$c_n = \sum_{k=0}^{n} a_k b_{n-k}. \tag{B.2}$$

We note that the sequence $\{c_n\}$ is also referred to as the *convolution* of $\{a_n\}$ and $\{b_n\}$. For each n, the term c_n is the sum of all products of terms of the sequences $\{a_k\}$ and $\{b_k\}$ whose indices sum to n.

The motivation for the definition comes from the desire to find a series product

$$\left(\sum_{n=0}^{\infty} a_n \right) \left(\sum_{n=0}^{\infty} b_n \right).$$

Were we to naïvely multiply the two series, ignoring any worries about convergence, we would obtain the sum of all possible combinations of $a_n b_m$ for $n, m \geq 0$. The same terms are obtained from summing $\sum_{n=0}^{\infty} c_n$, where we are first grouping product terms by the sum of their indices. Of course, we cannot actually ignore issues of convergence, and hence we prove the following.

B.15 Mertens' Theorem. *Let $\{a_n\}_{n=0}^{\infty}$ and $\{b_n\}_{n=0}^{\infty}$ be sequences of complex numbers with Cauchy product $\{c_n\}_{n=0}^{\infty}$. If $\sum_{n=0}^{\infty} a_n$ converges absolutely to a and $\sum_{n=0}^{\infty} b_n = b$, then $\sum_{n=0}^{\infty} c_n = ab$.*

Note that, under the circumstances of Mertens' theorem, the series $\sum_{n=0}^{\infty} c_n$ is also referred to as the *Cauchy product* of the series $\sum_{n=0}^{\infty} a_n$ and $\sum_{n=0}^{\infty} b_n$.

Proof. For any $N = 0, 1, \ldots$, we calculate

$$\sum_{n=0}^{N} c_n = \sum_{n=0}^{N} \sum_{k=0}^{n} a_k b_{n-k} = \sum_{n=0}^{N} \sum_{j=0}^{N-n} a_n b_j = \sum_{n=0}^{N} a_n \sum_{j=0}^{N-n} b_j$$

by recognizing that the middle double sums each represent the sum of all terms of the form $a_k b_j$ such that $0 \le k + j \le N$. If we write $s_n = \sum_{k=0}^{n} a_k$ and $t_n = \sum_{k=0}^{n} b_k$, then we see that

$$\sum_{n=0}^{N} c_n = \sum_{n=0}^{N} a_n t_{N-n} = \sum_{n=0}^{N} a_n(b - (b - t_{N-n})) = s_N b - \sum_{n=0}^{N} a_n(b - t_{N-n}).$$

Since $s_N b \to ab$ as $N \to \infty$, the proof will be complete once we show

$$\lim_{N \to \infty} \sum_{n=0}^{N} a_n(b - t_{N-n}) = 0.$$

Let $\varepsilon > 0$. Since the sequence $\{t_n - b\}_{n=0}^{\infty}$ is convergent (its limit is 0), it is bounded (Theorem 1.6.6). Thus there is $M > 0$ such that $|t_n - b| \le M$ for all n. Since $\sum_{n=0}^{\infty} a_n$ is absolutely convergent, there is $N_1 \in \mathbb{N}$ such that $\sum_{n=N_1+1}^{\infty} |a_n| < \varepsilon/(2M)$. If $\alpha > \sum_{n=0}^{\infty} |a_n|$, then $t_n \to b$ gives an $N_2 \in \mathbb{N}$ such that for all $n \ge N_2$, $|t_n - b| < \varepsilon/(2\alpha)$.

Let $N \ge N_1 + N_2$. Note that if $n \le N_1$, then $N - n \ge N_2$. Hence

$$\left| \sum_{n=0}^{N} a_n(b - t_{N-n}) \right| \le \sum_{n=0}^{N_1} |a_n||t_{N-n} - b| + \sum_{n=N_1+1}^{N} |a_n||t_{N-n} - b|$$

$$\le \frac{\varepsilon}{2\alpha} \sum_{n=0}^{N_1} |a_n| + M \sum_{n=N_1+1}^{N} |a_n|$$

$$< \frac{\varepsilon}{2\alpha} \alpha + M \frac{\varepsilon}{2M}$$

$$= \varepsilon.$$

This implies the desired limit. $\qquad\qquad\qquad\qquad\qquad\qquad\qquad\qquad\qquad\qquad$ $\square$

Leibniz's Rule

In this last section, we consider the condition under which the processes of differentiation and integration may be reversed that is key to our proof of Lemma 3.2.2, the first step in our proof of Goursat's theorem and our first encounter with the notion that an integral of a function about a closed contour can be used to determine the values of the function inside the contour.

As noted in Section 2.8, switching to a complex range has no impact on the result and requires no additional work. Indeed, in advanced real analysis texts, a complex range is a standard assumption. It is the move from a real domain to a complex domain that changes real analysis into complex analysis.

B.16 Leibniz's Rule. *Suppose that* $\varphi\colon [a,b] \times [c,d] \to \mathbb{C}$ *is continuous. Define* $f\colon [a,b] \to \mathbb{C}$ *by*

$$f(x) = \int_c^d \varphi(x,y)\,dy.$$

Then f is continuous. Furthermore, if $\partial\varphi/\partial x$ exists and is continuous on $[a,b] \times [c,d]$, then f' exists, is continuous, and satisfies

$$f'(x) = \int_c^d \frac{\partial\varphi}{\partial x}(x,y)\,dy.$$

In less precise terms, Leibniz's rule allows for differentiation to move from outside to inside the integral. Often this is written as

$$\frac{d}{dx}\int_c^d \varphi(x,y)\,dy = \int_c^d \frac{\partial\varphi}{\partial x}(x,y)\,dy.$$

Proof. We begin by establishing the continuity of f at an arbitrary $x_0 \in [a,b]$. Since $[a,b] \times [c,d]$ is compact, φ is uniformly continuous by Theorem B.13. Hence given $\varepsilon > 0$, there exists a $\delta > 0$ such that

$$|\varphi(x_1,y_1) - \varphi(x_2,y_2)| < \frac{\varepsilon}{d-c}$$

whenever $x_1, x_2 \in [a,b]$ and $y_1, y_2 \in [c,d]$ satisfy $\sqrt{(x_2-x_1)^2 + (y_2-y_1)^2} < \delta$. Accordingly, if $x \in [a,b]$ is such that $|x - x_0| < \delta$, then

$$|f(x) - f(x_0)| \le \int_c^d |\varphi(x,y) - \varphi(x_0,y)|\,dy < \int_c^d \frac{\varepsilon}{d-c}\,dy = \varepsilon.$$

Therefore f is continuous at x_0.

Now assume that $\partial\varphi/\partial x$ is continuous on $[a,b] \times [c,d]$. It is then uniformly continuous, and as above, given $\varepsilon > 0$, there exists $\delta > 0$ such that

$$\left|\frac{\partial\varphi}{\partial x}(x_1,y_1) - \frac{\partial\varphi}{\partial x}(x_2,y_2)\right| < \frac{\varepsilon}{d-c} \tag{B.3}$$

for all $x_1, x_2 \in [a,b]$ and $y_1, y_2 \in [a,c]$ such that $\sqrt{(x_2-x_1)^2 + (y_2-y_1)^2} < \delta$. Assume that $x \in [a,b]$ and $0 < |x - x_0| < \delta$. We will show that the modulus of the integral on the right-hand side of

$$\frac{f(x) - f(x_0)}{x - x_0} - \int_c^d \frac{\partial\varphi}{\partial x}(x_0,y)\,dy = \int_c^d \left(\frac{\varphi(x,y) - \varphi(x_0,y)}{x - x_0} - \frac{\partial\varphi}{\partial x}(x_0,y)\right)dy$$

is less than ε. By Theorem 2.8.8, it suffices to show that

$$\left| \varphi(x,y) - \varphi(x_0,y) - (x - x_0)\frac{\partial \varphi}{\partial x}(x_0,y) \right| < \frac{\varepsilon |x - x_0|}{d - c} \tag{B.4}$$

for all $y \in [c,d]$. By observing that

$$\frac{\partial}{\partial x}\left(\varphi(x,y) - x\frac{\partial \varphi}{\partial x}(x_0,y) \right) = \frac{\partial \varphi}{\partial x}(x,y) - \frac{\partial \varphi}{\partial x}(x_0,y),$$

we use the fundamental theorem of calculus to see that

$$\varphi(x,y) - \varphi(x_0,y) - (x - x_0)\frac{\partial \varphi}{\partial x}(x_0,y) = \int_{x_0}^{x} \left(\frac{\partial \varphi}{\partial x}(\xi,y) - \frac{\partial \varphi}{\partial x}(x_0,y) \right) d\xi.$$

But in the above integral, $|\xi - x_0| \le |x - x_0| < \delta$, and therefore taking the modulus of both sides and applying Theorem 2.8.8 gives (B.4).   □

Exercises for Appendix B.

1. Define the infimum in a manner analogous to Definition B.1 and state a theorem for the infimum similar to Theorem B.4.

2. For the following sets $A \subseteq \mathbb{R}$, determine $\sup A$ and $\inf A$. Give a justification using Theorem B.4 for the supremum and the analog developed in Exercise 1 for the infimum.

    (a) $A = [-1,3)$

    (b) $A = \left\{ \dfrac{-1}{n^2} : n \in \mathbb{N} \right\}$

    (c) $A = \{x \in \mathbb{Q} : x^2 < 3\}$

3. Suppose $A \subseteq \mathbb{R}$ is nonempty and bounded below. Show that the set $B = \{-x : x \in A\}$ is bounded above and $\inf A = -\sup B$.

4. Let $A, B \subseteq \mathbb{R}$ be bounded above, and define $C = \{x + y : x \in A \text{ and } y \in B\}$. Prove that $\sup C = \sup A + \sup B$.

5. The *intermediate value theorem* seen in calculus states that if $f : [a,b] \to \mathbb{R}$ is continuous, $L \in \mathbb{R}$, and $(f(a) - L)(f(b) - L) < 0$, then there is $c \in (a,b)$ such that $f(c) = L$. Prove this using the supremum of the set $\{x \in [a,b] : (f(x) - L)(f(a) - L) > 0\}$.

6. Assume the notation used in Definition B.10.

    (a) Fill in the details in Remark B.11. That is, verify that $\{\alpha_n\}$ and $\{\beta_n\}$ are monotone sequences that converge in $[-\infty, \infty]$.

    (b) Prove that $\liminf_{n \to \infty} x_n \le \limsup_{n \to \infty} x_n$.

    (c) Show that there is a subsequence of $\{x_n\}$ converging to $\liminf_{n \to \infty} x_n$ and a subsequence of $\{x_n\}$ converging to $\limsup_{n \to \infty} x_n$.

    (d) Prove that $\lim_{n \to \infty} x_n$ exists in $[-\infty, \infty]$ if and only if

$$\liminf_{n \to \infty} x_n = \limsup_{n \to \infty} x_n,$$

and hence are equal to the limit.

7. Let $f, g, h\colon \mathbb{C} \setminus \{0\} \to \mathbb{C}$ be given by $f(z) = 2z$, $g(z) = z^2$, and $h(z) = 1/z$. Show that f is uniformly continuous, but g and h are not.

8. Let $A \subseteq \mathbb{C}$, $f, g\colon A \to \mathbb{C}$ be uniformly continuous, and $c \in \mathbb{C}$ be a constant. Prove that the following functions are uniformly continuous on A.

   (a) cf

   (b) $f + g$

9. Let $A \subseteq \mathbb{C}$ and $f\colon A \to \mathbb{C}$ be uniformly continuous.

   (a) Show that if $\{z_n\}_{n=1}^{\infty} \subseteq A$ is a Cauchy sequence, then $\{f(z_n)\}$ is also a Cauchy sequence.

   (b) Prove that there is a continuous function $g\colon \overline{A} \to \mathbb{C}$ such that $g(z) = f(z)$ for all $z \in A$. The function g is called the *continuous extension* of f.

10. Let $A, B \subseteq \mathbb{C}$. If $f\colon A \to B$ and $g\colon B \to \mathbb{C}$ are each uniformly continuous, then prove that $g \circ f$ is uniformly continuous.

11. Show that if $a_n = b_n = (-1)^n/\sqrt{n+1}$ for all $n \in \mathbb{N}$, then the terms of the Cauchy product $\{c_n\}_{n=1}^{\infty}$ of $\{a_n\}_{n=1}^{\infty}$ and $\{b_n\}_{n=1}^{\infty}$ satisfy $|c_n| \geq 1$ for all n. What does this say about the hypotheses of Mertens' theorem?

12. Give an example in the spirit of Exercise 11 showing that $\sum_{n=0}^{\infty} c_n$ need not converge absolutely under the hypotheses of Mertens' theorem. Prove that if the hypothesis that $\sum_{n=0}^{\infty} b_n$ converges absolutely is added to the statement of the theorem, then we may conclude that $\sum_{n=0}^{\infty} c_n$ converges absolutely.

13. Let $a \in \mathbb{C}$ such that $|a| < 1$. Evaluate the following integral for all $x \in [0, 1]$ by first differentiating with respect to x using Leibniz's rule.

$$\int_0^{2\pi} \frac{e^{iy}}{e^{iy} + xa}\, dy$$

(Compare to Example 2.9.21.)

14. In a manner similar to Exercise 13, calculate the following integral for all $x \in \mathbb{R}$.

$$\int_0^{2\pi} \cos^2(xe^{iy})\, dy$$

15. Let $\varphi\colon [a, b] \times [c, d] \to \mathbb{C}$ be such that φ and $\partial\varphi/\partial x$ are continuous. Suppose that $g, h\colon [a, b] \to [c, d]$ are differentiable, and define $f\colon [a, b] \to \mathbb{C}$ by

$$f(x) = \int_{g(x)}^{h(x)} \varphi(x, y)\, dy.$$

Prove that

$$f'(x) = \varphi(x, h(x))h'(x) - \varphi(x, g(x))g'(x) + \int_{g(x)}^{h(x)} \frac{\partial\varphi}{\partial x}(x, y)\, dy.$$

(*Hint*: Make substitutions for expressions in the definition of f and apply the chain rule from multivariable calculus.)

REFERENCES

1. L. V. Ahlfors, *Complex Analysis*, 3rd ed., McGraw-Hill, New York, 1979.

2. T. Archibald, Analysis and physics in the nineteenth century: the case of boundary-value problems, in *A History of Analysis*, American Mathematical Society, Providence, 2003.

3. U. Bottazzini, Complex function theory, 1780–1900, in *A History of Analysis*, American Mathematical Society, Providence, 2003.

4. U. Bottazzini, Three traditions in complex analysis: Cauchy, Riemann and Weierstrass, *Companion Encyclopedia of the History and Philosophy of the Mathematical Sciences*, vol. 1, Johns Hopkins University Press, Baltimore, 2003.

5. J. W. Brown and R. V. Churchill, *Complex Variables and Applications*, 6th ed., McGraw-Hill, New York, 1996.

6. D. M. Burton, *The History of Mathematics*, 3rd ed., McGraw-Hill, New York, 1995.

7. J. B. Conway, *Functions of One Complex Variable*, 2nd ed., Springer-Verlag, New York, 1978.

8. J. D. Dixon, A brief proof of Cauchy's integral theorem, *Proc. Amer. Math. Soc.* **29** (1971), 625–626.

9. W. Dunham, *Euler: The Master of Us All*, Mathematical Association of America, Washington, D.C., 1999.

10. W. Dunham, *The Mathematical Universe: An Alphabetical Journey Through the Great Proofs, Problems, and Personalities*, Wiley, New York, 1994.

Complex Analysis: A Modern First Course in Function Theory, First Edition. Jerry R. Muir, Jr.
© 2015 John Wiley & Sons, Inc. Published 2015 by John Wiley & Sons, Inc.

11. G. B. Folland, *Fourier Analysis and its Applications*, Brooks/Cole, Pacific Grove, California, 1992.

12. T. W. Gamelin, *Complex Analysis*, Springer-Verlag, New York, 2001.

13. V. J. Katz, *A History of Mathematics, an Introduction*, 2nd ed., Addison-Wesley, Reading, Massachusetts, 1998.

14. J. L. Kelley, *General Topology*, Van Nostrand, Princeton, 1955.

15. M. Kline, *Mathematical Thought from Ancient to Modern Times*, Oxford University Press, New York, 1972.

16. J. Lützen, The foundations of analysis in the 19th century, in *A History of Analysis*, American Mathematical Society, Providence, 2003.

17. J. E. Marsden and M. J. Hoffman, *Basic Complex Analysis*, 3rd ed., W.H. Freeman, New York, 1999.

18. J. H. Mathews and R. W. Howell, *Complex Analysis for Mathematics and Engineering*, 4th ed., Jones and Bartlett, Boston, 2001.

19. O. C. McGehee, *An Introduction to Complex Analysis*, Wiley, New York, 2000.

20. M. Monastyrsky, *Riemann, Topology, and Physics*, 2nd ed., Birkhäuser, Boston, 2008.

21. Z. Nehari, *Conformal Mapping*, Dover, New York, 1975.

22. M. A. Pinsky, *Introduction to Fourier Analysis and Wavelets*, Brooks/Cole, Pacific Grove, California, 2002.

23. R. Remmert, *Classical Topics in Complex Function Theory*, Springer-Verlag, New York, 1998.

24. R. Roy, *Sources in the Development of Mathematics. Infinite Series and Products from the Fifteenth to the Twenty-first Century*, Cambridge University Press, Cambridge, 2011.

25. W. Rudin, *Principles of Mathematical Analysis*, 3rd ed., McGraw-Hill, New York, 1976.

26. W. Rudin, *Real and Complex Analysis*, 3rd ed., McGraw-Hill, New York, 1987.

27. E. B. Saff and A. D. Snider, *Fundamentals of Complex Analysis for Mathematics, Science, and Engineering*, 2nd ed., Prentice-Hall, Upper Saddle River, New Jersey, 1993.

28. E. Sandifer, e, π and i: why is "Euler" in the Euler identity, http://www.maa.org/news/howeulerdidit.html.

29. E. Sandifer, *How Euler Did It*, Mathematical Association of America, Washington, D.C., 2007.

30. R. Silverman, *Introductory Complex Analysis*, Dover, New York, 1972.

31. J. P. Snyder, *Flattening the Earth*, University of Chicago Press, Chicago, 1993.

32. G. B. Thomas, M. D. Weir, J. Hass, and F. R. Giordano, *Thomas' Calculus*, 11th ed., Addison-Wesley, Boston, 2005.

33. J. L. Walsh, History of the Riemann mapping theorem, *Amer. Math. Monthly* **80** (1973), 270–276.

34. A. D. Wunsch, *Complex Variables with Applications*, 2nd ed., Addison-Wesley, Reading, Massachusetts, 1994.

INDEX

257

Complex Analysis: A Modern First Course in Function Theory, First Edition. Jerry R. Muir, Jr.
© 2015 John Wiley & Sons, Inc. Published 2015 by John Wiley & Sons, Inc.